MAN THE BUILDER

Readers who take an interest in the progress of civilization and of the useful arts will be grateful to the humble topographer who has recorded these facts, and will perhaps wish that historians of far higher pretensions had sometimes spared a few pages from military evolutions and political intrigues, for the purpose of letting us know how the parlours and bedchambers of our ancestors looked.

MACAULAY

MAN THE BUILDER

Gösta E. Sandström

McGRAW-HILL BOOK COMPANY

NEW YORK   ST. LOUIS   SAN FRANCISCO   DUSSELDORF   LONDON   MEXICO

PANAMA   SYDNEY   TORONTO

TO A PATIENT WIFE

Contents

Foreword

Macaulay's comment on conventional history is as appropriate today as when he made it 120 years ago. Historical writing now as then is concerned with politics and war, or, as he puts it, "military evolutions and political intrigues." So indeed it has always been, from the beginnings of recorded history, because political and military events have a direct bearing on the affairs of men that cannot be ignored. But this emphasis on the workings of high policy tends to become repetitious and boring, which is perhaps the reason why history has come to be despised by the younger generation today.

Yet, politics and war are after all surface phenomena—the end results of hidden forces working in the deep and still poorly understood. Some no doubt derive from the hormone chemistry of man, his genetic programming, his deep-seated religious instincts, his environmental conditioning. Below the surface are the workings of economics, man's daily concern with physical survival. Exhausted land, droughts, floods and grasshoppers will force on migrations with accompanying military and political events that will shape the daily fortunes of men thousands of years later. Related to economics—indeed inseparable from it—is technology that helps provide a living in a harsh environment. It also forges the tools of conquest and empire building, and stokes up the pressures that bubbling up to the surface produce material for the establishment historian.

When through the efforts of merchants and *tecnites* men become adequately fed and clothed and sheltered and no longer fatigued from harsh labor, they begin to invent strange ideas of their position in the universe, about man's eternal soul, his god-given right to become master of his neighbor's property and to kill off all lesser breeds, to propagate the world and conquer the universe around him.

The Greeks had a word for it—*anthropocentric*—and the universities are largely concerned with its propagation. There is religious history, literary, art and architectural history, history of ideas—sane or insane—economic history, and much more besides. Then there is technological history which, however, for some obscure reason has not yet become generally accepted by the academic establishment. This has created a peculiar situation, to put it mildly. We live today, after all, in an environment shaped by technology, yet hundreds of thousands of university graduates, not to mention millions of secondary school students, finish their formal education without an idea of why or how the industrial society in which they have been raised and wherein they will have to make a living came into being. They have spent a vast number of man-hours listening to thousands of professors trying to explain the meaning of the writings of obscure local poets, but nobody has bothered to discuss with them the wider aspects of a power dam.

If a label is required, this book may perhaps be classified as an exercise in technological history, because it is primarily concerned with man in his capacity of builder. But it is not exclusively devoted to civil engineering in the narrow sense, because an attempt has been made to fit engineering into a general frame of reference. For example, it is all very well to give the known facts and figures on the Cheops pyramid, or for that matter the High Dam at Aswan. The pyramid—like the dam—makes sense only when viewed in context with its contemporary society—the totalitarian state, religious ideas, farm economics, the nature and behavior of the Nile, and so on. The same naturally applies to the ziggurats in the Euphrates valley, Greek temples, Roman aqueducts, medieval canal works and fortifications. They serve definite and different ends but express in their construction the social, mental, economic, political, and technical capabilities of their age.

In general, the structures chosen as representative for the engineering capability of a particular age belong to the great monuments of the past, some of which innocent Greek tourists named "the seven wonders of the world." Here, however, they merely serve as objects for discussion of the technology that brought them to fruition. Other examples, such as Westminster Abbey and the castles of Wales, have been included for special reasons, such as the wealth of detailed documented facts still available from the time of their construction. In sum, this book is not intended to be a catalogue of the engineering marvels of the past, nor is it concerned with anything that does not bear on the development of civil engineering in the western world.

When confronted with the magnificent works of the ancients, whether standing or in ruin, the academically conditioned mind seemingly inevitably conjures forth all sorts of explanations of how the structures were put together and the megalith members transported, about methods and know-how long since lost, or even more fanciful ideas, of which the latest tend to stray to the application of electric and even nuclear power.

Nothing could be more erroneous. In civil engineering, including building, there is nothing that cannot be accomplished by skilled men wielding hand tools: spade, hammer, pick and chisel. And nature provides the materials. After all, the vast majority of the structures still featuring our urban environment were put together by men working with a trowel. The principal tool used in building the railroads was the spade. So long as there are men willing to work for a sturdy meal and little more besides, a hand-tooled job is vastly preferable to anything accomplished by machines.

However, for the jobs discussed in this book something else is needed, namely organization ability. To an enquirer into ancient engineering, the real surprise is to find at what an early stage the abstract principles of organization began to be applied. Indeed, the art of management is found in full flower when the mists lift over King Menes' United Kingdom of Egypt. To put it briefly, the organization of an Egyptian mining expedition to Sinai anno 2800 B.C., a couple of centuries before the pyramids were built, could well have been lifted from a textbook in business management. There is the General Manager of the expedition accompanied by a staff representing the departments of the central government, including the Keeper of the Great Seal to make the venture legal, a senior officer commanding the security force, chief engineers in charge of mining and smelting operations, and then down the organization ladder to the foremen, each one bossing ten men.

There it is. Given the tools—pretty much the same as today—an adequate number of skilled and well-fed men, and this documented ability to organize and manage them, there is no mystery involved in erecting a pyramid or—more admirable considering the sophisticated hydraulic engineering involved—irrigating the desert along the Nile to feed the labor force engaged on these magnificent public works.

Greek and Roman construction is of course amply documented by a long and distinguished line of writers: Herodotus, Diodorus Siculus, Polybius, Strabo, Pliny, Frontinus, Vitruvius, and many others, who like all admirable writers unfortunately suffer from sloppy transcription and poor translation. However, anyone familiar with construction will have little difficulty recognizing the engineering facts in the garbled state in which they have come down to us through medieval monks and scholarly translators.

From the classical writings emerges a scene familiar to any senior engineer. There is the program stage, the heated political debates on financing, complementary reports and amendments of the original scheme, final surveys, drafting and estimating, planning and materials specifications, deployment of resources, and all the rest, including the final inspection and approval of the job by the engineering consultant. Indeed, a Roman aqueduct turns out to be a very familiar sequence of events. And why shouldn't it? There has to be an orderly pattern for such engineering accomplishments, and this pattern is seemingly eternal.

There is another pattern that emerges clearly and unmistakably from even a casual perusal of engineering history, and that is that the major technological innovations do not emanate from within the profession, but always from the outside. It is the outside amateur, a person wholly unfamiliar with the problems and difficulties with which the professional is acquainted, who comes up with the idea that sets engineering on a new track or lifts it up to a superior level. It was not a mason that saw the virtue of the pointed arch, but a churchman and politician; no canal builder in Lombardy or the Netherlands was aware of the possibility of bringing a canal over a summit—that idea came to a French tax collector; the admirable French engineering establishment could not conceive of a railroad, but a former English cowherder and tinkerer did. Mr. McAdam, a businessman, came up with the first economical way of building roads since Roman days; Bessemer, who confessed complete ignorance of the making of iron, instead hit upon volume production of mild steel; a mason found a way of making cement; and a Parisian gardener began reinforcing concrete with steel—just to cull a few examples from among the wealth of innovations that have brought engineering to its present state. Once the impact from the outside has made its impression it is taken over and further developed by the professionals, and before long the impulse that sparked the development is forgotten.

There is no doubt some deep moral in this. Any establishment, political or technical, tends to become set in its ways and will in the end perish, crushed under an excess of its own basic principles. Too much concern with security, regulations, standards, and formalities dulls the keen edge of enterprise. That is without doubt the reason why we look in vain for interesting engineering ventures in the modern welfare states. Whatever good they may accomplish, they fail to provide an environment conducive to innovation and one in which the odd genius can strike root and flourish. There, nothing is constructed but establishment-engineered slums.

Gösta E. Sandström

The Neolithic Revolution

For a hundred thousand years the face of Europe, or the better part of it, had been hidden under the Würm glaciers. It was a hard and difficult age; and vast areas were uninhabitable for plants, beasts, and men. But in the southwestern corner there were valleys sheltered from the Arctic gales and watered by the melting ice; here the grassy slopes teemed with animals that are now extinct outside the precincts of some zoos.

It was in these sheltered southern valleys that the piston action of nature and history, the response to a challenge, began to accelerate and produce distinguishable and distinguished results. For once, there was food enough to go around. It grew on the ground for Aurignacian and Magdalenian women to forage; the streams surged with salmon for their menfolk to spear; there were wild ponies and bovines for them to hunt. For some 20,000 years these favored people enjoyed a surplus; there was no need to grub every waking hour in order to keep alive; there was no need for everyone to participate in foraging, hunting, fishing. Some gifted individuals could stay behind in camp to invent and make new tools and, perhaps more important considing the evidence, prepare the ritual furniture required by the shamans of the clans; a few could engage in painting and fashioning out of rock and clay the deeply revered animals and the corpus of sex symbols needed to keep the fertility of men and beasts at maximum potential.

It was a time of awakening and innovation. The tools inherited by the Aurignacian hunters were simple and all-purpose, consisting of handaxes of different sizes. Then new methods of working flint were developed, first by blade flaking and subsequently pressure flaking, by means of which a whole array of special-purpose tools was made. The new tools and weapons—knives, burins, borers, lance heads, arrowheads, and so on—could be applied to the production of secondary tools made of other materials—of wood, bone, antler, ivory. Human muscle power was given increased leverage by means of the spear thrower, and the principle of concentration of energy was mastered with the aid of lever, wedge, the bow and arrow. Out of bone and antler were manufactured fishhooks, mattocks, and picks. Of perhaps more direct interest for subsequent building, the techniques of drilling, grinding, and polishing were also invented. By means of a bow drill—of a type still in use—provided with a stone bit with or without an abrasive, a hole was cored out of wood, bone, and antler, as well as hard rock. For example, the stone lamps used by the Magdalenian painters were hollowed out and shaped by means of such tools with the use of an abrasive, no doubt quartz sand.

By now many further details of the life of these paleolithic hunting clans are known and largely understood. There is no doubt but that these people possessed the brains, the creative ability, the tools and imagination to do practically anything. The age was charged with the potentials of what is commonly regarded a higher civilization. Why did they not take off into these higher spheres of technical and economic activity?

The answer seems quite simple. There was no need for it; the piston had come to rest at the end of a stroke. These European hunting people had everything they could possibly wish for. Their cave dwellings were excellent—cool in summer and warm in winter, dry and sheltered. Any way one looks at it—technically, esthetically, and above all economically—a limestone cave is an immeasurably better abode than anything built by human hands. They had all the food they needed; there is ample evidence that their creative and esthetic sensibilities were given free scope. They were attuned to their environment. Why should they want to change a well-nigh perfect manner of living? (For some details on *built* Paleolithic dwellings, see Paleolithic Houses, page 262).

Then came the rains

Nonetheless, in the end there was a change. For some inscrutable reason the climate changed. The cyclonic storms that for nearly a hundred thousand years had veered southward upon encountering the frigid air masses layered over Europe and released their watery burden over Africa returned to their old track. The ice melted, the glaciers retreated, and on their moraine tailings forests began to grow, nourished by the abundant rains.

That was the end of the good old times. The game vanished; the reindeer followed the retreating ice barrier to-

On this and some following pages are reproduced a few colored woodcuts from the Nürnberg Chronicles, published in 1493. This book is, after the Gutenberg Bible, the most ambitious work to come out of the early printing presses. The author was Hartmann Schedel, town physician in Nürnberg; and the illustrations, numbering some two thousand colored woodcuts, were made by or under the personal supervision of the two greatest artists of southern Germany, Hans Pleydenwurff and Michael Wohlgemuth. The latter had among his pupils a young man by the name of Albrecht Dürer. The Chronicles contain a large number of topographical town views, most of them purely imaginary, but some based on sketches made by Erhard Reuwich who in the 1480s took part in a major expedition to Palestine. His work was published as a pilgrim guide printed in Mainz in 1486, and can therefore be regarded as reasonably authentic. The woodcuts, here taken from the original edition, naturally possess no technological value but are nevertheless of great indirect interest. The above woodcut shows Noah supervising the building of the Ark, naturally in the form of a fifteenth century cog built with contemporary timberman's tools. The idea of the Ark has always fascinated mankind, or at any rate the ark as depicted in the Bible, chiefly for religious reasons, but the great flood catastrophy around 4000 B.C. is of great interest to the technological historian because the recovery of the flooded areas in the Euphrates valley and the rise of Ur into its status as the first metropolis mark the beginnings of technology in the narrow sense.

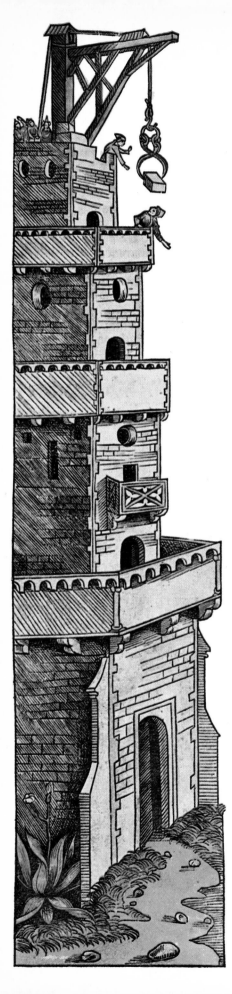

The Tower of Babel as visualized by Wohlgemuth is a strange confection of nonrelated building elements. The crane atop the tower is an interesting detail and so is the lewis tong, used by Greek builders. To the right is a view of Jericho, another artistic fantasy, although certain details of topography and flora suggest that a sketch made by somebody (Erhard Reuwich?) on the spot inspired the artist.

King Matthias of Hungary in his campaign against the Turks in 1462 ran across an armed camp, known as Sabatz in the Chronicles. Aside from the plaited palisade with the sharpened stakes and thorns that recall Neolithic defense, the picture is a concoction of puerile artistic fantasy. The Turkish barracks have become half-timbered houses like those lived in by contemporary wealthy Nürnberg burghers and never found on the Balkan plains. The tall gate leading to the camp is touching in its naïveté.

ward the north; ponies and bison trotted east. The Magdalenian hunters became engulfed in the forest wilderness, and when their descendants subsequently appeared around the gloomy lakes and bogs in Schleswig-Holstein, they were a sorry ragtail lot whose total energy was consumed in keeping alive. With the Magdalenians gone, nothing of original interest was to happen in Europe for the next 10,000 years.

It was elsewhere, in the Mount Carmel region in Palestine, that an interesting pattern began to unfold at about this time. The Natufian hunters, apparently direct descendants of hominids that had lived in the caves on the slopes of Mt. Carmel for over 100,000 years, gradually changed their diet. Their women began to gather the wild grasses still growing in the Galilean hills, the oat and barley and einkorn wheat, and ground their seeds into a meal. Eventually, they found it more convenient to plant the seeds in a small plot near their camp spring.

Because of the fortuities of archeology, it has in recent decades been fashionable to refer to Jericho as the original farming center, the takeoff point for the Neolithic Revolution. Jericho is situated near the Dead Sea some 1,000 feet below sea level and is for most of the year a smoldering pit scorched by the cruel sun. It is—or at any rate used to be—a lethal valley for anything but scorpions, but in Jericho there was and still is a spring of sweet waters flowing from the western hills. This was part of the Natufian hunting territory, and around the spring their ancestors had camped for untold millenia. Here some time in the eighth millenium B.C. (carbon dating gives 7800 B.C.) they established a water shrine, after which their seasonal camp became a permanent settlement, because it was necessary for the women to irrigate their plots of land to bring in a harvest. Their men devised a new tool for them, a reaping knife made of flint microliths set in an ornamented bone. By then they had also invented stone querns on which to grind the grain. Their early flimsy huts have turned into dust at the bottom of the Jericho tell.

But obviously there were other places where women at this time tended their small seeded plots. Indeed, Jericho's claim as the original diffusion center of farming has recently been challenged by another Natufian settlement at Eyhan, also in Palestine, which has been carbon-dated to 8850 B.C. Here flint-bladed reaping knives used to harvest wild or perhaps cultivated cereals have also been found. This gazelle-hunting clan lived in cave shelters and in circular huts 25 feet in diameter and partly sunk in the ground.

At Jarmo, a tell on a spur in the Kurdish hills in the Tigris drainage area, a permanent community engaged in food production appears to have been established some time around 6500 B.C., as determined by carbon dating. The Jarmo farmers lived in multiroomed rectangular houses made of compacted clay, so-called pisé, and sometimes set

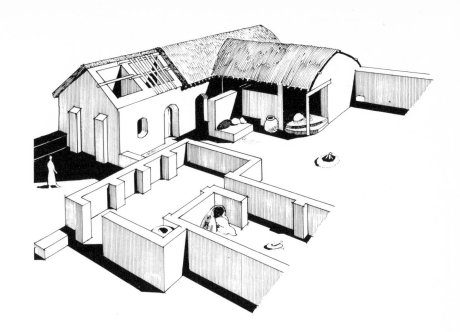

A Neolithic farm of 4500 B.C. might have looked like this sketch of its reconstruction. It was excavated from a tell at Hassuna, on the west bank of the Tigris south of Nineveh. The buildings were constructed of pisé blocks, and the grain was stored in large pots or pits in the ground lined with plaster and insulated with bitumen.

on stone foundations. The houses had basins and baking ovens sunk in the floor. The Jarmo villagers, like those in Jericho, made their household vessels of soft stone. They used the polished stone ax and adz, the characteristic tools of the new age, from which indeed it gets its name. Since bones of sheep, goats, bovines, pigs, and dogs have been found in abundance, it can be assumed that these animals had become domesticated. Each house had its divinity, the Mother Goddess shaped of clay.

At Jericho, Jarmo, Hassuna, and numerous other places in prehistoric economic geography, hitherto of interest only to field archeologists, we encounter the early use of the most common building material ever employed—pisé. It consists of soil, preferably clay, tempered with chopped straw or dung mixed with water and well trodden. When timber is available, the material is rammed into walls between plank shutters, but usually it is made up of rough blocks or molded into bricks and left to dry in the sun. A pisé house will last no more than a couple of generations, after which it disintegrates. Repeated building with pisé on the same building site will result in a mound, or *tell,* which grows into considerable height as time passes. At Jericho the Neolithic levels were 44 feet thick.

In Jericho, after the first rough settlement, a new generation of beehive houses built of sun-baked pisé blocks had evolved, surrounded by a masonry wall and a round tower. Nonetheless, the fortified farming community was

11

not able to withstand attack, and the 12-foot walls were scaled by some hill people, descended from the Natufians, who took over Jericho's wonderful oasis. This second pre-pottery occupation of the town—because that is what it had grown into by now—produced a new type of rectangular house approached through a courtyard enclosed by storage rooms and subsidiary buildings—in other words one resembling the houses still being built in the Near East. In the main house the walls were lined with plaster and painted. The plastered floors were polished to a high sheen and covered with mats of pleated rush. Door frames were made of wood. The material used for walls was sun-baked brick shaped like flattened cigars. It has been estimated that the population of Jericho during the early part of the Seventh

Millenium was around 3,000, but it can be doubted whether all the Jericho families lived in such fine houses.

Although the space standard of "Tahunian" Jericho appears like a dream to the crowded tenants living in the flats provided by contemporary socialistic welfare states, the household amenities were still rather elementary. The bowls and dishes were of soft limestone, the storage vessels of wood and leather. Tools were made of imported obsidian —a volcanic glass—and local flint. The principal cultivation implement was still the digging stick weighted with a perforated stone. The bones of deer and other wild animals and the large number of arrowheads found suggest that the inhabitants had not altogether abandoned their ancient hunting economy.

The diffusion of Neolithic farming

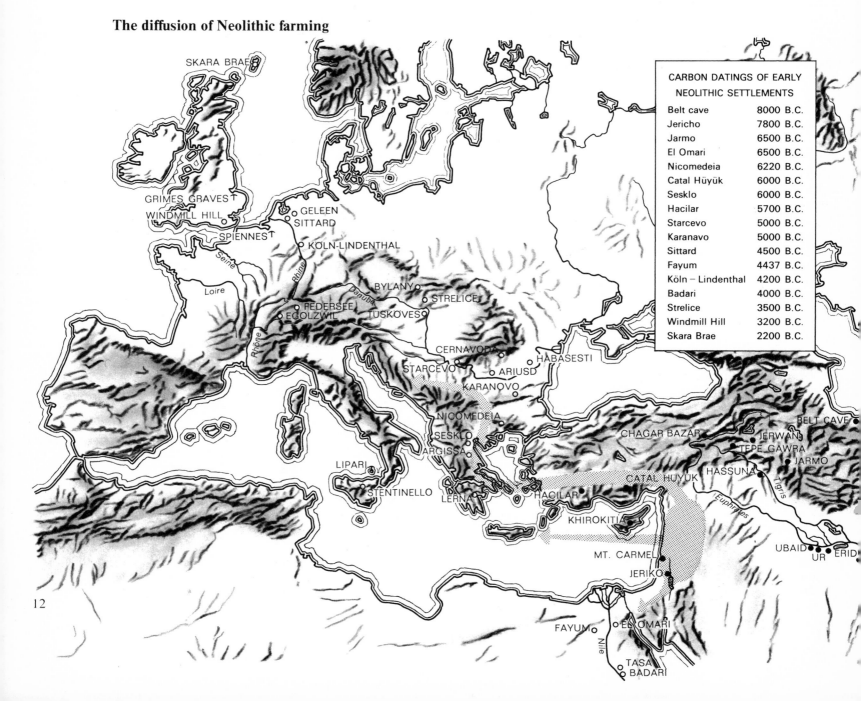

CARBON DATINGS OF EARLY NEOLITHIC SETTLEMENTS	
Belt cave	8000 B.C.
Jericho	7800 B.C.
Jarmo	6500 B.C.
El Omari	6500 B.C.
Nicomedeia	6220 B.C.
Catal Hüyük	6000 B.C.
Sesklo	6000 B.C.
Hacilar	5700 B.C.
Starcevo	5000 B.C.
Karanavo	5000 B.C.
Sittard	4500 B.C.
Fayum	4437 B.C.
Köln – Lindenthal	4200 B.C.
Badari	4000 B.C.
Strelice	3500 B.C.
Windmill Hill	3200 B.C.
Skara Brae	2200 B.C.

The Neolithic Revolution

What had happened at Jericho, Jarmo, the Belt cave, Hassuna, et al.—to wit, the ability to raise a harvest, to have herds of domesticated animals on call—proved to be of revolutionary importance for the further development of building technology, and more besides. The entire transition except the domestication of animals had been the work of women, but now gradually men took over.

Southward, eastward, westward marched the Neolithic Revolution. Let us consider for a moment the westward one. The takeoff point for the spread of the new farming economy appears to have been Syria, but strangely enough it was on such islands as Crete and Sicily, as well as others in the Mediterranean, that, judging from an early type of pottery, the first peasant settlements developed. Whatever means these Neolithic pioneers used to reach these distant islands, they had to bring their seed corn with them, because emmer and the other grain grasses grew wild only in the Near East—Palestine, Syria, Mesopotamia, and Persia. That is the reason why cereal cultivation could begin only in this area.

The overland diffusion via Anatolia—"The Royal Road to the Aegean"—is easier to follow. In Anatolia farming was begun in the early Sixth Millenium, and by 5000 B.C. the new economy flourished in numerous villages throughout Greece and Thessaly. As in Anatolia, the farmers in Greece and Thessaly made a prosperous living on mixed farming of wheats, barley, and millet; they grew vegetables, fruit, and grapes, and raised animals besides—cattle, sheep, goats, and pigs.

From Greece the Neolithic pioneers hacked their way through the Macedonian forests and ran into snow and cold. No doubt the so-called Sesklo people, bearers of the new ideas into sunny Greece, were stopped in their tracks by the chilly blasts from the north. But their ideas were absorbed by the hardy Vardar-Moravian people who became instrumental in bringing farming into Europe.

The further advance along the Danube valley into Belgium and northern Germany can be easily traced. The trek followed a corridor along the lightly forested loess land, excellent for simple farming but rapidly exhausted. After a generation, sometimes less, the light soil ceased to yield, but it did not matter greatly because there was always plenty of land beyond the horizon. So the Danube peasants kept pushing along, leaving exhausted land behind them as they moved into the fertile regions yonder. Their peculiar behavior was watched with incredulity by the scattered Mesolithic hunting tribes inhabiting the area since of old; but the colonization seems to have been carried out without bloodshed because the peasant villages along the Danube valley were left unfortified. No doubt some natives saw the benefits of the new way of life and climbed on the Neolithic bandwagon. By 4220 B.C., as determined by carbon dating, the colonizers had reached western Europe, to the neighborhood of Magdeburg, southwest of Berlin.

Thus, by 4000 B.C. a broad stretch of Europe had been settled by peaceful peasants minding their own business and devoting themselves to growing einkorn and emmer wheat, beans, and lentils in small hand-hoed plots. They kept a few pigs and sheep and cattle and had apparently abandoned hunting and fishing entirely. They appear not to differ much, if at all, from the peasant farmers who are still grubbing for a living in isolated valleys on the European continent.

Communal houses

Naturally, there are some differences between then and now. During the first stage of the agrarian colonization, the Neolithic pioneers built splendid communal houses, rectangular timber structures measuring up to 160 feet in length. The buildings had steeply pitched roofs—the ridgepole was supported by a row of posts and, in order to carry the long rafters, flanked on both sides by two lines of timber columns. Such a house was divided into two parts: one with a raised floor had walls of split logs; the other was less solidly constructed with wattle and daub for the walls. It appears that the timbered one was used for human beings and the other for animals.

Later, the Danubians in the Alpine region abandoned their large communal buildings in favor of a small two-room megaron type of house, which when built on soft ground was carried by a framework of timber raised on a timber platform. Each room had a hearth; in the inner room the hearth was placed near the partition, and in the right-hand corner was a raised bed. The outer room with its clay oven served as the kitchen. In front was a planked terrace. By about 3000 B.C., the layout of the peasant house had become strictly standardized. The time, skill, and labor required to build such a house must have been considerable. It is interesting to note that birch bark was often used to insulate the houses, just as it has continued to be used within recent generations.

It should also be noted that nowhere in Europe has there yet been found a ritual building or a house of this period set apart for a chieftain or ruler. Nor were the villages fortified, at least to begin with.

The large Danubian village at Köln-Lindenthal is probably a typical case of how a community fared during the long peace that reigned in Europe before the arrival of the Indo-Europeans. When first settled, the village was left wholly unprotected; later it was fenced in to keep out animals. Köln-Lindenthal appears to have had 25 "households" at the most, and was resettled seven times, each time for a duration of 10 years, after having been abandoned for an interval of 50 years. It was during the last settlement, ca. 4200 B.C., that the village was surrounded by a tall palisade and a steeply inclined earthern wall resembling a glacis.

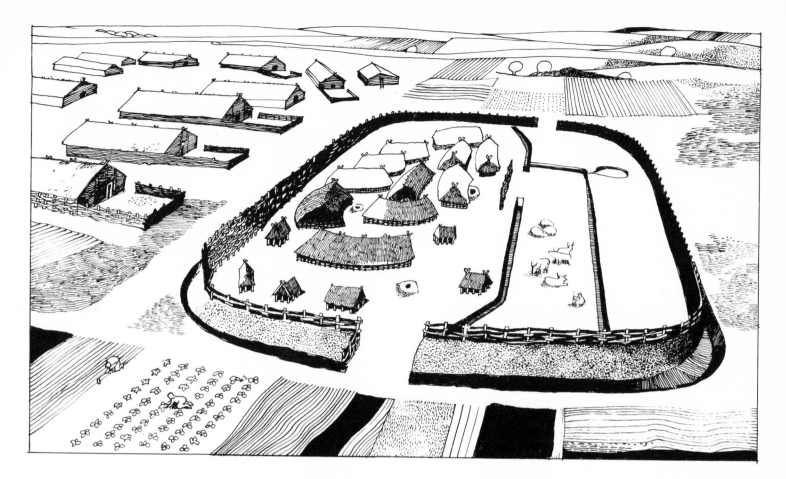

The Neolithic village Köln-Lindenthal (in the neighborhood of Cologne) consisted of 115 × 65-foot-long houses erected with poles stuck in the ground and plaited with withies and smeared with clay. The reconstruction shown here depicts an occupation of the site around 4200 B.C. when the village was enclosed by a plaited palisade and a trench as a protection against animals. During the last occupation, stronger defense works were erected along the perimeter.

The history of Köln-Lindenthal, as revealed by excavations, spans 370 years and accentuates a characteristic feature of Neolithic farming, namely, the rapid exhaustion of the light loess lands in central Europe. After 10 years or so of cultivation, the land had to be abandoned and left fallow for about 50 years before it could be used again.

It has been estimated that, under primitive conditions of farming, an acre of land gave about 7 bushels of consumption grain, and that about 10 bushels was required to maintain a man for a year. These figures suggest that the 100 to 150 villagers of Köln-Lindenthal would have to farm 150 to 200 acres. Had they stayed permanently and cultivated the land by medieval methods, leaving one-half to two-thirds of the fields fallow, they would have to work a farm area of 300 to 600 acres, or 12 to 24 acres per family, clearly an impossible task without ploughs and draft animals.

Hence, the Neolithic peasant had no alternative but to exhaust the soil and move on. This method was employed in northern Europe until a few hundred years ago and is proceeding apace in the Congo today. A settler puts fire to the forest and keeps taking crops from the burned-off land until the plant nutrients contained in the ashes are used up, whereupon he moves on and puts the torch to another tract of timber.

It can be clearly seen that under such primitive farming conditions there will never be a surplus; the best that can be expected is to salvage enough seed and corn to maintain the cycle. There will never be enough to fuel a population increase, and since the labor invested is wasted in getting only a few crops, there will never be accumulated energy for breaking out of the closed circuit and taking off on a spiral leading to technological developments and all that follows in their train. No civilization ever spired in the ashes of burned-off land.

The furthest outpost—Skara Brae

The revolution reached Britain somewhat before 3000 B.C. when the Windmill Hill farmers shipped over the Channel in boats carrying their livestock, hoes, and seed corn. These farmers arrived along the southern route, via Spain, up the Rhone valley, and along the roundabout way of the coast. Their encounter with the natives was apparently not altogether peaceful because they had to fortify their hilltop villages from the outset. The newcomers grew bread wheat and barley, and kept sheep, goats, and pigs. But they favored cattle and bred a longhorn stock, probably with the aid of the local aurochs.

This Neolithic train diagonally across the continent of Europe can be explained and understood, but it is not easy to understand what drove these peasants further along to the utmost reaches of their long trek, to the Orkney Islands. Why did they not return to more hospitable regions when they encountered these treeless, storm-swept, inhospitable islands in the northern sea?

In any case, to the Orkneys they came, and rather fortunately as it turned out. At Skara Brae they established their truly Neolithic economy—although by now it had become obsolete around the Mediterranean, and built a village which, considering the formidable difficulties, is one of the most admirable ever constructed in the north, in its accommodation to a vicious nature.

The 20 × 18-foot Skara Brae houses were built of stone, on a square plan with rounded corners. The roofs were corbeled, and the opening in the center appears to have been closed with rafters of whalebone. The door, also of stone, was 4 feet high and pivoted on a stone bearing; and in the center of the hut was a square hearth on which peat was burned. The village contained seven such huts connected by a roofed-over alley, and under them ran a stone-lined sewer. The little community was completely sheltered from

A sketch of the interior of a 20 × 18-foot Skara Brae house on the Orkneys as it remains today. In the rear is a bed, and to the right a two-shelved dresser, both of them made of stone. In the foreground is a water or food storage pit sunk in the floor. The Skara Brae village consisted of seven such houses connected by a roofed-over alley and served by a stone-lined sewer.

the raging Atlantic gales, and the inhabitants did not even have to leave their sheltered security to use the privy, because each house had one, directly joining it, and like the living quarters buried under sand, refuse, and ash.

But the wonderful thing about Skara Brae is the furniture, which was made of readily split stone, since the islands were, and still are, lacking in timber. So here, at Skara Brae, we can learn what cannot be found out elsewhere, namely, how the houses of Neolithic peasants were furnished. Everywhere else, the furniture has turned to wooden dust; here it remains.

On one side of the hut were beds made of stone slabs, each with stone posts carrying a canopy of skins, and a mattress of heather. Just above the bed were shelves fastened to the walls; and against the rear wall was a two-shelved dresser, also made of stone. Water and/or provisions were stored in rectangular containers sunk in the floor and lined with stone slabs with finely luted joints.

Except for the luxury of the canopied bed and the furs, perhaps also the dresser and the ready convenience of stores sunk in the floor, this is the way small peasant farmers have lived until now in isolated mountain valleys. Indeed, life in Europe some 5,000 years ago no longer holds any particular mysteries.

The gray grind of European farmers

For thousands of years life went on as usual. The European peasant got stuck in a circular groove of hoeing, planting, harvesting, threshing, tending his animals. He and his family were self-sufficient, or nearly so. Now that the diet was based mainly on cereals, he had a need for salt that had to be brought to him, he did not know quite from where. In numerous settlements not favored by local deposits of suitable rock for toolmaking, the peasant's need for tool stone had also to be supplied from somewhere.

There were individuals, or groups of individuals, living in localities with convenient access to these desirable raw materials: lava for querns, green schist, flint nodules, glassy hard obsidian. They worked the surface deposits, and when these gave out, they followed the outcrops into the ground. It was found that flint nodules mined underground were technically superior to the weathered stone aboveground; they were sounder and behaved better when worked. The need for tools inspired the local peasants in favored localities in France, Belgium, England, Denmark, Poland, and Bohemia to spend their idle time mining tool rock. They became adept at working this up into ax and knife blanks at the mine and traded the tools over long distances.

The simple laws of economics directed the ensuing developments. Specialist flint miners, like those in Grimes Graves in eastern England or at Spiennes in Belgium, contented themselves with mining the nodules from ever more complex underground workings and sold the desirable

The sketch shows the reconstruction of a village with 38 one-family houses excavated at Aichbül, on the shore of the Federsee, a one-time lake that has turned into a bog, in Würtemberg. It was occupied during the third and last Neolithic period in Europe and has been dated to ca. 1100 B.C. Here the houses were erected with standing timber or with wattle and daub. All houses had steeply pitched roofs. Futher investigation is likely to turn it into an early Celtic village.

stones to fabricators who roughed them out to blanks by a series of expert blows. They sold their semimanufactured goods to peddlers, who took them on the road to trade to villagers, who ground and polished them into finished tools. But there is also ample evidence of ready-made tools turned out by specialist manufacturers.

So the gray Neolithic life in continental Europe had its brief glimpses of excitement with the arrival, at the end of the harvest, of the salt and flint peddlers. Along the coasts a few daring souls braved the sea for fishing or trade in coracles made of skins stretched over a flimsy framework of wood. But with these exceptions, not much of interest to a technological historian occurred in Europe during these dull millenia. Civilization as we know it did not begin in this peaceful but gray world of European clodhoppers. It began elsewhere, in the riverine valleys and deltas of the Nile and Euphrates, the Paradise of the Bible, where the hot sun and the fertile slime combined to accelerate the economic, technical, and also the political developments, where—to put it poetically—"the woman saw that the tree *was* good, and that it *was* pleasant to the eyes and a tree to be desired to make one wise . . ."—where exciting but not always pleasant things began to happen. (For the arrival of Indo-Europeans, see note on European Prehistory, page 265.)

The Garden of Eden settled

Here, as elsewhere, the Neolithic Revolution began in a slow and unobtrusive way, although quite differently in the Tigris-Euphrates valley from in the Nile valley. Whereas previously the silt carried by the Tigris and Euphrates had been borne off by the currents in the Persian Gulf, now it began to accumulate and formed a flat, topographically uninteresting but tremendously fertile delta. The channels and lagoons swarmed with fish, wild fowl, and game; date palms grew wild on the occasional elevations.

To this newly created and no doubt unhealthy paradise —the Eden of the Bible—some time in the Fifth Millenium, came migrants from highland farm centers to the east or north. One of the settlements of these proto-Sumerians gradually grew into a tell, the one of al Ubaid, from which the settlers have been called the Ubaid people. Owing to local circumstances, from the beginning they had to unite their individual labor to master the formidable difficulties confronting them. To begin with they had to drain the marshes; and later develop an elaborate system of perennial irrigation, with all that is entailed in the construction and maintenance of dams, sluices, canals, and ditches—an eternal labor on a huge scale, far beyond the capacity of one family, or clan, or even tribe. The recovered soil gave a tremendous surplus. When Herodotus visited this area some 3,500 years later, he reported that the Euphrates delta gave two-hundredfold or more of wheat. Unlike the loess lands of Europe, which became exhausted after a couple of decades, the Euphrates valley and delta continued to yield large harvests. Year after year, generation after generation, century after century, millenium after millenium, the Tigris and Euphrates descended into the lowlands, choked with fertile sediments (five times more than the Nile) of clays and sands, of phosphates and organic slime. There was a fortune to be made in this hothouse where each harvest brought a huge surplus, but it was assuredly not one suited for rugged individuals. Here everybody had to cooperate to produce wealth, to provide the surpluses needed to convert luxuries into necessities. Or to put it in a nicer way

"Behold, now everything on earth rejoined . . .
The fields produced much grain, the harvest
of the palm groves and vineyard were fruitful,
It was heaped up in granaries and mounds . . ."

The wheat heaped high, the surplus of the early Sumerian farming effort. What happened to it? Well, of course, there were the priests, waiting to take their share. The Ubaids were a religious people and regarded their fertile land as a blessing from the gods. Their hunting ancestors had been accustomed to sacrifice the first kill of the season as a first-fruit offering, and the idea of making contributions to the gods for their good fortune came naturally to Sumerians and other prosperous people.

The early surplus went into the erection of a shrine and a suitable building to enclose it. At Eridu in the south and at Tepe Gawra in the north they erected rectángular sanctuaries, each provided with an altar and offering table. The temple was raised on a platform and built of prismatic mud-brick walls, heavily buttressed. But such a ceremonial building required experts in ritual; indeed the initiative for the building and the mobilization of labor required for the work, as well as the continuous maintenance of the shrine, needed an establishment possessing more power than the old-fashioned shaman of the hunting peoples. These early temples of the Ubaid people were the work of priests, whose religious duties also included the collection of taxes. And to keep track of the tax collections some method of recording was required, which gradually led to the development of writing. The priestly scribes freed from the necessity of grubbing for a living could devote their time to intellectual pursuits, such as the early development of the sciences: arithmetic, geometry, calendrical astronomy, and so on, all of which found practical application in construction. But that came later. Before the Sumerian temple state reached its summit of civilization and political power, the Ubaid people led by their priests kept replacing their Eridu temples with larger and more elaborate structures as the pisé buildings collapsed, always taking care to keep the original site of

the shrine intact. In time, not less than 16 temples had been raised in succession on the debris of the preceding ones, until the temple mound rose 50 feet above the surrounding flatlands.

Then came the Flood. The Tigris, always wild and unpredictable in its behavior, suddenly rose; and a curtain of swirling water inundated the lowlands of the delta and swept everything before it. Of the many small agricultural towns nestling around their local temple mounds, only one survived, that of Ur; but here too the waters lapped against the foundations of the temple atop the mound.

We are not here concerned with the details of Sumerian history. Suffice it to say that the land left desolate by the Flood was settled by the Uruk people residing higher up the valley, and the newcomers carried on and brought to fruition such revolutionary innovations as writing, the invention of the wheel, and the use of metals.

Origin of building elements

Of greater interest to an inquirer into ancient methods of construction is the building heritage left by the original settlers of the reedy Euphrates-Tigris delta. Long before their labors were swept away by the Flood, they had invented and perfected a method of house building that is still being used by the marsh Arabs living in the area. The material employed was reeds, the strong 30-foot-tall reeds growing in the delta. The settlers formed the reeds into thick round bundles, or fascines, which they placed as uprights spaced equidistant from one another in a straight line. To these vertical members they lashed two, three, or more thinner bundles of reeds horizontally, whereby they obtained a sturdy framework for a wall. To complete the wall, the spaces between the vertical and level fascines were filled in with reed matting. By placing two such walls parallel to each other and some distance apart, the long walls of a house were obtained. The two gables were constructed in a similar manner, with upright and level reed fascines.

The final phase in building a Sumerian reed house involved the construction of the roof. This was done in the simplest fashion: the tall vertical fascines were bent inward and lashed together where they met. Horizontal reed bundles were tied to these rafters, and matting was placed over the roof frame. By bending the tall reeds the roof could be formed either into a semicircular vault or, by not stressing the reeds too much, into a pointed arch.

The Sumerian reed house constructed in this manner gave satisfactory shelter while it lasted. But it was a drafty house when the wind rose, and the next step was to plaster the roof and walls with mud. Also, in this more advanced state, the house had a serious defect—its fire hazard. An ember from the fire might ignite the dry reeds, whereupon the house burst into flame instantly. However, when the house was wholly plastered, it frequently happened that the shell

17

In the Euphrates delta a strong and tall reed is still being used as a building material. (There is a project afoot to turn it into paper pulp.) The reeds are bunched into fascines for use as uprights, to which are lashed thinner bunches of reeds horizontally, whereby a sturdy framework is obtained. The spaces between the structural members are filled with reed matting. The roof is constructed by bending the tall vertical fascines inward and lashing them together where they meet. They can be bent to a circular or a pointed arc, as demonstrated on a pre-Flood water trough from about 4000 B.C. and shown on the next page. The reed house was usually plastered with mud. This house, which contains all the structural members required for building a Romanesque or Gothic cathedral, appears from Sumerian legends to have been fully developed long before the Flood washed away an unknown number of them.

remained intact after the combustion of the reed members. From such common experiences, it is believed, the idea developed of building houses of mud alone, without a framework of reeds, and thus the *terre pisé* houses of the ancient world developed.

Be that as it may, it is worth noting that in this pre-Flood reed house are found all the major elements of ancient and modern buildings: the large vertical reed fascines carrying the houses are of course equivalent to columns, or (perhaps more technically accurate) buttresses; the horizontal reed bundles are equivalent to beams and/or girders; the matting to wall paneling; the bent-over roof fascines are the origins of the ribs required to construct an arch, semicircular or pointed. Thus, long before major construction in stone or brick began, the structural elements required had been invented and developed into a satisfactory state of technology. These developments took place before 4000 B.C., when the Neolithic Revolution had run its course in the Euphrates valley.

Early happenings along the Nile

It remains now for us to discuss briefly some of the early developments in a riverine area destined by nature to become another major center of civilization, namely the Nile delta and the narrow river valley stretching linearly to the south and hemmed in by desert.

The change in climate which brought about an end to the last glaciation in Europe, and as a consequence straightened the tracks of the Atlantic cyclones and furthered forest growth on that continent, had an altogether different effect farther south. The grassy highlands on both sides of the Nile, which like the central African plateau previously had

teemed with steppe animals were no longer watered by rain and began to desiccate. The water table sank and the wadis emptying into the valley, hitherto crowded with vegetation and animal life, also felt the inroads of the desert. On the other hand, the formerly swampy and uninhabitable river valley began to dry up and invited settlement by early farmers. In the north, the reedy and swampy delta also reached a state inviting human habitation, at least along the edges.

Among the first to arrive were, not unexpectedly, Natufian colonizers from the Palestine region. A Natufian settlement has been recognized as having existed near Helwan. There, at El Omari, one of the graves investigated contained a carved wooden baton resembling the Ames scepter, later the insignia of Lower Egypt kingship. From this find the conclusion has been drawn that these early Neolithic colonizers were ruled by chiefs.

But the best-known site, whether settled by Natufians from the east or by Libyans from the west, is the Fayum depression to the south of the delta and to the west of the river. Here, around the lake, the early Neolithic farmers grew emmer, wheat, and barley; kept domesticated bovines, goats, and pigs; grew flax and wove it into linen; made good coiled pottery and excellent baskets. Their reaping knives were straight like those of the Natufian, except that the hafts were made of wood. They used ground and polished stone axes and hunted hippotami with them; their barbed fish spears were Natufian in form. They lived in flimsy huts of which nothing remains except sunken hearths and a few grain storage pits originally lined with matting. They appear to have begun farming during the second half of the Fifth Millenium, or at any rate the carbon datings of the two oldest grain pits give 4145 ± 250 B.C. and 4437 ± 180 B.C.

The Fayumese were a great deal more sophisticated than the Moravian peasants on the loess lands along the Danube. They painted their eyes and ground the malachite pigment on rectangular palettes of alabaster. They decked themselves with shells imported from distant shores, from the Red Sea and Indian Ocean. They differed from the Natufian farmers in Palestine by burying their dead in cemeteries outside the villages; having their ancestors hanging around the house did not appeal to them.

Except for the care taken to orient the dead, there is nothing in these early Egyptian farming settlements to indicate or forecast the elaborate funeral rites and monuments that came to characterize the historical Egyptian civilization. These obviously derived from another source; indeed, the economic, social, religious, and political forces shaping the ancient Egyptian civilization, as we know it, derived not from the delta but from the south, from Upper Egypt.

The Neolithic Revolution as it was played out at Badari (some 20 miles upstream from present Asyut) had considerably more lasting effects on human history. On a desert spur at Tasa have been dug out what are still regarded as the oldest farming communities in the Nile valley. The African peasants living here in flimsy shelters grew emmer and barley and kept sheep and goats. They ground the grain in large saddle querns, made rough pottery, and loomed textiles. They wore ivory beads and bangles, painted their eyes, and when they died they were wrapped in skins and enclosed in straw coffins.

Their descendants, the Badarians, who occupied a much larger territory stretching southward, continued to hunt with bow and boomerang and fished with the same shell hooks as the Tasians, but they also bred cattle and stored their wheat in mud-lined huts. Apparently, they did not need to irrigate their plots. They were satisfied to live in simple huts made of reed matting, but otherwise they wanted more out of life than their Tasian ancestors.

From all accounts, the Badarians carried on an extensive foreign trade. Besides the usual seashells, they took home malachite from Sinai, cedar and juniper wood from Syria. They also navigated the Nile in simple boats made of papyrus bundles lashed together.

In their burial rites the Badarians began to show a rough pattern which was later to be developed to its bloated perfection. Their dead were wrapped in skins like their ancestors and buried in pits, with pots containing food and drink, and occasionally accompanied by female statuettes of ivory or clay. Cattle and sheep were also ceremoniously interred with them.

From then on the economic pace accelerated. The Amratians, apparently direct descendants of the Tasians and Badarians, began an extensive cultivation of the flood plain of the Upper Nile, kept cattle for dairying, used the ass as a beast of burden, and developed the high-prowed Nile boats so well known in later times. Their flint tools reached a state of perfection; their carved ivory combs, ladles, and stone vases were beautifully designed and skillfully executed; the ubiquitous pigment palettes were given the shapes of animals, such as fishes or hippopotami. Pictures of animals, crocodiles, scorpions, and so on, suggest that the permanently settled villages along the river were occupied by totem clans.

The Amratians made use of native gold and copper which they hammered into personal adornments, but copper was also turned into thin sheets from which harpoon heads and pins were made. The Amratian cemeteries contained a few thousand graves, and the pits were much more richly furnished than those of previous clans. The dead were provided with the usual tools and weapons, ornaments and pots, but also with models of women and servants. On the pots and slate plaques buried with the dead were inscribed pictures of their worldly possessions.

By about 4000 B.C., then, the people living in Upper and Lower Egypt stood on the threshold of history and the Age of Metals. Gone forever was the free but hazardous life of the hunter and the peaceful unpretentious existence of the Neolithic farmer and husbandman; about to go was the Mother Goddess, perhaps not always so gentle in the personification of a sharp-tongued maternal grandmother but nevertheless an expression of the enduring life-giving force of man and beast, "of every foule of the ayre, and to every thing that creepeth upon the earth . . ." Gone, too, was communism in its true sense, because except for a mild acquisitiveness finding expression in personal adornments and the obvious joy in the possession of a fine pot, all the means of production and, in Europe at least, the houses also, were property shared equally by the members of a clan. From now on man began to take over, to shape the world in his own image. It may be worth recalling these aspects of communal life when we go to explore the man-made structures to which the following chapters will be devoted.

A water trough from Mesopotamia, now in the British Museum. It has been dated to ca. 4000 B.C.

19

River Control and Irrigation

In the previous chapter, mention was made of the fertile burden of clays, sand, and organic slime carried by the Euphrates and Tigris which kept the lowlands in a permanent state of high production. But these natural blessings were not obtained without a huge input of human labor, and the city states of Sumer and their imperialistic heirs were, from an early date, organized to deal with the problems arising from the behavior of the rivers.

Unlike the Nile, these two rivers rise without warning. The Tigris usually starts off in early March, followed by the Euphrates a week later. Sometimes they rise simultaneously, and when that happens the lowlands are flooded; one such prehistoric inundation which put an end to the original settlers of the delta still echoes in the Bible.

Some system of flood control, by means of dikes, was therefore required from early times, because another serious difficulty was the time and the duration of the inundation period. After rising in spring, the waters receded in June, leaving the ground parched under a burning sun which raised, the average summer temperature to 120°F. A third and, in the long term, even more serious trouble was the salt carried by the rivers; since gradual salination of the soil would have made it useless for cropping.

This bundle of problems presented by the twin rivers had to be solved in order to keep the lowlands fertile. The solution was found in a system of perennial irrigation, incorporating structures for flood control and water storage, i.e., dikes and dams, as well as canals and ditches, to conduct water over the fields in such a manner that the ground would not absorb too much of the salt carried by the water.

Method of perennial irrigation

It is no longer possible to trace even in rough outline the early Sumerian irrigation works. Canals and ditches have a habit of silting up if not maintained; dikes and earth-filled dams are swept away or simply disintegrate. Moreover, since the days of the early Ubaid settlers, the Euphrates has changed its course repeatedly, through natural causes, deliberate action by men in times of war, or sheer neglect.

Nevertheless, it is possible to visualize the early means used for the control and utilization of the water. To live in the inundated valley it was necessary to surround the village communities with dikes, but as the central authority grew in power, it eventually became possible to contain the river itself between embankments.

In this manner the bed of the Euphrates was raised above the plains, and it kept on rising with accumulating silt deposits. Thus the source of the irrigation water was at a considerably higher level than the fields. Although this difference in elevation facilitated gravity distribution of the water, the dikes were a constant source of worry and had to be maintained at all times to prevent a catastrophe. From the river, lateral canals were dug to conduct the water to the area farmed by the temple city. These in turn were joined by a number of feeder canals supplying a network of ditches surrounding the fields.

By breaking the ridge wall of the irrigation ditch, water was admitted to cover a small plot of land to a height of a few inches. The land was not permitted to soak up all the water for fear of salinization, and after a short period of inundation the surplus was drained off and fresh water led to the next plot until in the end the whole field had been watered.

Each irrigation canal had its name, sometimes after the man who bossed the work of digging it, sometimes after the village that was responsible for its upkeep. The canals were also transport lanes for barges which brought the produce into the growing towns and cities.

In this way the peasants of Sumer and their successors succeeded in raising several harvests a year, which accounts for Herodotus' statement of twohundredfold yields. Theophrastus (ca. 287 B. C.) speaks of hundredfold harvests, which seem more realistic. The Babylonians sowed their seedcorn thickly, about $\frac{4}{5}$ bushel to the acre, and obtained a return of 30 bushels per acre. The perennial irrigation permitted two or even three such harvests a year.

But it follows from the above outline that the maintenance alone of such an extensive hydraulic system required an immense input of labor. The canals and ditches silted up fast; the dikes and barrages eroded quickly and had to be raised and strengthened. It was also inevitable that as the works along the river became more extensive, friction arose about water rights between neighboring city-states, which kept them in a latent state of hostility that frequently flared up into open war. There is still a word for this ancient strife over water rights: *rivals* and *rivalry* were first applied to holders of water rights along a river or irrigation channel.

Hammurabi and corvée labor

Long before the enforced unification of downriver Sumer with upriver Akkad was accomplished, the old rules of water rights had been codified into law, and when in 1760 B.C. the entire Tigris valley as well as the Euphrates valley as far as Mari was conquered by Hammurabi, the whole river system was brought under central control and administered under one law. Hammurabi's Code was concerned, not unexpectedly, to a large extent with irrigation. Each farmer was responsible for keeping his dikes and ditches in good repair, and in default thereof, he was charged with any damage to the neighboring land caused by his faulty dikes.

The major hydraulic works were maintained by the central administration in Babylon, from where a stream of orders and directives was issued on cuneiform clay tablets to the local governors. One such directive to the governor of Larsa has survived: "Summon the people holding fields to the west of the Danau Canal to free it from silt. Work to be finished before next month."

By this time the logistics in canal and dike construction were fully mastered and applied in routine fashion. The governor of Larsa would know the volume of dirt that could be lifted by one man in a day and the number of man-days needed to carry out the order. Tablets with definite instructions—"work orders"—were sent out to the peasants subjected to corvée and living to the west of the Danau Canal. Each peasant would receive a properly signed and dated clay tablet summoning him to perform a specified task and stating the duration of the job. He could protest and become exempted upon the payment of a certain sum of money. Hence, under the Code, corvée labor was regarded as a tax-in-kind that could be converted to cash payment.

Under a powerful central government ruthlessly directing the resources of a large wealth-producing area, much larger public works could be undertaken than by the small city-states. In Babylonian times, the area between the lower Tigris and Euphrates was crisscrossed with canals which irrigated the land enclosed by them, as well as serving as efficient transport lanes. The largest and most famous of them was the 400-foot-wide Nahrwan Canal which left the Tigris somewhere between Tekrit and Samarra and ran for a distance of 200 miles along the right bank of the river to irrigate the country to the west of it.

Considering the nature of the Tigris, the construction of this large canal involved ancillary works on a huge scale. There must have been a large barrage thrown across the Tigris backing up and storing the wild floodwaters; there were regulating sluices, and at least in two places the canal was lifted on superstructures above torrential rivers. Upon completion of the works detailed centrally issued directives were required to maintain maximum and minimum water

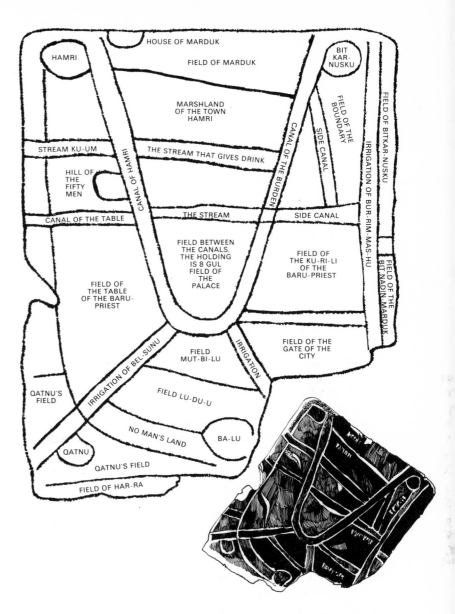

Map of an Assyrian irrigation area in the neighborhood of Nippur, drawn from a cuneiform clay tablet dated ca. 1300 B.C. (lower right). The area belonged to the town of Hamri and obtained its irrigation water from the parabolically shaped "Hamri Canal" which enclosed the Hamri reed common from which reed was taken for the building of houses and construction of dikes.

levels in order to keep the canal viable for both irrigation and navigation. Unfortunately, little is known today of this magnificent canal.

Methods of waterproofing

The civil engineering know-how of the Babylonians and their Assyrian heirs left nothing to be desired. The "Hanging

21

Gardens" in the royal palace of Nebuchadnezzar caught the fancy of Greek travelers, which is the reason why they became famous in history. They consisted of vaults carrying a platform over a cellar containing a well worked by wheels which raised the water to the gardens planted above. The deck was carefully waterproofed by means of alternate layers of bitumen and reed mats, stone slabs, brick and mortar, plus sheet lead. On this was a thick layer of earth on which trees and shrubs were planted.

In waterproofing the structures, the Mesopotamian builders from an early date made liberal use of bitumen which was native to the area. The Greek word naphtha derives in fact from the Babylonian *napu,* a verb meaning to "flare up," to "blaze." The heavy fractions of napu, such as the bitumen, were used in building construction, and deliveries of as much as 265,000 pounds of bitumen are known from 2400 B.C. This particular shipment derived from Magda, but later the chief source of supply was Hit, on the Euphrates 200 miles north of Babylon.

The Hit bitumen derived from seepage and was free from adulterants. It was sold by volume or weight, and the price of $3\frac{1}{2}$ shekels of silver per 2,200 pounds was maintained for several centuries. Related to grain or cattle, the ancient Babylonian price for bitumen was just about the same as the price for the modern refinery product.

The bitumen was applied as a mastic; i.e., it was melted and mixed with fillers consisting of sand and a fibrous material. The mastic was shaped into cakes in molding baskets holding a standard volume. In the early days of Ur, the price of the finished mastic was 21 shekels per 2,200 pounds; subsequently it dropped to 18 shekels, and this is the price usually quoted in the many building contracts that survive.

The mastic varied in composition according to its end uses: there was masonry mastic, floor and wall mastic, and a special irrigation mastic used to coat hydraulic structures as well as barges.

The mastic that has survived in ancient buildings contains 25 to 35 percent bitumen, the rest being made up of loam and lime, or quarry dust, chopped reeds, and straw. The fibrous matter amounts to about 15 percent by weight. The builders used tables stating the weight of bitumen needed to compound a 25 percent mastic for a given floor area, with a standard thickness of 2 cm. It was sparingly employed in walls, seldom in pisé walls; but in temple and palace buildings erected with kiln-fired brick, mastic was used as a mortar. As such it has never been improved upon, because it is now virtually impossible to separate two ancient bricks bonded with mastic. Other uses included waterproofing the brickwork of drains and sewers, dikes, embankments, quay walls, bridges, canal bottoms, and processional roads.

The use of bitumen as a building material ceased with the ancient builders in the Near East. Greek and Roman builders either did not know about it or ignored it in favor of waterproofing methods that were far inferior.

Came the Assyrians

To the corvée-subjected farmers of Mesopotamia neither time nor politics brought any perceptible change in their off-season labor on dikes and ditches. Nature rather than politics dictated their fate: kings and dynasties came and went, the long chalcolithic period ended, the Bronze Age had its brief flurry in the valley, to be succeeded by the Age of Iron. Then they became subject to a new set of rulers, the Assyrians, the first empire builders whose clanging war engines were heard over the entire Fertile Crescent, and whose iron-clad battering rams knocked over any mud-brick wall in their way. Although this northern hill people was chiefly occupied in carrying war to all corners of the ancient world, there were peaceful interludes between the expeditions when the Assyrian kings devoted themselves to such pursuits as palace building and canal construction.

The Assyrians did not contribute anything to the building technology of the subject people but were content to carry on their traditions. They extended the irrigation schemes into their home country and initiated vast irrigation systems, always accompanied by the planting of trees and shrubs that had caught their fancy during campaigns in distant lands. Tiglath-pileser (1115–1102 B.C.), the first Assyrian king to reach the Mediterranean, brought home with him 42 varieties of fruit trees and shrubs, as well as cedar, sycamore, and other timber trees, and tried to acclimate them in his gardens and parks.

But the Assyrians' principal hobby while resting between campaigns was canal construction, and Nineveh, their capital built at the confluence of the Tigris with its small tributary the Khosr, was in the end provided with 18 canals bringing fresh water to the capital. But again, little is actually known about their construction, except for the canal built by Sennacherib (705–681 B.C.), son of Sargon II. This 65-foot-wide canal, which ran a distance of 50 miles from Bavian in the northern foothills to the capital, was built in the surprisingly short time of one year and three months. Of this tremendous works there was not a trace until 1932, when at former Jerwan in Iraq a Danish archeologist found the remains of an old man sitting on a stone bearing an Assyrian inscription that was deciphered as "bridge." Subsequent excavations uncovered both the aqueduct and the canal.

The "bridge" at Jerwan was found to have consisted of a stone aqueduct which was 900 feet long, 70 feet wide, and 23 feet high. It was carried over a steep ravine on corbeled arches, and about two million stone blocks each measuring $20 \times 20 \times 20$ inches were used in its construction. They were quarried in the Bavian hills but dressed on the site.

Despite the short time taken to finish the work, there was assuredly nothing slipshod about the construction of the Bavian canal. The bottom consisted of a 16-inch bed of concrete poured on a 1-inch bed of mastic and carrying a stone pavement with fine joints. The bottom had a gradient

At Jerwan in the eastern part of present Iraq the Assyrian king Sen-nacherib had constructed a 920-foot-long and 66-foot-wide viaduct spanning a valley which lay across his canal from Bavian to Nineveh.

A river was taken under the viaduct through pointed corbeled vaults. The 50-mile-long Bavian canal was constructed in one year and three months according to inscriptions in the rock close by the source.

of 1:800, and the concrete used contained one part lime, two parts sand, and four parts limestone aggregate.

However, all these ancient achievements in hydraulic construction were, not surprisingly, surpassed by King Nebuchadnezzar (604–561 B.C.) with the building of a high dam 5 *beru* (16 miles) in length, joining the bank of the Tigris with that of the Euphrates. On the huge lake thus formed, the waves rose high "like on the sea." From the King's own description, it was an earth-filled dam lined with kiln-fired brick set in mastic, and the chief reason for the works was to provide the capital with a system of water defenses.

This was civil engineering on a grand scale, not even approached later by the Romans, and it was to take something like 2,600 years before anything like it would be undertaken again.

The end of the ancient construction

One may wonder what became of this huge and technologically advanced hydraulic civilization in Mesopotamia. Of course, it would not take long before the canals if left unattended would choke with silt and the embankments crumble, permitting the water to flood the land and turn it into a reedy swamp. Such a system has to be maintained at all times and requires a tremendous input of labor to repair and keep repairing the eroding effects of running water. There were times in the past when barbaric people descended into the lowlands and in their ignorance temporarily permitted the structures to fall into disuse. But from about 200 B.C. onward it appears that the political rulers of the valley, regardless of their origins and policies, understood the necessity of keeping the system intact, and devoted

their peacetime energies to expanding it. Under Persian, Hellenic, Roman, and Arab hegemony, the wealth-producing irrigation system was kept in efficient order for the duration of more than 1,700 years. Then in 1258 A.D. the valley was invaded by a Mongol army commanded by Hulagu, younger brother of Kublai Khan. He burned Bagdad to the ground and in scourging the land he had his cavalry ride along the soft canal banks.

When he left, the caliphs were no longer able to restore the crumbled banks, the entire system silted up, and a few centuries later nothing remained of the hydraulic civilization of the ancient valley. Its back was broken, and not even its tremendous subterranean wealth of *napu,* when it at long last began to be exploited, proved adequate to repair the damage. A rehabilitation scheme, financed by the surplus produced by the oil was indeed put through, the Tigris was again being dammed as a necessary preliminary, but the mental resources of the people had also become desiccated, and the attempt aborted in a bloody revolution that brought on further chaos. Civilization is a consuming flame that devours both natural and human resources. A region that has carried a civilization for a few thousand years ends in a wasteland beyond recovery.

Basin irrigation in Egypt

In Egypt, the regular behavior of the Nile directed the development of an irrigation technique along different lines. To the peasants living in the Nile valley the approaching flood was announced by the river itself. At the beginning of July the water turned green owing to its burden of vegetable scum picked up by the White Nile during its sluggish passage through the swamps in southern Sudan. Two weeks later the river turned red with its burden of unorganic silt carried by the Blue Nile from the Ethiopian highlands, and the water began to rise. It kept on rising until by September the arable land was inundated to a height of 6 feet, sometimes more. The depths of the annual flood varied along the 605 miles of the river within Egypt. In ancient times, a favorable inundation measured 28 royal cubits at Elephantine, 24 at Edfu, 16 at Memphis, and 6 in the delta. But the river did not always behave according to the rules, and occasionally there occurred severe water shortages or catastrophic floods. After reaching its maximum level, the water rapidly subsided and by the end of October the river returned to within its banks.

Before the Aswan Dam was completed in 1902, the alluvial burden of the Nile was estimated at about 100 million tons per year. An analysis of the solids suspended in the Nile water at the turn of the century has been reported as follows, in percent:

Organic material	15.02	Sodium	0.91
Phosphates	1.78	Clays	20.92
Calcium	2.06	Sand	55.09
Magnesium	1.12	Carbonates and	
Potash	1.82	other solids	1.78
		Total	100.00

There exists no estimate of the land area covered by inundation and hence benefiting from this annual accretion of fertile deposits that the kind river not only brought along but also distributed uniformly over the narrow valley between the escarpments. However, an almost unbroken series of Nile records since the Arab conquest shows that these deposits have raised the floor of the valley by about 8 feet during the past 1,300 years. This figure suggests that in the Old Kingdom the level was about 20 feet below the present valley floor.

With basin irrigation, only one crop per year could be raised; and since only gravity distribution of water could be used, beyond marginal narrow strips that could be watered by means of such simple lifting devices as *shadufs* (or balanced scoops), the topography of the valley put an upper limit to the area that could be irrigated and hence cultivated. This, in turn, put a limit to the population that the arable land could support. It appears that through the long history of the valley, one acre of basin-irrigated land could support one man in an adequate manner, which suggests that the maximum population of Egypt could never exceed 7 million people. This upper limit seems to have been reached in ancient times, on those occasions when the country enjoyed a good administration. A period of misrule, or internal strife accompanied by plague, could well reduce the population by a third, which indeed had occurred when Napoleon invaded Egypt in 1799.

Wheat planted in the fertile slime grows fast in the mild

A painting in tomb No. 52 Nakht from the Eighteenth Dynasty (1580—1340 B.C.) gives a schematic but technically excellent description of contemporary Egyptian farming. The ground was worked with a well-developed plow provided with a pole to which was hitched a pair of oxen. The yoke was split and placed around the oxen's horns. The painting shows clearly that the sowing was done after the ploughing, whereas in the Old Kingdom the seed was broadcasted and trampled down by driving a flock of animals over the seeded plot.

winter climate and is harvested in April. This was the way the Neolithic farmers grew their winter crops of wheat, barley, and flax. But as the population increased, the irrigation engineered by nature became inadequate, and land not previously reached by water had to be irrigated by man-made measures. From early attempts to retain the flood water by a simple barrage of stones and mud there was developed a system of basin irrigation, whereby the entire valley was provided with earth banks running parallel to the river banks and crossed by dikes running at right angles to them. Such a checkerboard of dike-enclosed areas varied in size from a few thousand square feet to 40,000 acres. Canals, enclosed by dikes when necessary, conducted water into higher areas that were not reached by the natural inundation.

Within the dike-enclosed area the water was conducted by means of sluices into one basin at a time, where it was held until it had saturated the scorched mud, whereupon it was admitted to the next basin, and so on, until the surplus water was drained off at the lowest level and eventually returned to the river via a canal.

Egyptian water administration

The Pharaonic system of basin irrigation must have been well developed by the middle of the Fourth Millenium because the "cutting of the dikes" had become a ceremony conducted by the King, as shown on the macehead of a predynastic king of the Scorpion clan (ca. 3200 B.C.). King Menes, conqueror of Lower Egypt and the first dynastic pharaoh, was also concerned with the administrative and ceremonial aspects of the irrigation system.

In dynastic times the principal duty of the governors of the nomes was the maintenance of the dikes and canals. The peasants of the nomes were subject to corvée, just as in Sumer, although much more stringently. In high flood the dikes had to be patrolled day and night to detect cracks or seepage, because once saturated, the mud banks would rapidly give way and flood the villages and towns.

A nome administration included tax inspectors and cadastral surveyors to check on the boundary stones which were liable to be lost or moved during the inundations.

This enlarged detail of a macehead (lower left) belonging to an unknown predynastic king (commonly referred to as the "Scorpion King" because of the scorpion in front of his face) shows him engaged in the ceremonial cutting of a water dike. The miniature engraving suggests that the state water administration had already become established before 3000 B.C.

The willful shifting of such a stone was of course a serious misdemeanor subject to severe penalty. But, obviously, in a monolithic state the central administration kept growing at the expense of the nomes, and in the end the national water administration became centered in the "Labyrinth," a huge building containing 3,000 rooms erected during the reign of Amenemhet II (ca. 1850 B.C.) on the west bank of the Nile south of Memphis. Both Herodotus and Diodorus Siculus visited the place at an interval of 400 years, and the accounts of these ancient reports agree so closely that it cannot be dismissed as a myth. Of this tremendous structure nothing remains; but the site, near the pyramid of Hawara, was found and surveyed by Flinders Petrie in the 1880s.

Of the three great sources supplying information on the material aspects of ancient civilizations, namely, Herodotus, Diodorus Siculus, and Strabo—all of whom were dismissed as nonsensical by nineteenth century scholars—Herodotus can be characterized as a roving reporter intent on filing a good readable story, while Diodorus is a more serious writer, and Strabo writes like the secretary of a reporting committee and therefore was never read by his contemporaries and remained unknown and unrecognized until a copy of his works was found in Constantinople some five hundred years ago.

In their treatment of the Labyrinth, however, the three agree rather closely, but since Herodotus was the first one on the spot and his account of what he saw is much more fun, it will be quoted here in preference to the other two. Herodotus visited Egypt some time around 460 B.C. and describes the Labyrinth in the following way:

"Moreover they (referring to some kings) resolved to preserve the memory of their name by some joint enterprise; and having so resolved, they made a labyrinth, a little way beyond the Lake Moeris and near a place called the city of Crocodiles. I have myself seen it, and indeed no words can tell its wonders; were all that the Greeks have builded and wrought added together, the whole world would seem to be a matter of less labor and cost than was this labyrinth, even though the temples at Ephesus and Samos are noteworthy buildings. Though the pyramids were greater than words can tell, and each one excels the monuments built by Greeks, this maze surpasses even the pyramids.

"It has 12 courts, with doors over against each other, six face the north and six to the south, in two continuous lines, all within the outer wall. There are also double sets of chambers, 3,000 altogether, 1,500 above and the same number underground. We ourselves viewed those aboveground, and speak of what we have seen; of the underground chambers we were only told; the Egyptian guards would not show them, there being, they said, the burial vaults of the kings who first built the labyrinth, and the sacred crocodiles. Thus we speak only from hearsay of the lower chambers; the upper ones we saw for ourselves, and they are creations greater than human. The outlets of the chambers and the

25

many passages hither and thither through the courts were an unending marvel to us as we passed from court to apartment and from apartment to colonnade, from colonnade again to more chambers and then into more courts. Over all this is a roof, made of stone like the walls, and the walls are covered with carven figures, and every court set round with pillars of white stone most exactly fitted. Hard by the corner where the labyrinth ends stands a pyramid 40 fathoms high, whereon great figures are carved. A passage has been made into this underground."

It seems quite likely that this administrative monster grew from the early necessity of knowing beforehand the scale of the annual flood. Although the Nile normally behaves in a gentle manner, rising and falling with predictable regularity, it does happen that it goes on a rampage when some of the tributaries of the Blue Nile fill up with snow water at the same time. The last time this occurred was in 1946 when Egyptian irrigation engineers took a calculated risk of holding back the crest of the flood behind the Aswan Dam. The worst flood in recent memory occurred in 1879 when the river broke through the banks and submerged all of Lower Egypt. Upstream, the river rose to the level of the Karnak temple and undermined the columns in the Hypostyle hall and caused their collapse. During the inundation in 67 B.C., water reached 24 inches high on the walls of the Luxor temple.

As a measure against such contingencies, a hydrographic service was established at an early date. The service installed and maintained a number of nilometers—in the end there were 20 of them—along the river from Elephantine in the south to the island of Roda, to the south of Cairo, in the north. Every important temple was required to take daily readings from its nilometer, which consisted of a scale marked in royal cubits and fractions thereof, engraved on the wall of a stone quay or temple building fronting the river. At Elephantine, the existing nilometer consists of markings on the wall of a staircase leading down to the lowest winter level of the river; at Edfu the scale is marked on the walls of a deep well, the bottom of which is connected to the river.

Obviously, the further upstream the nilometers were placed, the more valuable were the readings because they enabled the service to gain a few days in forecasting the rise of the flood downriver. Thus the southern expansion of the Egyptian empire into Nubia came to be marked with nilometers. The most southerly one, that at Semna at the second cataract, began to be manned at the time of the Twelfth Dynasty, but the most important one during the long history of Egypt, including the recent Arab hegemony, was at Elephantine, the viceregal administrative center in the south. From here runners brought daily readings by relays to the Labyrinth where they were recorded and compared with the previous records. If anything abnormal was noted in the early Elephantine readings the nome service along the river were alerted. The system worked in principle like the ones now employed in any well-regulated river system, whether used for power generation, transport, or irrigation.

Conduction of drinking water

The water on which so much thought and labor was expended was not fit to drink. It was—and still is—lethal in character. It would be interesting to know the mortality in these ancient riverine valleys due to malaria and other diseases propagated in stagnant water. Alexander came through unscathed from his innumerable campaigns, including desert marches that left his veterans gasping for water, but he succumbed to a fever contracted on a pleasure excursion on an irrigation canal south of Babylon.

Good healthful drinking water has to be obtained from sources other than irrigation water, and if a community is to be kept healthy the two should be separated from each other, from the source to the place of consumption.

Of course there were and are numerous exceptions to this simple rule: water may first be used by human beings, the overspill by animals; and then the waste or surplus used for irrigation. This was the custom in Rome at a later date, and in ancient times this was common practice in desert oases, such as Dakhla and Kharga, where local wells supplied water for drinking as well as for irrigation.

The wells of these two oases to the west of the Nile valley are sufficiently interesting from the point of view of technological history to warrant a brief excursion from the mainline developments. When they were first discovered by European travelers, the two oases were watered by *artesian* wells sunk 230 feet into the sandstone. Among the local people the visitors found men who were specialists in rope drilling and families whose male members were trained to be divers. They supported themselves by working under water when an artesian well was to be repaired.

The rope drilling method that travelers had occasion to observe in the Kharga oasis as recently as 50 years ago began with the digging of a 6 × 6-foot well through sand and eroded sandstone. The well was lined with timber, down to where surface water was encountered, in the Dakhla oasis usually at about 100 feet. Then a 14-inch tube made of joined acacia staves was placed in the center of the well. The tube was built in short sections joined together. As each section was securely joined to the one below, the well was backfilled with clay. When the well had been backfilled to the top of the ground, a timber horse was placed at a suitable distance from the pipe. A rocker beam was supported in a bearing atop the horse, and through a one-sheaf block at one end of the beam ran a rope carrying a heavy iron rod with its lower end forged into a broad bit. The drill rod was sufficiently heavy to pull down the

An early papyrus in the Cairo museum gives a schematic idea of the layout of a basin irrigation scheme, with its supply and drainage canals, dikes, and so on. The principle is illustrated by the sketch to the right. In order to cultivate land above the level of a normal inundation of the Nile, a supply canal was dug or enclosed by dikes from a high-level junction with the river at a considerable distance upstream. The area between the supply canal and the river was divided by a system of dikes that enclosed basins on different levels. By cutting the upper-level dikes the basins were filled with water, and when the soil of subsequent basins had been thoroughly soaked, the surplus water was returned to the Nile.

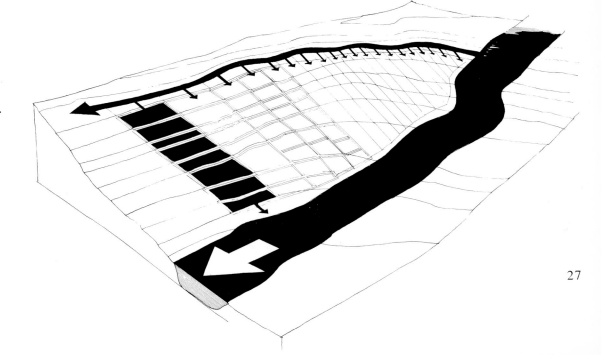

27

Since many of the walled-in cities in the Near East frequently obtained their water from a spring outside the wall, the population had to fetch its household water from a deep well. The shaft was therefore stepped along the sides. A later development resulting from a better mastery of subterranean geometry was to sink an oblique shaft, either straight or angled, leading to the well. This was done in Jerusalem, Meggido, Lashish, and other cities in Palestine.

rocker beam, which was lifted by 10 men tugging ropes attached to the other end of the beam.

The drillmaster stood at the hole and turned the rope and hence the bit a few degrees of arc, whereupon the men let go of the rope, permitting the heavy bit to drop against the rock. In this way an experienced drill team could achieve 20 to 30 impacts per minute. When the drilling became impeded by the accumulation of cuttings at the bottom of the hole, the bit was replaced with a spoon which was dropped down and scooped up the cuttings.

With this method of impact drilling, a 160-foot hole could be sunk in five months, but frequently it took a year to reach water which owing to the local geological conditions, i.e., the presence of an impervious cover of rock, was under sufficiently high hydrostatic head that water was ejected through the pipe up to and beyond ground level. An artesian well sunk in this manner would deliver 260 gallons per minute of water for several centuries, i.e., enough to irrigate about a hundred acres of desert land.

Although early travelers and writers complained of the total lack of evidence, documentary or archeologic, some of them were inclined to assume that the rope drilling method found in these isolated oases was the continuation of an ancient Egyptian technique developed in the Middle Kingdom. Some even insisted that a prerequisite for the colonization of the two oases must have been the mastery of rope drilling in order to get the irrigation water needed to make the colonization economically viable. They found proof for their assumption in a statement by Olympiadores that the wells in the Kharga and Dakhla oases were at least a couple of thousand years old.

The likelihood is that these writers fell victim to a semantic trap. Olympiadores speaks of deep wells, and his English translator uses the word *bore* in the manner applied also to shafts and tunnels without in any way indicating the method employed to obtain a well, shaft, or tunnel. When the early visitors ran across these artesian wells and also found a fully developed rope drilling method, they took for granted that Olympiadores referred to these drilled wells when he actually spoke of deep wells sunk by digging and chiseling at the time of the Middle Kingdom.

The origin of these artesian wells in the western desert is altogether different. Rope drilling is a Chinese innovation, and the earliest mention of it in Chinese literature is dated to A.D. 1013. The Arabs adopted the method, and the wells found in the Kharga and Dakhla oases were drilled in medieval times.

The earliest artesian wells in Europe were sunk in and around Modena and Bologna as described by Cassius in 1660. At that time, the arms of the former city carried two augers crossed. The water table around Modena is situated below an impervious cover of *argilla,* or clay, and the practice developed by the local well diggers was to sink a well by digging down to the clay and placing a wooden pipe in the center of it, whereupon the well was backfilled with clay. An auger was then put into the pipe, and a hole drilled through the clay to the water table that was usually found at 63 feet. Owing to the pressure under the argilla cover, the water shot up through the pipe to a height of some 5 feet above the ground.

These auger-bored wells in the Modena district can no doubt be called artesian, but the first rope-drilled artesian well in Europe was sunk at Grenelle, near Paris, as late as 1832. Here, water under tremendous pressure was found at 1,795 feet.

Water logistics

The Egyptian state required gold, and the acquisition of gold was a never-ending concern of successive Pharaonic administrations. But owing to the laws of nature, the gold could only be mined in the desiccated wadis beyond the first cataract, and subsequently also in the gold-bearing quartz reefs higher up the wadis, in the desert region between the Nile and the Red Sea. The quest for fine building stone also drove succeeding administrations to sustained efforts to dig and maintain surface wells along the march routes to the mines and quarries. There is a record of such an expedition into Wadi Hammamat in 2000 B.C. when a force of 3,000 men was put to the task of sinking wells as a preliminary to the opening of a communication link to the mineral deposits in the wadi.

It appears that keeping open these desert routes to the golden wadis taxed the Memphis administration to the utmost, and in times of political unrest the routes disappeared altogether. During the reign of Ramses II (1292–1225 B.C.), the route to the gold mines in Wadi Allaqi in Nubia had been lost, owing to the sanding of the wells. The Allaqi mines had been used also as a place of imprisonment where the Pharaoh put away those who for some reason or other had invited his displeasure, in addition to common criminals; and therefore it held high priority in the southern policies pursued by this imperialistic king.

But Wadi Allaqi proved a hard logistic nut to crack. "If some gold washers went thither," writes a chronicler of Ramses II, "only a few arrived, for the rest died of thirst, together with their asses. There could not be found for them the necessary water to drink going up and coming down the wadi, from water in their water skins." The King's lieutenants managed eventually to find water at only 20 feet, whereas expeditions sent out by Seti I (1313–1292 B.C.) had failed to strike water at 200 feet and had to abandon the wadi.

The caravan routes and military roads also required wells; those along the 9-day-long march route to Palestine, apparently dug during the reign of Seti I, were fortified. Palaces and towns were provided with deep brick-lined wells which had to be extended upward as the elevation of the sites grew owing to the accumlated debris. But getting water out of a well, although of vital importance to the people in need of it, is wholly inadequate for the development of a civilization with some pretensions to scale. To provide an expanding city with water requires altogether larger sources of supply, and the main centers of the ancient world obtained their drinking water by means of freshwater canals extending over a distance of some 50 miles or more.

Storage dams and cisterns

For an ancient city lacking local sources of water there were other ways than costly and labor-demanding canal building to secure an adequate supply of water. One method that did not prove successful in Egypt and was not repeated was to build a dam across a wadi to contain the rain water coming down it as flash floods during a few weeks in winter. Such a dam was built in Wadi Gerrawe (El Kofaro) near present Helwan some time during the Old Kingdom. The abutments of the dam still remain, from which it may be deduced that the dam must have been 361 feet wide and 39 feet high. It was constructed as two separate stone barrages placed 118 feet from each other and each 79 feet wide at the base. The space between was filled with erosion debris. If the site measurements are to be believed, this modest dam would have had a base of altogether 276 feet. Upstream the dam was laid with limestone blocks with a slope of 3:4.

The obvious purpose of the Gerrawe dam was to catch the flash floods of the wadi and store something like 600,000 cubic yards of water for the benefit of the hundreds of quarrymen employed in the nearby alabaster works. The idea was sound, but the unknown Old Kingdom designer was not quite up to the job. During the first flash flood the water rose over the rim of the dam, and the whole structure collapsed owing to the combined action of erosion and pressure. The quarry workers had to be supplied with water in some other way.

Better luck was had with the Orontes barrage built ca. 1300 B.C. during the reign of Seti I. The site chosen for the 500-foot-long and 20-foot-high dam was the northern end of a long flat valley. It was a stone-fill dam faced with basalt blocks. It raised the water level 17 feet and formed the Lake of Homs, 6 miles long and 3 miles wide. The barrage and lake still remain.

The other more common and successful method of storing winter rains was by means of cisterns cut into the rock. The Arabian peninsula still abounds with rock-cut cisterns, and the city of Aden alone has 50 of them with a storage capacity of 30 million gallons. At Jericho there is an ancient cistern with an area of 6 acres, and 11 miles outside Jerusalem are the "Pools of Solomon" which supplied the city with water. To prevent evaporation the cisterns were placed underground, chiseled out of the rock.

Conducting drinking water in open canals is feasible only when the volume is large enough to permit wastage owing to evaporation. But there were occasions when the topography between the source and the place of consumption precluded canal building. This was true of "the canal that brings abundance," i.e., the one that supplied the old Assyrian capital of Nimrud built around 1240 B.C. It was "cut to the chord"—i.e., in a straight line—and had to be brought through tunnels cut in rock through the intervening hills to join the open canals in the valleys.

The qanaat builders

This manner of conducting drinking water through tunnels subsequently became a standard Assyrian practice, resorted to in difficult country and when the volume required was rather modest, say, on the order of a few gallons per second. The Persians adopted the old practice and perfected the *qanaat,* an underground conduit that during their hegemony became standard all over the ancient world, from North Africa to India.

The qanaat tunnel evidently derives from the mining tunnel, or adit into a hillside, pioneered by Armenian miners; and it seems likely that the early qanaat builders were recruited from the ancient Armenian mining region. The underground excavations required to drive a qanaat from a water source on sloping ground, such as an alluvial fan, to the place of consumption in the plains called for a great deal of hydraulic and mining know-how. The length of the water tunnel varied; the existing Aleppo qanaat is 7.5 miles, and the Arbil qanaat in Kurdistan is 3.4 miles long. The tunnel was excavated from shafts sunk on line at a distance of 65 to 165 feet from each other. Depending on the topography of the intervening country, the shafts were sunk to a varying depth that occasionally could reach 300 feet. At intervals, the qanaat was provided with an oblique and stepped shaft used for inspection and the removal of accumulated silt.

The capacity of a qanaat is not large, usually on the order of 100,000 gallons per 24 hours, although the Arbil qanaat fed by three branches delivers 250,000 gallons. There are reports of others with a capacity of up to 3 million gallons per day.

The driving of a qanaat is an excellent example of the clever application of simple engineering principles that tend to be lost as technology becomes more scientific and abstract. When confronted with an ancient Persian qanaat, a scientifically educated engineer will immediately inquire about the instruments or other survey aids employed for alignment and, beyond all, grade. Actually, these underground structures were built without any measuring aids whatsoever.

The manner in which these ancient tunnels were driven

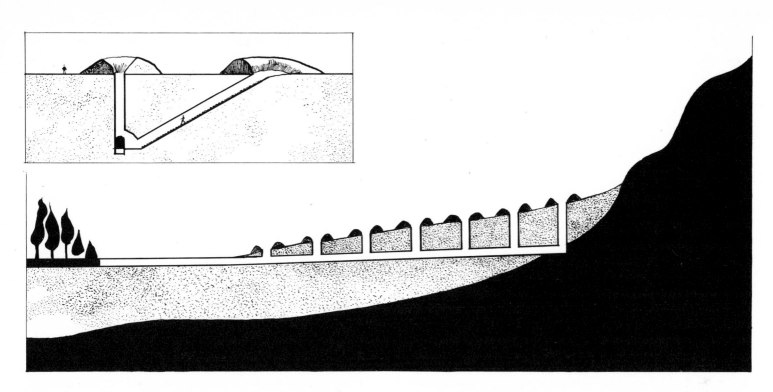

Many old cities in the Near East, and elsewhere in the ancient world, still get their water through a qanaat, a small tunnel leading from a hillside source to a deep well or surface adit within the walls. The qanaat appears to have been an Assyrian innovation adopted by the Persians, who introduced it wherever they went. Hence, qanaats are found in areas that at one time were under Persian administration. The insert shows an intersection of a qanaat and an oblique stepped inspection shaft. After 3,000 years Iran has about 22,000 qanaats with a total length of 168,000 miles, supplying altogether 730 cubic yards of water per second.

was as follows: Having selected a spot on geological indications, on the spur of a hill or an alluvial fan, the qanaat master sank a shaft until he reached the top of the water table. He then lined up the shaft with the distant place of consumption, and along this usually straight line, he sank shafts at a distance varying from 65 to 165 feet from each other, depending on the topography. With a trickle of water from the original shaft used as reference, the rest of the shafts were sunk to the level of the water table as encountered in the reference shaft. Hence, the bottom of the tunnel was kept at a gradient slight enough to permit water to trickle from one shaft to the next. The qanaat driven in this fashion terminated either in a well or most frequently aboveground, and was then bricked over. The end result of the labor at this stage was merely a trickle of water, suggesting a near-level grade from the reference shaft to the exit. Then followed a second stage. The qanaat was extended backward into the rising water table, and deepened to increase the flow.

In the Kharga oasis a captain in Darius' navy by the name of Scylax, in 518 B.C., supervised the driving of an interesting variation of a qanaat. He sank 5 × 2½-foot shafts into the sandstone to a depth of 174 feet, from which 2 × 5-foot egg-shaped headings were driven. From Um

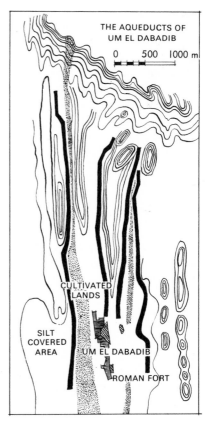

THE AQUEDUCTS OF
UM EL DABADIB

0   500   1000 m

When the oases in the desert west of the Nile were under Persian hegemony, work began in 518 B.C. on four qanaats in the hills to the north of Um el Dabadib for the purpose of collecting water from the fissures in the sandstone. The total length of the tunnels is 9 miles, and they were driven from the bottom of 150 shafts sunk at a distance of 65 feet from each other to an average depth of 70 feet. The volume of the rock excavated—at a temperature in excess of 90°F— came to 26,000 cubic yards. The work was supervised by a captain in Darius' navy.

CULTIVATED LANDS

SILT COVERED AREA

UM EL DABADIB

ROMAN FORT

31

el Dabadib north of the oasis he drove four such qanaats running roughly north-south along the sides of three separate valleys. The longest tunnel measures 3 miles, and the total length of the system is 9 miles. One of the tunnels was driven from 150 shafts, and when recovered in 1900 A.D. it gave 34 gallons of water per minute.

But the interesting thing about these Kharga qanaats is that none of them ever tapped a spring. They were cut through fissured rock with the aim of interrupting water percolating through the fissures during the advance. A similar method was used by Italian engineers to gather water along the hillsides in the Dolomites to eke out the water supply for postwar power generation.

The Siloam tunnel

The technical excellence of the Assyrian and Persian qanaat builders is brought out by comparing their work with the amateurish attempt to drive a rock tunnel through the Ophel hill to the southwest of present Jerusalem, as ordered by Hezekiah, King of Judah, in 700 B.C. Indeed, this tunnel which is still well preserved—each sweep of the miners' picks is clearly shown by the scars left on the walls—displays most of the errors that can be committed by inexperienced miners.

For all that, or perhaps because of it, the Siloam tunnel is one of the most interesting examples of ancient underground construction that can be seen in the Middle East today. Moreover, it is actually mentioned in the Bible, although the author evidently had not the slightest idea what he was talking about. This is how this particular passage reads in II Chron. *32*:2–4:

> "And when Hezekia saw that Sennacherib was come, and that he was purposed to fight against Jerusalem, He tooke counsell with his Princes, and his mighty men, to stop the waters of the fountaines, which were without the citie: and they did helpe him. So there was gathered much people together, who stopt all the fountaines, and the brooke that ran thorow the middest of the land, saying: Why should the Kings of Assyria come, and finde much water?"

It requires a great deal of imagination to interpret these verses as referring to a water tunnel, but that is indeed the historical fact behind this garbled account.

At any rate, the mining gangs put to work by Hezekiah and his "Princes and mighty men" at both ends of the Siloam tunnel, in the north close by the Gihon well outside the wall of the old Sabatean town and in the south from a pit inside the wall, were confronted with the job of excavating 366 yards of rock. Each team had to advance but 183 yards before breakthrough.

In actual fact, the length of the Siloam tunnel became 583 yards, since the miners proved unable to keep line. The

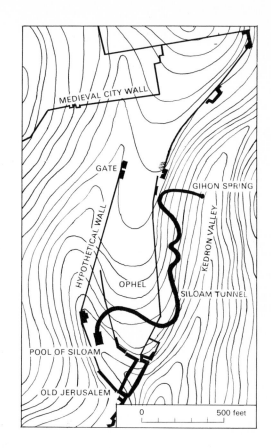

The old Jerusalem—"the City of David"—on the Ophel hill to the south of the present city, obtained its water through a tunnel driven in 700 B.C. during the reign of Hezekiah when threatened with a siege by the Assyrian king Sennacherib. The tunnel joins the Gihon well outside the northern wall to a well—"the Pool of Siloam"—inside the wall. The linear distance between the portals is 366 yards, but owing to the inability of the miners to keep line, the tunnel meanders through the limestone rock, and its length became 583 yards. The area of the tunnel is 2.3 × 5.6 feet.

tunnel meanders and twists in the limestone rock, and the miners obviously lost their way from the outset. When they realized that they had become lost, they had to make raises to the surface to see where they were. They did not keep grade; the bottom of the tunnel dips and rises by as much as a couple of feet. Worse, from the evidence still available, the team at the southern end, now the Pool of Siloam, attacked the rock something like 20 feet above grade and had to keep picking downward to find proper grade. The job was a mess from beginning to end.

But the two teams eventually met and on a tangent. The exact place of the breakthrough can still be seen—after nearly 2,700 years—by the curved scars meeting on the walls. The breakthrough was an important event because a local poet was commissioned to write some nice verses appropriate for such an occasion. For reasons unknown, they were not included in the Holy Writ but were engraved on a stone tablet mortared into the wall just inside the southern portal. There they remained until 1880 when the tablet with its inscription in the earliest known cursive of the Phoenician alphabet was removed and placed in a museum in Constantinople. This is what it says:

"This is the tunnel, and this is the story of the tunnel
whilst yet the miners were lifting up the pick, each
towards each other, and whilst there were
three cubits to be cut through there was heard the
voices of men
calling to each other, for there was a split in the rock
on the right hand and on the left hand And on the day
when they met the miners struck, each to meet his fellow,
pick upon pick; and there flowed
the waters from the pool for two hundred and a thousand
cubits, and hundred cubits was the height of rock
above their heads"

Such are the fortuities of history. Of all the great works accomplished in ancient times we know hardly anything except what can be deduced from random archeological diggings. But a miserable little amateurish job remains practically intact, and one phase of it is recorded in detail.

The Red Sea Canal

Compared to the vast hydraulic works previously discussed, the Red Sea Canal was a rather modest affair. Its principal interest lies in its character of being a so-called ships' canal, used only for navigation, and as such it was unique in the ancient world. However, the odd thing about the canal is that despite detailed accounts of it by ancient writers, its existence has been hotly denied by many modern historians. But after finding five steles along the route, even the most skeptical will have to admit that after all there might have been such a canal.

Indeed there was. The Red Sea Canal was 150 feet wide and 16 feet deep and ran from ancient Bubastis, on the eastern arm of the Nile in the delta, along Wadi Tumilat to the Bitter Lakes and then south to Arsinoe (now Suez) at the northern end of the Red Sea (or the Arabian Gulf as it was then called). The total distance was about 125 miles.

Of the three sources, Herodotus states that Pharaoh Necho (609–595 B.C.) began the work on the canal, but after losing 120,000 diggers, he abandoned the project because of a prophesy that "he was toiling beforehand for the Barbarian." The Barbarian, as it subsequently turned out, was Darius the Persian (521–485 B.C.) who finished the work. Darius' canal was so wide that two triremes could pass each other, and the journey from Bubastis to Arsinoe took four days.

Diodorus Siculus also attributes the canal construction to Necho and Darius, but adds that the latter abstained from finishing it for fear that the Red Sea, being on a higher level than the Mediterranean, would flood and devastate Egypt. Ptolemy II (ca. 285 B.C.) had no such qualms, and after installing a cofferdam, he made the final cut and joined the canal with the sea.

Strabo, finally, states that the canal was originally cut by Sesostris "before the Trojan War," as he puts it; but he also refers to Necho as having begun the work which was continued by Darius and finished by Ptolemy.

Thus the three accounts are in close agreement, at least in outline. It seems quite possible that the early canal had been built by Sesostris, or one of the Senwosret pharaohs of the Twelfth Dynasty (2000–1785 B.C.), but that the cut had sanded in and was opened again by Necho and his successors. However, the archeological evidence clinches the ancient argument in favor of Darius as being the one who at any rate finished the work, because this is the reading on one of the five steles found along the canal:

"I am a Persian. From Persia I captured Egypt. I commanded the canal to be built from the Nile which flows in Egypt, to the Sea which comes from Persia. So was the canal built, as I commanded, and ships passed from Egypt through this canal to Persia, as was my purpose."

This Roman-Egyptian terra-cotta relief shows a Negro slave tramping an Archimedean screw, then commonly used for raising water.

The Megalith Builders

In the following notes on the building and construction activities in the chalcolithic centers of civilization in Sumer and Egypt, the esthetic, i.e., the architectural and artistic, qualities of the structures will be largely ignored in favor of the basic reasons why they were built and how they were built. Inevitably, they also shape into a study of *hubris,* as the Greeks named the wanton arrogance arising from reckless disregard of the moral and physical laws of restraint, which always seems inescapable upon mastering a new technology. The chalcolithic structures, regardless of time and place, share common characteristics of exaggerated size and a ruthless and extravagant use of material and labor. The builders had just learned to quarry stone and move it to the building site. Rapidly, within the course of a century or two, they were no longer content with working stones of a size that lent themselves to convenient handling and piling them into a building conforming to human scale.

The contrary was true; men and beasts had to be strained to the limit and beyond, in quarrying and transporting huge monoliths weighing several hundred tons, to be incorporated in gigantic structures rising high into the sky, in ziggurats, pyramids, palaces, tremendous "pentagons" housing the administration of rulers who had turned themselves into gods and lost all touch with the realities of human life. A thousand years later or more, in distant Britain and elsewhere on the barbaric marches of the unknown world, the same newfound ability to quarry and transport monoliths was turned into similar dehumanized exercises, such as Stonehenge and the passage graves. Revolting as the social and political processes leading to these building excesses may appear, it cannot be denied that at any rate in Egypt building in a purely technical sense reached such excellence that nothing achieved since can compare with it. What masons other than the Egyptian have managed, for example, to join monolithic stone slabs with the accuracy of a tenth of a millimeter or even less that will stay put for 5,000 years or longer? It was a stupid exercise; but as building performance it was sublime.

Rise of the tribute state

What is less admirable was the means used to erect the buildings. As mentioned in the previous chapter, the Neo-lithic farmers in the Euphrates and Nile deltas were content to turn over some of their farm surplus to the god to whom their land belonged. The tribute was collected and administered by the priests, serving as the god's representatives. The accumulated surplus was channeled into temple buildings and the maintenance of a hierarchy of priest-scribes, but it also aided indirectly the development of specialists, potters, masons, and, during the age considered, merchants, coppersmiths, and other artisans shaping the metal into artifacts.

Out of the nonlaboring temple hierarchy eventually emerged powerful and ruthless personalities who claimed suzerainty in the name of the city's divine Lord over all residing within it and the land farmed by the citizens. The more ambitious and warlike of these local god-kings did not rest content with what they had, and on some pretext or other, in Sumer commonly conflict over water rights, set out to subdue the territory of a neighboring god and his subjects and reduce them to vassalage. Eventually, beginning around 3500 B.C. there arose the tribute state wherein the god-king and his high priests and scribes could appropriate whatever the peasant masses produced, beyond their personal consumption. The peasants were made subject to corvée and could be pressed into service by the ruling hierarchy, to labor on the dikes and ditches, and as transport workers on temple and palace construction.

These, then, were the conditions leading to and providing the means for megalith construction. They were admittedly somewhat different in Sumer and in Egypt. In the latter country, after the merger of the two kingdoms the state had become monolithic and totalitarian in character, whereas in the Sumerian city-states the citizens still retained a measure of freedom and certain limited rights, in some ways resembling the tax-ridden citizens of the modern welfare state. But they had in common a ruthless appropriation of the rapidly increasing surplus produced on the fertile lands by a large bureaucratic hierarchy obtaining its powers from the divine lord and charged with the duty of channeling the surplus into public buildings to impress the people with the might and glory of their god and king.

Now, nothing remains of the splendors of ancient cities

except a pimple on the desert floor. For archeologists these tells mark the whereabouts of an ancient village, town, or city buried under the shifting sands. Digging into a tell won't give much to old-fashioned treasure-hunting archeologists. It contains nothing but layers of rubbish, the remains of sun-dried brick or pisé houses which, after serving as shelter for a few decades, collapsed, whereupon the site was used again for another house. Rubble and pottery shards are about all that the average tell has to offer.

The city of Ur appears to have had a population of 360,000 by 2000 B.C., with a density of 44 persons per acre. Outside the Temenos, the walled-in temple compound, the city was a jumble of houses, winding streets, and narrow alleys, as shown by this map of an excavated housing area. The names of the streets are those given by the excavator Sir Leonard Wooley.

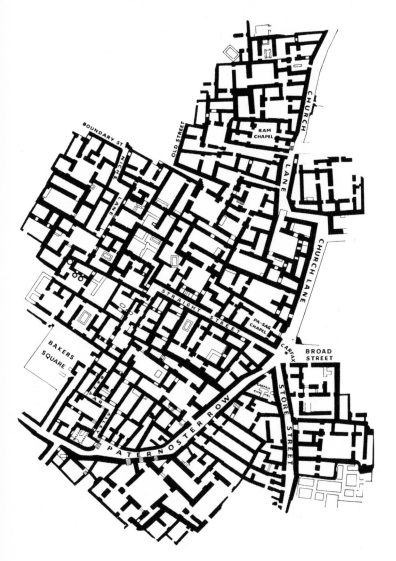

But to the modern science of archeology the tailings of ancient civilizations are of greater importance than the show pieces on display in the museums.

The very fact that a tell exists is significant to begin with. The shards,—the shape, decor, and hardness of the one-time pots—speak of a pattern of communal life, indeed of the mental quality and capability of the people who lived there some 6,000 years ago, of their passions and dreams, their political and economic ambitions, and how they succeeded in realizing their ambitions. In distant laboratories, thousands of miles from the site, bits of bone, the charcoal remains of fires quenched long before the first seed of emmer was planted in Europe, bits of metal and skeletal remains of ancient householders are all subjected to analysis which reveals the realities of life that existed in a living community now dead and buried under the sand and fills in the pattern revealed in outline by the work on this or some other site.

Beyond classical documentary sources the following notes are based on the interpretation of archeological material published in recent decades.

The first metropolis

In Sumer, Akkad, and elsewhere along the twin-river valley, the population lived in walled cities, of which Ur although bigger is typical of the others. Like all Sumerian cities Ur began as an administrative center of an irrigated district in Ubaid days, but on the mound formed by the rubble of its predecessors, the city changed its character and became an industrial and commercial center of the fertile delta. The excavated city of Ur derives from the Larsa period (1920–1800 B.C.), where at one time a family resided which was destined to have a remarkable influence on a large part of the then unknown world thousands of years after the city of Ur had been abandoned and turned into a tell. This was the household of Terah, whose son-in-law Abraham came along when the family for some unknown reason packed up and left the sophisticated metropolis to return to its native town of Harrar in the Anatolian foothills. Both places were subject to the same divine lord Nanner, the Moon God, in whose honor the major temples were erected.

The Larsa city, or if the somewhat pretentious label be tolerated, the "city of Abraham," had at the time a population of 300,000. It was a crowded city; the density of the population appears to have been about 250 persons per acre. The layout of Ur was an irregular oval with a maximum length of ¾ mile and a maximum width of ½ mile. The city was surrounded by a mud-brick wall about 25 feet high with a steeply sloping glacis. Along the western wall ran the Euphrates, and along the eastern wall a broad canal joining the river upstream from the city. The layout of the residential section suggests no system of planning; the city had been permitted to grow freely from the primitive

35

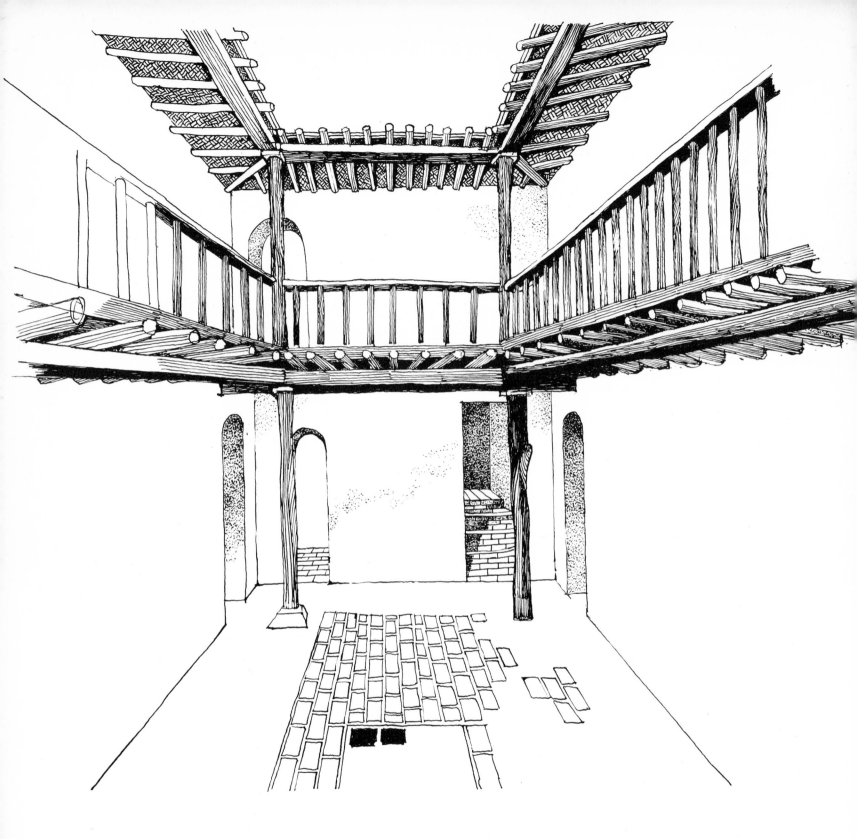

The private houses in Ur at about 2000 B.C. were of the so-called atrium type, i.e., the living quarters were grouped around an open court. An overhanging second-floor balcony supported on palm logs was also a characteristic feature. The walls were built of kiln-fired brick up to the second floor, and above that of mud brick, and showed a plain undecorat- ed face to the outside. On the bottom floor was a kitchen and rooms for servants and guests, and under the stairway a water privy. The second floor was occupied by the owner's family and duplicated the layout of the rooms below. This was the way the "lower middle class" lived.

The Lion Gate at Mycenae served as the main entrance to the citadel. It was placed between a megalithic curtain wall and a bastion (to the right) also built of roughly worked Cyclopean blocks. The gate posts consist of $10 \times 6 \times 2\frac{1}{2}$-foot breccia megaliths supporting a $15 \times 6\frac{1}{2} \times 3$-foot lintel. The opening was closed by double gates swinging on pivots with a diameter of 6 inches. Above the lintel, the wall stones are corbeled to obtain a triangular space, a typical Mycenean feature, which some art historians and archeologists still insist on calling a "relieving triangle," no doubt because at one time it was believed to serve a static function. A sculptured $13 \times 11 \times 2\frac{1}{2}$-foot megalith of hard limestone is fitted into the space. The sketch to the left shows clearly that the sculpture rests entirely on the lintel and that the corbeled stones only push up against it. The heads of the lions were made of steatite and fastened to the bodies with dowel pins. They were removed and lost long ago.

37

In the Aswan quarries there still remains a 138-foot-long megalith weighing 1,200 tons. It was to be shaped into the largest obelisk ever, but when the trenches along the sides were sunk, a couple of cracks were found and the block was abandoned. Practical trials using the dolorite balls left in the quarries have proved that roughing out the block, including the trenching, could be accomplished in nine months. Below is shown the transport in 1768 of a 1,500-ton granite float block that was used as a base for the statue of Peter the Great in St. Petersburg. The block was moved a distance of 2.2 miles in 213 days. The engraving is contemporary, and in common with all artistic representations it is highly unreliable because the French artist on the spot was unable to resist the temptation of exaggerating the size of the block.

village that it originally was. The houses, most of them two stories high, were jumbled together. Awnings were stretched over the narrow streets lined with open booths. There is really nothing strange about Ur; it resembled the bazaar section of any existing Middle East city, except possibly for the many small private chapels hemmed in between the houses.

The houses, large or small, were very much alike. They turned an undecorated wall built of brick toward the street. A house was built around an open central court lined with a balcony carried by palm trunks. The roof sloped inward to slightly beyond the second floor balcony and left a large opening admitting air and light to the house. Gutters from the roof projected rainwater to the center of the brick-paved court where it was collected in a drain pit or well.

The rooms were placed around the court—kitchen and service rooms, sleeping quarters for the domestic staff, a workroom, and, in the rear, a chapel and burial vault. On the second floor, reached by a stairway the lower flights of which were of hard brick and the upper of wood, the rooms were laid out similarly to the ones below. Under the stairway was a lavatory. The doorways were sometimes arched, sometimes topped with a lintel.

The houses of Ur were well adapted to the climate, and just about the same plan is still being used in Basra and Bagdad, the difference being that whereas now they are occupied by well-to-do Arabs, in Ur they appear from documentary evidence to have been owned by people of the lower middle classes: shopkeepers, tradesmen, merchants, lower-echelon scribes, and the like. The temple slaves lived in slave lines situated at the opposite end of the city, to the northwest of the temple area.

The ziggurat compound

The Temenos, or temple and palace compound, was situated on a raised terrace built of mud brick and enclosed by a wall, also of mud brick. The rectangular enclosure measured 270 × 190 yards; and in the west corner, raised on a higher walled-in terrace, stood the huge staged tower —the ziggurat—reaching a height of 86 yards. In front of the sacred ziggurat compound was a large courtyard around which were the storerooms and offices where the tenant farmers came to pay their land rent and taxes to the god's publicans, the only time they were admitted within the walls. The "Hill of Heaven" dominated the entire city and could be seen from a distance of 25 miles.

The Ur ziggurat—or more accurately Ur-Nammu's ziggurat—was built during the Sumerian revival under the Third Dynasty (2100–1950 B.C.) when Mesopotamian brick construction culminated, in the size of the buildings as well as in the techniques used. It should be noted, however, that the ziggurat represented the continuation of temple building

that by then had progressed continuously since before the Flood (ca. 4000 B.C.). There was, for example, the prehistoric temple at Tepe Gawra (ca. 4000 B.C.), built of sun-dried mud brick, with articulated recessed walls deriving from the reed houses, as discussed in the previous chapter. The Eridu temples were raised on a solid platform of mud brick approached by a steep ramp and enclosed by a wall of limestone. Considerable refinements had been attained at an early date. Wall surfaces exposed to weather were protected by a momumental mosaic consisting of terra-cotta cones painted red, yellow, and black, stuck into the soft mud plaster and set in mastic. Beyond all, these early temple builders understood how to relieve the brick surfaces with recesses and reinforcement buttresses, half-columns divided by vertical T-shaped grooves.

The Ur-Nammu's ziggurat had the form of a stepped pyramid three stories high, of which only the lowest stage remains in good condition. This stage measures 200 × 150 feet and is built of kiln-fired brick set in mastic. The walls have a prominent batter, the converging lines leading to the shrine on the top stage, and they are relieved by shallow buttresses to accentuate their height. To counteract the appearance of weakness in a huge structure built of such small elements as brick, the architect gave the sides a slight outward curve of 1:125, while the faces of the brickwork were given an *entasis* (as the Greeks called this trick of compensating an optical illusion of weakness in long straight lines) of 1:100.

In the Ur ziggurat the interior of the solid structure was filled with mud brick, and to prevent shrinking and uneven settling the fabric was interlaced with a layer of reed matting every fifth course. Prestressed cables of tough reeds running from face to face of the structure and at right angles to one another also aided in tying it together.

The *raison d'etre* of this tremendous pile of brick was to carry the shrine of Nanner, the Moon God. It consisted of a single bedchamber which according to Herodotus contained "a great and well-covered couch and a golden table is set hard by. But no image has been set up in the shrine, nor does any human creature lie therein for the night, except one young native woman chosen from all women by the god, as say the Chaldeans who are priests of this god. These same Chaldeans say that God himself is wont to visit the shrine and rest upon the couch, but I do not believe them," concludes Herodotus.

It is tempting to pursue Herodotus' line of enquiry into the details of the mysterious rites that went on in "the High Place." But that is not the concern of common people, now as well as then. By the Third Dynasty of Ur, Nanner was the absolute autocrat of the state. His priesthood formed a court headed by the ministers of state who kept the god unapproachable to the taxpayers, except during the spring festival when he went in solemn procession to consummate the rites that ensured the fertility of the earth.

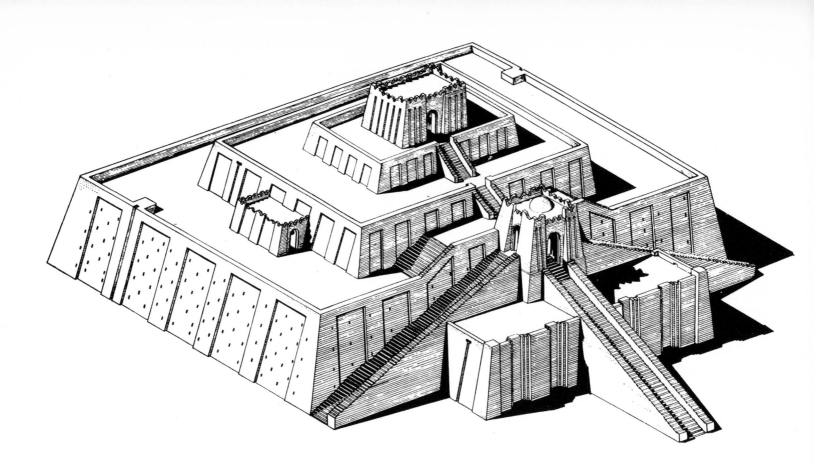

Ur-Nammu, first king of the Third Dynasty of Ur, erected a ziggurat in 2100 B.C. in honor of the moon god Nanner. Only the lower 200 × 150-foot stage of the three-stage structure remains in good condition, but what is left of the rest permits a satisfactory reconstruction. The height was 85 feet, and the walls consisted of an 8-foot revetment of kiln-fired brick set in bitumen. They had a prominent batter and were given a horizontal entasis of 1:125 and a vertical one of 1:100 to present an illusion of strength. The deep red shade of the brick differs altogether from the color of the brick used in contemporary buildings, and remains to be explained. Perhaps a special clay was reserved for important structures.

The end of the Tower of Babel

Owing to Nanner's royal status, the citizens were prevented by the priesthood from gaining access to their god, and they had to vent their religious urge on more accessible deities. It was one of these that the Terah family brought along with them when they left Ur: "the God of Abraham and the God of Nahor (Nanner), the God of their fathers." Four hundred years later Moses identified him as Yahweh who alone among the vast semitic pantheon managed to survive, to shape the mores in distant and as yet nameless lands, and incidentally to inspire an unmeasurable building activity in his honor.

Yahweh, as every Jew and fundamentalist Christian knows, is a jealous god that does not tolerate the existence of any other diety. Indeed, he does not take kindly to such manifestations of the human hubris as the building of ziggurats and the organization of labor required for building them.

"Behold, the people is one, and they have one language, and this they begin to doe, and now nothing will be restrained from them which they have imagined to doe; Goe to, let us goe downe, and there confound their language, that they not understand each other's speech. So the Lord scattered them abroad from thence upon the face of all the earth: and they left off to build the city."

This is one way of explaining the reaction that inevitably set in to the hubris that found expression in the erection of such huge and labor-demanding structures. But the ancient author of these verses in Genesis failed, no doubt in his ignorance thereof, to provide a better and much more dramatic account of the destruction of the Tower of Babel. The summit of another ziggurat, that at Birs-Nimrod and locally known as the "Tower of Babel," shows unmistakable signs of having been damaged by a violent stroke of lightning. Large chunks of masonry were torn out of the fabric, and the bricks were subjected to such tremendous heat as to become vitrified. It seems strange that such a cataclysmal bolt of hot anger from on high has left no traces in the religious traditions of the East. Surely, the unknown northern sibyl who conjured forth Voluspa would not have failed to turn such celestial fireworks into a universal conflagration ordered by Yahweh in his frustrated anger at the wanton wickedness of man, whom after all he had created in his own image.

Yet, tremendous in scale as they were, there was nevertheless a limit set to the size of the Mesopotamian ziggurats. Brick, being a man-made material, has its limits: it will crush under its own weight. Fuel, labor, and transport will also determine the supply of brick available at any given time and place. A ziggurat has its built-in brakes which set its upper limit, beyond which it was not possible to go, even for the most ruthless god-king.

Building in stone, on the other hand, has no limit except that set by human ruthlessness and organization ability. Given enough men and beastly overseers, food to feed them, and an unlimited supply of rock, there appears to be no theoretical limit to what can be accomplished. Piling stone upon indestructible stone provides such a durable structure that it requires something approaching geological time to bring about its demolition. A pyramid will withstand earthquakes and bolts of lightning, heat, frost, and wind; the largest ones will even stand up against human greed in attempts to quarry them. In the end, of course, weathering will level them to the ground; after 5,000 years a great many Egyptian pyramids have been ground into dust by the action of wind, water, frost, and heat.

With the building of the Gizeh pyramids the megalith hubris reached its zenith. As structures, they were from the outset a useless exercise in simple solid geometry frozen in stone that did not serve the intended purpose of protecting the body of the exalted mortal who ordered the monstrous work to be performed. However, as an exercise in building, the Great Pyramid has not yet been surpassed.

The pyramids of Egypt

The pyramids of Egypt hold many secrets, and chief among them is the organization of the resources of the state only a few centuries after the dawn of history, after the unification of the two kingdoms, the one of Lower Egypt and that of Upper Egypt, by the falcon clan, "the followers of Horus." In later days Egyptian tradition attributed the tremendous and rapid upswing of the Amratian settlers in Upper Egypt to people coming from the east, and recent excavations also prove that the newcomers brought with them the superior technology developed in Sumer. It also appears from archeological evidence that these settlers came from the Red Sea and via Wadi Hammamat struck the Nile valley at Koptos. Present opinion leans toward the idea that they were not actually Sumerians but rather a seafaring people who participated in the Oman copper trade, perhaps the same seafarers from which the Phoenicians also claimed descent.

A monolithic state nurtured with blood

The newcomers found a narrow strip of land rich in farming potential but poor in everything else. There were no metals, no timber, nor a great many other necessities required for the new way of life. They set about to find these, and for the first time the clash of arms was heard in the formerly peaceful valley where the previous settlers had rested content with their own labor. They pushed northward downriver and eventually—led by King Menes, the chief of the clan—they defeated the King of the Delta in a battle lost in the mists of prehistory. All that remains as a reminder of this dramatic event is a slate palette showing the conqueror on a battlefield strewn with corpses. Egypt had been united and the seeds of a superior civilization sown on a field manured with blood. The King of the Followers of Horus became the "Lord of the Two Lands." By 3000 B.C. the fusion was complete; and the conquerors had built a new capital, "The White Hall" at Memphis, as a seat for the new government. The First Dynasty was secure in power, and a monolithic state established.

This was an entirely different state from the autonomous city states of Sumer where at least the early rulers were the tenant farmers of the local god. Here Pharaoh himself was god, from the very beginning; and although the local nomes, administered by his chieftains, had their own deities, no provincial gods could dispute the authority of the supreme Pharaoh. No Egyptian ever became a citizen: he remained for thousands of years a serf to the *corvée,* obliged to perform the work as ordered by the delegated authority of the distant Pharaoh. The Egyptian peasants and artisans were not slaves in the legal sense of the word, but they were tied to the land and always suppressed. So long as they were needed to till the fields and reap a harvest, they were left alone, but when the harvest was gathered and the tax on it collected, they were called to labor for Pharaoh: to dig ditches, clean canals, repair dikes, and do the unskilled porterage work in Pharaoh's quarries and on his building sites.

One may wonder how it was possible to subdue so easily, so early, and for so long such a gifted people living in one of the richest countries of the ancient world. The answer is no doubt found in the topography of Egypt. The settled valley is narrow, hemmed in by desert on both sides, and a central government could keep the population under control simply, effectively, and economically by patrolling the river. Wooded regions, deserts, or mountains have always been the best shields of freedom against suppression by a central authority. The poor Egyptian peasantry has always lived exposed and unprotected against ruthless suppressors. A barge with archers or a paddle-

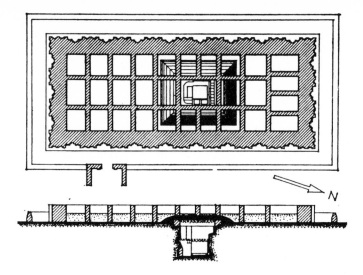

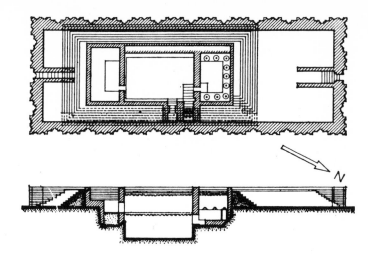

An early mastaba tomb from about 3100 B.C. still had a mound of desert sand heaped atop the burial shaft, as a reminder of the original burial pits in the desert. This sand naturally proved a poor foundation for the superstructure and eventually brought about its collapse.

The next step in the development of the mastaba tomb was to enclose the sand in a shell of brick and reinforce all four sides of the structure around the burial shaft by piling on mud brick in stepped revetments. These tombs have remained intact for nearly 5,000 years.

wheel gunboat is all that is needed to keep the narrow valley, to the first cataract, in submission.

Being a god, Pharaoh obviously could not accept death; a god lives forever. So, for that matter, do his relatives, the higher officers of his court and, since such ideas have a way of rubbing off, also the common people, provided of course that they could afford the luxury of a third-class embalming.

As mentioned briefly in a previous chapter, the burial customs during the Amratian period were simple and dignified. The dead were buried in pits dug in the sand enclosed in skins or a reed mat, together with their personal belongings: a few necklaces, their hunting arms, some pots containing food and drink. Those who could afford it liked to line the pits with boards lashed together with leather thongs at the corners.

This was the burial custom that the Followers of Horus brought along with them on their march to the north. But there was a weakness attached to this traditional burial method: the shifting sand eventually exposed the bodies to jackals and buzzards. To the Followers of Horus this end to their corporeal remains was too revolting to contemplate. Having now the means to do so, they lined their burial pits with brick and provided them with a solid superstructure of sun-baked bricks. Out of these early protected burial shelters developed the so-called mastaba tombs, and from these modest "benches" in the sand grew ultimately the fantastic pyramid tombs.

But before that happened, the mastabas had grown in size and pretension. They became copies of the noblemen's houses or royal palaces and, as is to be expected from a rising ruling class, the members could not bear the thought of having to fend for themselves in the afterworld. Accordingly, household and personal servants were buried with them. Some Egyptologists dislike the evidence of a household establishment being buried alive with the dead master and prefer to think that they were drugged or poisoned to death before interment.

Old customs are slow in dying. Although in the end the mastaba burial palaces had become poles apart from the simple Amratian burial pits in the sand, the two nevertheless had one feature in common. In the elaborately designed and decorated mastabas of the First Dynasty there was on the ground floor—immediately above the central burial pit holding the corporeal remains of the exalted person—a mound of sand overlaid with a protective covering of brick. This represented the mound thrown up over the simple Neolithic pit grave. Such is the conservatism of human customs.

The simple personal adornments and the pots filled with food and drink accompanying the Amratian peasant on his last journey were imitated in a similar fashion, and consequently the mastaba tombs became treasure caves attracting robbers. It became a contest between tomb builders and grave robbers to outwit each other. The burial pit with its gold treasure was made as difficult of access as

From mastaba to pyramid

The sketches on this page illustrate the evolution of the stepped pyramid from the mastaba tombs of the First Dynasty, as described in the text. Originally, a mastaba consisted of a low brick structure erected above the burial shaft and was provided with a large number of storage cells. In subsequent improved construction, the cells were moved underground to make it more difficult for grave robbers to reach them. At the same time, the structure was reinforced with stepped revetments of mud-brick.

Pharaoh Zoser's tomb began as a conventiontional mastaba on a square plan with a 207-foot side and a height of 26 feet. The mastaba was oriented in the four cardinal points. But instead of building it in mud brick, the architect Imhotep used stone and lined the entire structure with polished limestone slabs. This, no doubt, was one of the greatest innovations in building.

The further developments can be followed with the aid of the figures in the sketch at the top right. When the stone mastaba (1) was finished, its sides were extended 14 feet to a height of 24 feet whereby a 2-foot step was obtained along the four sides (2). Then the east side was extended 28 feet, which made the plan rectangular (3). But before the walls had been lined, the plan was changed again, and all four sides were extended 9½ feet. Stone was piled on top of the mastaba, changing it to a four-staged pyramidical structure (4). On the northern side, a funeral temple began to be built, but before it was completed the north and west sides were lengthened (5). After all four sides were extended once more, the height was increased to a six-stage pyramid and lined with polished slabs of Tura limestone.

The middle sketch shows the development of the substructure: a 23 × 23-foot shaft (6) was sunk into the rock and a sloping tunnel was driven northward for a distance of 66 feet, from where it continued as an open trench for a further 70 feet (9). The burial shaft was now sunk to 92 feet, and the grade of the adit was sharpened so that it intersected the shaft 40 feet above its floor. A funeral chamber of red granite was built on the bottom of the shaft, which after the funeral was backfilled with stone.

About 70 feet from the bottom chamber and parallel to its side were driven four galleries. A stepped oblique shaft led to the east and west galleries (11,12). But then the plan was changed, and after the mastaba had been extended (2), 11 shafts were sunk 108 feet along the east wall, and galleries driven from them in a westerly direction. However, the subsequent extension of the east side (3) covered over the shafts, and the galleries could be reached only by a long stairway on the eastern side. Up to the time of the final extension (5), access to the subterranean structure was by means of the open trench and the northern ramp (9). This trench was backfilled when the pyramid was extended toward the north (5), and a new adit (10) driven from the funeral temple replaced the old entry ramp. The temple (13) is thought to have been a copy of the palace in Memphis. A statue of the departed king was set up in the serdab (14).

The sketch below shows how the compound was surrounded by a 33-foot wall with a perimeter of one mile. The wall is built of stone and lined on both sides with slabs of polished limestone. It is provided with buttresses every 13½ feet, and 14 of them have been made extra wide and shaped into blind gates. The wall is thought to be a copy of the "White Wall" built by Pharaoh Menes around Memphis, capital of the united kingdoms.

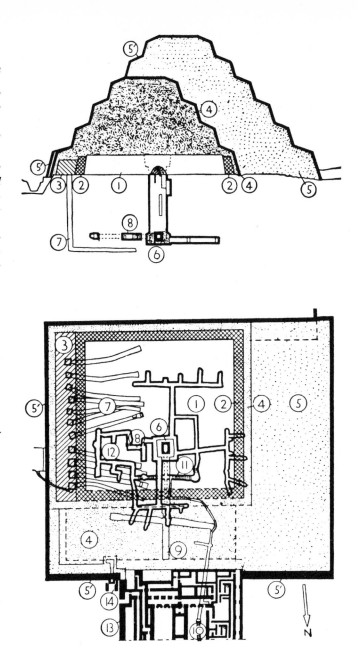

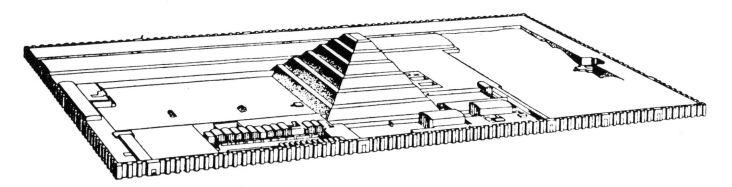

43

human ingenuity could devise. Shafts and tunnels, blind or real; ramps, staircases, and portcullis gates consisting of one, two, or more solid slabs of stone and carefully sealed and camouflaged in numerous ways to prevent detection; all these devices were used to protect the corpse and the treasures buried with it from robbers.

King Zoser and his architect Imhotep

But the irony of it all was that no matter how clever the means of hiding the illustrious body, grave robbers sooner or later found their way to it. And in the few cases where ancient grave robbers failed, modern archeologists succeeded. But that is a minor point. More serious from a religious point of view was the fact that these elaborate and costly burial palaces failed to accomplish what came natural to the simple burial pit in the sand, to wit, preservation of the body. In the simple burial pits in the Sudan excavated before the flooding by the backed-up waters of the High Dam, bodies of men buried in 2000 B.C. have mummified muscles stretching over their well-preserved bones. But the mummified flesh rapidly turns into a brown powder when shaken slightly in the air. In the splendors of the mastaba the body rapidly decomposed and thus defeated the purpose of the burial: to preserve the living dead for time without end. Eventually, this dilemma was resolved by chemical means: the body was embalmed, and if Herodotus is to be believed, the higher the status of the beloved departed, the more costly and more enduring was the chemical treatment of the body.

We are not here directly concerned with Egyptian religious and burial customs, and what has been noted above will have to suffice as a somewhat rough outline of developments leading to pyramid construction. The first pyramid, that of Pharaoh Zoser of the Third Dynasty, began as a mastaba, but before the royal architect Imhotep, the first builder to be recognized by name, was through with his commission, he had conjured forth something entirely different. Sophisticated inquiry has revealed how his mind worked, step by step, while he was occupied with this monument intended to be the eternal dwelling of his lord and master.

It has been made clear without doubt that what Imhotep set out to do on a site measuring 597×304 yards on high ground at Saqqara overlooking the capital of Memphis was to reproduce exactly the palace compound of the capital. To that end he enclosed the burial compound by a wall 33 feet high, made with an inner core of masonry and faced with Tura limestone (so-called because the stone derived from the underground Tura quarries situated on the east shore escarpment immediately to the south of present Cairo).

The walls were deeply recessed with rectangular bastions spaced $13\frac{1}{2}$ feet apart along the 1-mile perimeter. Four-teen were considerably larger and provided with faked gates carved in the stone. The actual entrance was situated near the southeast corner. There is now no doubt but that the wall was a copy in enduring stone of the "White Wall" built in mud brick overlaid with gesso, as erected by Menes around his new capital of Memphis.

Inside the compound was a large open court surrounded by numerous buildings serving different purposes, some of which were duplicated, since the ceremonies performed in them had to be repeated by the Pharaoh in his capacity of King of Upper and Lower Egypt. The buildings at the southeast corner deserve mention because they served a vital function in prolonging the reign of Pharaoh. A king to be any good must possess full physical vigor, and when he grew old or in other ways decrepit, it had formerly been the custom to kill him and crown a young man as his successor. But by this time the Egyptian priests had succeeded in resolving this unpleasant dilemma of an aging Pharaoh by substituting a religious ceremony—the *hebsed*—whereby the reigning Pharaoh after a certain number of years was able to regain his vigor through the exercise of magic. The hebsed ceremony involved the reenactment of the coronation when Pharaoh in stately procession led by his priests would enter the chapels along the hebsed court housing the nome gods of Upper Egypt and obtain their consent to his continued reign. He would then be placed on a dais and crowned King of Upper Egypt. The ceremony was repeated, in order to get the approval of the nome gods of Lower Egypt.

Having obtained this approval—or perhaps before which would seem logical—he would have to run a fixed course to prove that he possessed the physical agility so necessary for preserving the fertility of the land. The hebsed ceremony had to be repeated ad infinitum also after the body was interred within the cemetery compound.

Having built these important but architecturally subsidiary buildings, Imhotep began the work of erecting the tomb. He started in the traditional manner and built a mastaba with a square plan, each side measuring 206 feet, and gave the structure a height of 26 feet. The boxlike tomb was constructed of local stone and lined with slabs of Tura limestone.

But from all evidence he was not satisfied with what he saw—few architects are—so he added 14 feet to all sides and lined the extended structure with dressed limestone. Then he changed his mind again and added 28 feet to the east side, whereby the square became a rectangle. But before this new extension was lined the architect had an entirely new idea.

He began piling rock on the finished mastaba, using it as the first stage of a stepped pyramid rising in four stages. Toward the north he began work on a mortuary temple. Imhotep then once more used the prerogative of women and architects: he changed his mind again. He extended the structure toward both the north and west and began piling rock in a stepped formation until he reached the level of the fourth step, or the top of the original pyramid. By that time he had a better idea: now he extended all sides of the bottom stage and, after the sixth attempt, finally completed the structure which by then had reached the height of 204 feet. It was no longer a square; the base measured 411 feet east-west and 350 feet north-south. The whole structure was cased with dressed limestone from the Tura quarries.

The substructure of the stepped pyramid

Zoser's pyramid was a solid pile of rock containing nothing, but the rocky site under the original mastaba is honeycombed with underground excavations which, considering that they were ordered by one of the outstanding architects in history, show as little sense as the building history of the superstructure. But that may possibly be due to the work of robbers who have been as busy as moles tunneling toward the buried treasure, and so it is no longer possible to distinguish the original excavations from the galleries driven after Imhotep's time.

The main outline of the substructure seems clear. Inside the mastaba, a 23-foot-square shaft was sunk into the limestone foundation to a depth of 28 feet, and from this shaft a tunnel was driven northward for a distance of 66 feet, or beyond the northern line of the original mastaba and four-stepped pyramid. It continued as a sloping trench for a further 70 feet. The shaft was then sunk to 92 feet, and the grade of the tunnel was sharpened so that it joined the shaft 40 feet above the bottom.

At the bottom of the shaft was the burial chamber, measuring 9 feet 9 inches × 5 feet 6 inches and lined with pink Aswan granite. The roof had an opening at one end to admit the body, after which the tomb was sealed with a 6-foot granite slab weighing 3 tons. The sealing slab was stored in a chamber above the tomb until after the funeral, when the entire shaft was filled in with rubble.

This substructure would have been adequate for the purpose of providing a secure tomb for the exalted Pharaoh. But the architect had other ambitions, and so he began work on galleries running parallel to the sides and at a distance roughly 70 feet from the chamber. Access to the four galleries was by way of flights of steps leading from both sides of the adit. Apparently the idea was to line the walls of the galleries with glazed tiles in patterns resembling the reed mats covering the walls of Zoser's palace. At any rate, remains of such tile panels have been recovered in the east gallery, as well as some reliefs of the king performing religious ceremonies, possibly associated with the hebsed.

This subterranean work was done while the original mastaba structure was being erected. After the first enlargement of the superstructure, the architect had 11 shafts sunk to a depth of 108 feet into the rock along the eastern line of the mastaba. At the bottom of each shaft a gallery was driven westward, i.e., toward the center of the tomb. Apparently a chapel was meant to be erected over each one of the shafts and associated galleries, to serve as tombs for the members of the royal family. (Two alabaster coffins have been found at the end of the middle gallery.)

But the idea was abandoned, and as the mastaba was enlarged the second time, all 11 shafts became buried under the eastern extension, and access to the northernmost shaft had to be arranged by excavating a long and steep stairway terminating in an entrance building, just outside the east line of the final pyramid.

The open trench leading to the main entrance gallery also became filled in and covered by rock as the pyramid base extended northward. It therefore was necessary to arrange another adit to the main gallery leading to the tomb shaft. Beginning some distance to the north of the pyramid, a sloping stepped shaft was sunk into the rock, and from the bottom of the stairs, a tunnel was excavated, running roughly parallel to the former trench for a short distance, after which it curved to the east to join the gallery just where it began under the northern wall of the first mastaba. This was a stupid tunneling job, and one would be inclined to dismiss it as the work of robbers but for the opinion of experts that this tortuous gallery, interrupted by two separate flights of stairs to reach grade, was indeed the main entry from the outside.

The substructure of the original mastaba is a rabbit warren of tunnels, serving no sensible purpose except possibly as exploratory excavations by ancient grave robbers. Their underground work is almost as impressive as the original excavations. After finding their way into the subterranean structure, the robbers cleared the large burial shaft of its rubble fill, after first having underpinned the large stone which closed the opening of the shaft and carried the weight of the superstructure. They plundered the royal tomb thoroughly of all its contents and emptied the 11 family tombs as well, except for the two alabaster coffins. It was of course primarily the metallic treasure they were after, because they left thousands of vases and vessels made of alabaster, porphyry, quartz, and rock crystal for recovery by modern archeologists. Indeed, a lot still remains in the royal pit tombs, some of which are heaped full from floor to ceiling with beautifully executed stone vessels.

The Stepped Pyramid of Zoser has long been regarded as the first major stone building ever erected. No links have been found between the poor attempts at using stone roughly dressed as paving blocks in the tombs of the First and Second Dynasty kings and the mason's art springing up suddenly fully developed in the magnificent Zoser structure. Perhaps it is true, as has been suggested, that the revolutionary change was brought about by the individual genius of Imhotep. Nonetheless, it should be recalled that by the time Imhotep kept changing his mind, groping for architectural solutions satisfying his ambition (ca. 2700 B.C.), there was already a large body of craftsmen capable of shaping beautiful vessels out of the hardest rock. Indeed, the traditional skills of stoneworking extend much farther back into Neolithic times than the art of potting. Shaping a vessel in porphyry is a far more intricate and difficult job than shaping a column in granite or polishing a slab of marble. The skill was there and highly developed; it awaited the call from somebody with the imagination to put it to work for other ends. When Imhotep decided to execute his structures in durable stone instead of the traditional materials of mud brick and wood, there was an ample supply of stoneworkers to do this. It was up to him to tell them what he wanted done.

Conventions frozen in stone

As previously noted, Imhotep was not at all sure of what he wanted done. The whole structure, as well as its details, shows that he was groping his way, never quite sure of what he was doing except executing traditional building elements in stone. He had reproduced in stone the ornaments, patterns, and shapes previously developed for use in wood, reed, and brick. The columns derived from reed bundles or palm trunks; the fluted ones recall the grooves made by metal tools on tree trunks; the ribbed ones imitated bundles of reeds used as fenders to protect exposed corners of friable brickwork. The stone columns were painted red except for a narrow white band 2 feet above the base and a black one at the bottom. These details were also reminiscent of conventional construction, the narrow band representing a reinforcement ring of copper, and the black the protective leather placed around the bottom of wooden columns. The stone ceilings were likewise retrospective; the blocks were laid on edge with their lower edges rounded to give the appearance of tree trunks put side by side.

The architect evidently had his doubts about the structural strength of the stone columns and refrained from using them as detached free-standing roof supports. They are always incorporated with the masonry, either attached to the main wall faces or at the end of flywalls jutting out from them.

The building stone used also differs from the cyclopean blocks subsequently employed in Egyptian dry masonry. The stone was quarried and dressed in small blocks that could be handled without mechanical devices. The blocks were dressed so as to obtain an accurate joint to a depth of only an inch or so from the face; to the rear and unseen was a widening unbonded gap between adjacent blocks. In other words, structural solidity was sacrificed for external appearance. This is not the only instance of Egyptian disregard for elementary structural requirements, and the construction during the New Kingdom shows numerous examples of how structural soundness was ignored for theatrics.

Zoser's Third Dynasty successors continued the fashion originated by him. Pharaoh Sekhemket built himself a seven-step pyramid in a compound, also at Saqqara and immediately to the southwest of the Zoser monument, but it was not finished owing to his short reign. Another differently constructed pyramid was erected at Zawiet el-Aryam, between Giza and Saqqara, and believed to have been ordered by Khaba, an obscure king of the Third Dynasty. Farther south, at Serta, Zawiet el-Mayitin, Naga, and El-Kula (between Luxor and Aswan) there are also four small stepped pyramids of which little is known.

The really remarkable developments in pyramid building had to wait another century, upon the crowning of King Cheops, second king of the Fourth Dynasty.

King Seneferu's three pyramids

The transition from the stepped to the true pyramid occurred in the building of three pyramid structures, all of them attributed to Pharaoh Seneferu, founder of the Fourth Dynasty in 2680 B.C. The first one, at Meidum 30 miles south of Memphis, began as a small stepped pyramid which was enlarged to a seven-step structure with the sides inclined about 75°. The next phase consisted of raising the top 45 feet and widening the structure all the way to the base. The exposed faces were lined with dressed Tura limestone.

Then, for some reason not understood, the unknown architect filled in the steps with local stone to form a pyramid the faces of which were cased with dressed limestone from the Tura quarries.

No change was made in the original substructure which consists of a $19\frac{1}{2} \times 8\frac{1}{2}$-foot tomb chamber partly built into the center of the superstructure and partly excavated in the rock. The roof of the chamber has a corbeled vault. From the tomb a vertical shaft descends into the rock where it is intercepted by a level gallery provided with recesses on each side for storing the stone slabs used for sealing the approach to the tomb. After 31 feet the gallery angles 28° upward and continues on this gradient for 190 feet where it pierces the northern face slightly above the first step of the original structure. Against the east side was a small mortuary chapel joined by a causeway down to a valley building situated at an elevation barely reached by the inundation of the Nile in flood.

The pyramid was investigated in 1882, and from inscriptions—so-called graffiti—scribbled on the walls by visitors during the Eighteenth Dynasty (ca. 1550–1350 B.C.) the structure was attributed to King Seneferu. Nothing was found in the tomb, and it is doubtful whether it had ever been used.

To the right is a pile of Cyclopean blocks used in the wall around the citadel of Tiryns, contemporary with Mycenae. Below is a woodcut from the Nürnberg Chronicles purporting to be Babylon, a conglomerate of nonrelated elements. The column with the kingly statue is frequently used in the town views and is likely to have been inspired by some picture from Rome. This is how medieval scholars and artists visualized a distant and to them unknown world, and is one reason why it is more difficult to obtain an accurate idea of medieval engineering than, for example, of Egyptian, Greek, or Roman work. Not until the sixteenth century did technically useful illustrations become available north of the Alps.

Immediately beyond the wall in the northeastern part of the citadel of Mycenae is the Perseian well, an underground cistern that supplied the citadel with water. A staircase with 16 steps led to a landing 9 feet belowground. Here, three steps turn the stairway in a different direction, whereupon follow 54 steps down to a 5 × 2 × 16-foot basin topped by a V-shaped roof 13 feet above it. The basin is situated 38 feet below ground level. The water derives from a terra-cotta pipe placed in the roof. The picture shows the second stairway section with its 20 steps.

In the southeast corner of the curtain wall of Mycenae and well shielded from the outside is a postern built by corbeling uncut megaliths. The length of the postern is 23 feet, or the same as the thickness of the wall; the width varies from 3½ to 4 feet, and the height is 8 feet. The construction is the same as at Tiryns. The postern was never closed by a gate and was defended by a small force stationed in a 40 × 105-foot court inside the wall. Through the postern the garrison could make sorties and attack an enemy threatening the Perseian well in the flank.

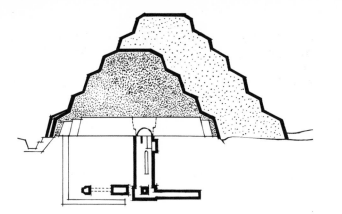

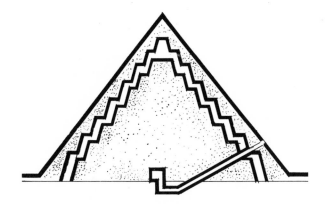

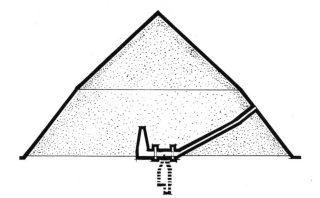

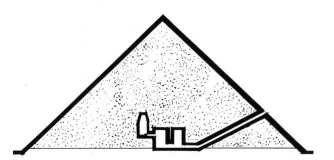

Evolution of the pyramid

The development of the pyramid during the Fourth Dynasty, particularly under the reign of Seneferu, is shown in the four sketches to the left.

On top is Pharaoh Zoser's stepped pyramid from 2680 B.C. Then follows the Meidum pyramid, probably built for Pharaoh Seneferu, whose Horus name was Neb-maet. It was originally constructed as a stepped pyramid with seven stages, after which the sides were extended and the height increased by 45 feet, whereupon the structure was lined with polished slabs. The adit was on the north side where the gallery slopes 28° for a length of 190 feet. It then runs level for 31 feet to a raise, leading to an 18.5 × 19.5-foot burial chamber, the floor of which is level with the ground.

The third pyramid in the series is the so-called Bent Pyramid at Dahshur which has a square plan with a 620-foot side, as against 473 feet for the Meidum structure. It was built from the outset as a true pyramid, but somewhat above half the height the angle of the slope changes from 54°31' to 43°21'. The break has been explained by the necessity (for some reason or other) of finishing the work as quickly as possible, and the masonry above the break is of poorer quality than that below it.

The pyramid has two adits, one on the west side and one on the north side. The latter is situated 39 feet aboveground, and the 241.5-foot gallery slopes 24°24' to a 16-foot level vestibule provided with a corbeled roof 4.5 feet above the floor. The hall leads to the lower of the two chambers that measure 20.5 × 16.5 feet with a height of 57 feet. The corbeled roof has a 3 × 1-foot capstone.

The west exit is situated higher up on the side, and the gallery slopes at first 30°09' and then 24°17' for a distance of 211 feet, where it reaches ground level and continues horizontally 66 feet, ending in a chamber with a corbeled roof.

The pyramid was never used, except for the burial of one owl and five bats interred in a wooden box sunk in the floor of the upper gallery. However, since the Horus name of Seneferu has been daubed in ocher on a couple of stone slabs, it is assumed that the Bent Pyramid was erected during his reign.

Immediately to the north of the Bent Pyramid is a third one having a square plan with a 710-foot side. The sides slope 43°36' instead of 52°52', which is the right batter for a geometrically correct pyramid. The adit is situated on the north side, and the gallery through the masonry has a 27° gradient and ends in three chambers placed in line, one of which is positioned exactly on the vertical axis of the structure. The chambers measure 31 × 12 feet and have corbeled roofs. The third chamber has its floor 25 feet above the ground which constitutes the floor of the other two. From stone markings it has been deduced that this pyramid, too, was erected for Seneferu.

Normally, such evidence would have been accepted without doubt, were it not for the fact that 28 miles north of Meidum, at Dahsur, there are two pyramids of which the southern one was most definitely built by Seneferu and probably also the northern one. The southern pyramid was built on a square plan, each side measuring 620 feet, and its original height must have been 336 feet. It is the best preserved of all surviving pyramids, owing principally to the fact that the stones were laid with an incline toward the interior, whereas in other pyramids they are laid in level courses. No other pyramid has retained so much of its Tura limestone casing. Its unique feature is that at slightly above half of the total height the angle of incline changes abruptly from 54° to 43°, and this break has given the structure the name the "Bent Pyramid." It looks as though the builder had finished off the work in a hurry; indeed, the masonry above the bend was found to have been laid with less care than that below it.

The substructure contains two chambers, one within the pyramid and the other excavated in the rock and provided with a corbeled vault at ground level. Two separate sloping adits connect the northern and western sides. Structural weakness must have developed during the building because the chamber within the pyramid has had to be trussed up with large cedar timbers, and cracks in the walls have been filled with plaster. The "Bent Pyramid" started out as an attempt to build a true pyramid, but for some reason or other it aborted. Nonetheless, the pyramid and its associated buildings have revealed a wealth of details of the utmost interest to Egyptologists and other experts.

Of perhaps greater interest in this context is the northern pyramid at Dahshur, built on a square base. This is the first structure that was built as a true pyramid, although the inclination of the sides is only 43°, as against 52° for those that were to follow. An adit in the northern face has a sloping gradient of 27° and leads to a series of three burial chambers, each one measuring 31×12 feet, with the floor at ground level. The chambers have corbeled vaults which in the third one rise to a height of 52 feet.

Judging from the red ocher quarry markings on the casing blocks, which have been read as Neb-maet, the Horus name of Seneferu, this pyramid too was built during his reign. (Each block delivered from the quarries was marked in red ocher by the quarry gang that had worked it, such as the "Scepter Gang," "North Gang," "Vigorous Gang," and so on.) Assuming that this is true, the three pyramids attributed to the first king of the Fourth Dynasty have been estimated to contain 9 million tons of rock, excluding all ancillary structures such as funeral temples, causeways, and enclosures. Erecting piles containing 9 million tons of rock within the reign of one Pharaoh in order to provide three alternate tombs for his exalted body makes no sense even when transgressing the border of the murky realm of Pharaonic megalomania, and the intriguing problem of these three pyramids will have to be left for future generations to resolve. They have been touched upon here merely to suggest that when the magic of the priests failed to put old Seneferu through his hebsed and he was put away — only his grave robbers knew where—his son and successor Khufu, better known under his barbaric Greek name of Cheops, had no need to engage in development work when as his first enactment he ordered his eternal abode to be built. The true pyramid, the only tomb fit for a chalcolithic ruler, was there, for his architects to copy.

The Great Pyramid at Gizeh

*"But Cheops brought the people to utter misery
. . . he compelled all Egyptians to work for him."*
HERODOTUS

Cheops chose as a site for his tomb a limestone plateau on the western edge of the desert about 5 miles to the west of the river and 10 miles north of Memphis. His son and successor Chefren and his grandson Mycerinus also had their tombs built nearby, slightly to the south. In an outcrop left between the quarries that supplied the stone to the Great Pyramid, Chefren has his portrait fashioned, the head of the Giant Sphinx being in his likeness while the rest of the sculpture is a recumbent lion. Together, the four form without a doubt the most magnificent group of monuments ever produced.

Here, however, the interest will be centered on the Great Pyramid which stands at the apogee of not only pyramid construction but of building generally. There have been subsequent structures, such as power dams, that compare with and exceed the Great Pyramid in sheer volume, in earth excavated, and concrete poured; but in skill and accuracy of building nothing can touch this pyramid. By the time modern dams are gone, the great pile of rock erected by Cheop's builder will still be standing. It is indeed a fantastic structure, the foremost megalith in history, and this chapter will therefore be concerned with it.

The dimensions of the Great Pyramid are known with great accuracy, having been measured in 1925 by G. H. Cole of the Egyptian survey service, with the application of geodetic survey methods. They are as follows:

Geodetic survey of the Cheops Pyramid

Side	Length, meters	Azimuth	Angle (calculated)
North	230.253	89°57′32″	90°03′02″
East	230.391	359°54′30″	89°56′27″
South	230.454	89°58′03″	90°00′33″
West	230.357	359°57′30″	89°59′58″
North		89°57′32″	
Diagonal SW-NE	325.699	44°56′55″	90°00′08″
Diagonal SE-NW	325.868	134°57′03″	
N-S axis	230.374	179°56′00″	90°01′48″
E-W axis	230.354	269°57′48″	
General mean	230.364	3′06″	

The original height can no longer be determined with this degree of accuracy since the top is missing, but it has been calculated to 481.4 feet. The sides incline at an angle of

Cheops' was the first geometrically true pyramid built. It has a square plan with a 755-foot side. The original height was 481.4 feet and the batter of the sides was 51°52′. The entrance is on the north side about 55 feet from the ground, and from there a gallery sloping 26°31′ leads into the rock under the pyramid for a distance of 345 feet to an abandoned chamber in the vertical axis of the structure. From this 3-foot 5-inch × 3-foot 11-inch gallery, a raise with the same dimensions has been driven 60 feet from the portal, running at an incline of 26°2′ for a distance of 129 feet and ending in the 153-foot-long Grand Gallery leading to the 34-foot 4-inch × 17-foot 2-inch burial chamber dressed with pink Aswan granite. The nine granite slabs forming the roof weigh 400 tons and above them are four compartments with flat roofs and a fifth with a pointed roof. The purpose of this construction was to relieve the weight of the rock piled on top of the burial chamber.

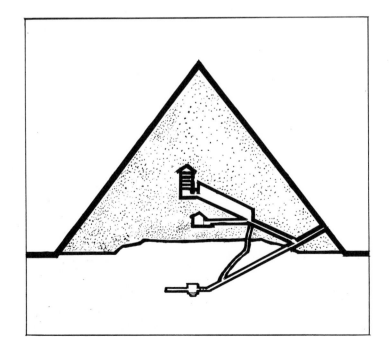

51°52′ to the horizon; i.e., they differ by only one angular minute from the theoretical batter. The area of the base is 13.1 acres.

To build this structure required 2.3 million stone blocks, each block averaging 2.5 tons in weight, although some weigh as much as 15 tons. The granite slabs of the "King's Chamber" weigh 50 tons apiece. The total weight of the structure would be about 6 million tons.

The pyramid as it stands today has suffered greatly from having been a quarry for building stone since the Middle Ages. It had remained intact, although robbed, until the ninth century A.D. when the son of Harun el-Rashid ordered his men to break into the structure to see what it contained. Since then the fine limestone casing has been removed; it can be found incorporated in buildings in Gizeh and Cairo.

In its original state the Great Pyramid was cased with polished limestone laid with such unbelievable accuracy that the joints between neighboring slabs measure 0.01 inch, a precision used in machine components. The tremendous area of the four sides, from the bottom to the top, diagonally or from the side—no matter how viewed—presented an unbroken flat surface. The top consisted of highly polished granite sculptured with reliefs and gilded. It was without doubt an impressive sight, particularly when seen at sunup and sunset.

But there was more to it. Close to the east side of the pyramid was a 171 × 132-foot mortuary temple consisting mainly of an open court with a row of square granite columns and walls decorated with low reliefs, and probably one or more statues of the king placed in recesses. From the east wall of the temple began a walled-in causeway leading to a Valley Building which was located near the line of the highest inundation with the Nile in flood. The Valley Building and the Mortuary Temple served separate important functions in the funeral ceremonies. In the former the exalted body was embalmed, in the latter it was subjected to the final rites.

The pyramid compound was surrounded by a high wall, and outside, to the east and west of it, was a large number of mastaba tombs, the eastern ones for royal relatives, the western for court dignitaries.

The galleries in the structure

The internal arrangements of the galleries and chambers in the Great Pyramid differ radically from those in the previous structures, indeed from those of any other pyramid. The first design was conventional in concept: from a portal on the north face, about 55 feet above the base, a gallery was driven into the structure with a sloping gradient of 26°31′23″. It measures 3 feet 5 inches in width and 3 feet 11 inches in height, and extends on this gradient for 345 feet, where it becomes level for a distance of 29 feet and terminates in a chamber measuring 46 × 27 feet 11 inches, with

a height of $11\frac{1}{2}$ feet. It is only possible to descend the gallery in a crouched position, and getting in the Royal Body enclosed in its three layers of coffin could not have been a very dignified operation.

The large burial chamber was never finished and provides therefore an excellent illustration of the methods of mining used at the middle of the Third Millenium B.C. On both sides of the chamber a trench descends from the ceiling to the floor, just wide enough for a man to work in. On the top, too, rock has been excavated leaving a space below the ceiling for a man to work in a crouched position. The excavations left a bench to be wedged out, but this work was never executed.

The north side of the unfinished chamber is in line with the vertical axis through the apex of the pyramid. In the south wall, opposite the adit, is a narrow tunnel extending for about a hundred feet and then abandoned. A square pit or shaft in the floor in front of the bench suggests that the architect had a plan for deepening the chamber, but nothing came of that idea because the entire layout of the substructure was abandoned.

Instead it was decided to place the burial chamber within the superstructure of the pyramid. The miners were ordered back to about 60 feet from the portal and from here they struck out on an ascending grade of 26°2′30″. By the time this work started, the building of the superstructure had progressed to above this level, and the miners had to chisel their way through the masonry already laid. This ascending gallery has the same dimensions as the descending adit for a distance of 129 feet. The interesting thing about the gallery is that while the lower part is cut through stones with irregular joints, the masonry in the upper part has close-fitting joints. In other words, somewhere along this gallery the tunnelers broke out in the open and caught up with the masons erecting the pyramid. The rest of the gallery was built, not tunneled.

This ascending gallery also reveals an important feature of the construction of the pyramid: at intervals of 17 feet 2 inches, the gallery pierces a different type of masonry which the early investigators called "girdle stones" but which actually consist of internal casing slabs placed at an angle of 75° to the base and obviously serving to interrupt lateral movements in the large structure where the stones are laid in level course.

The ascending gallery ends in a level approach to a chamber measuring 18 feet 10 inches × 17 feet 2 inches, having a pointed roof with a maximum height of 20 feet 5 inches. The chamber is situated right on the vertical axis of the pyramid and is traditionally known as the "Queen's Chamber." In actual fact it was never finished as the scheme was abandoned before even the floor was paved.

Instead, a much more grandiose plan was carried out. The ascending gallery was continued on its gradient for a further 153 feet, but was widened to 5 feet 2 inches and the

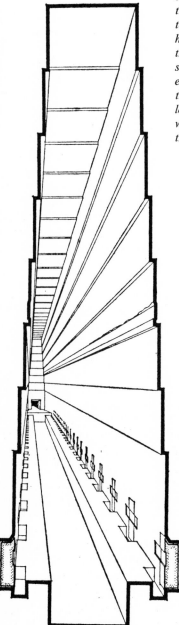

The Grand Gallery in the Cheops pyramid forms a 153-foot-long ramp with an incline of 26°2'. The walls consist of polished limestone slabs placed so that a course extends 3 inches beyond the face of the one below, producing a width at the roof of about 3 feet 5 inches. The roof slabs rest in indentations cut in the top course so as to prevent the load of the slabs from accumulating along the slope. In the fourth course, there is a continuous, 1-inch-deep scoring that may have been used to anchor the timbers carrying the sealing slabs before the funeral. The two side ramps appear to have been used for hauling up the heavy coffin against the steep grade to the funeral chamber. The original plan seems to have been to backfill the entire gallery with stone blocks, but in the end only the chamber and the lower end of the gallery were closed with accurately fitted sealing slabs, three at each end.

height was increased to 28 feet. This so-called "Grand Gallery" is lined with polished limestone, each of the seven courses being 7 feet 6 inches in height. Beginning with the second one, each course corbels about 3 inches, and the 3 foot 5-inch-wide space at the top is spanned with roofing slabs laid in such a way as to avoid transmitting the total weight of the slabs along the sloping ceiling. The weight of each slab is taken up by an indentation at the top of the wall.

The Grand Gallery ends in an antechamber which has three sides lined with Aswan granite. The east and west walls are provided with four slots intended for portcullis slabs for closing the passage to the King's Chamber beyond.

The King's Chamber is lined with large granite slabs and measures 34 feet 4 inches from east to west and 17 feet 2 inches from north to south. The height of the rectangular room is 19 feet 1 inch. The entire chamber lies slightly to the south of the vertical axis of the pyramid. Near the west wall stands a lidless granite sarcophagus which no doubt at one time contained the coffin holding the king's embalmed body. The sarcophagus is too wide to have been brought through the galleries and must have been placed in the chamber when it was being built. For some reason or other it was never finished, and its surface still shows the marks of the tools used in making it.

All this work was obviously done in the open while the level courses surrounding the chamber were put in place. Having finished the chamber with nine slabs of granite, each weighing nearly 45 tons, the architect had qualms about its structural endurance. After all, these slabs had to carry a lot of rock extending all the way up to the top of the pyramid. Had he known better, he could have bridged the span in a number of ways, but since he did not know what he was doing, he built a series of five compartments above the ceiling slabs, apparently with the idea that they would relieve the weight. The first four had flat ceilings, while the fifth was given a pointed roof.

There is one more strange feature that should be noted in connection with the chamber. In the north and south walls and about 3 feet above the floor are the openings of two small oblique shafts, the southern one extending at an angle of 31° and the northern one at 45°, piercing the entire structure and terminating about halfway up the two faces of the pyramid. The shafts are too narrow to have been driven and could only have been built. Just what their purpose was has never been explained.

However, there can be no doubt about the purpose of a crooked and poorly executed raise joining the adit with the bottom of the Grand Gallery. This was used by the funeral laborers and the supervising priest after they had closed the tomb with its three portcullis slabs and plugged the opening of the oblique shaft leading to the bottom of the Grand Gallery. Without this subsidiary raise they would have shut themselves up inside the pyramid. When they had slid down the raise supported by a rope, they sealed the bottom openings of the two raises with stone slabs, one of which fell down when Caliph Ma'mun's men forced their way into the structure in the ninth century.

Construction of the Great Pyramid

The first important consideration preliminary to the construction of a pyramid was the choice of a proper site. For political reasons it had to be close to the capital, for religious reasons to the west of the Nile. For structural reasons the site had to be capable of supporting a pyramid weighing several million tons. Only a rocky site free of faults would be capable of carrying such a giant structure. Logistics also played an important part; there should be suitable building rock for quarrying reasonably close by.

The limestone plateau to the west of Gizeh met all the requirements. There was plenty of good rock on the site, and across the river, less than ten miles distant, were the Tura quarries with their fine-grained limestone free from fossils.

Once Royal approval of the Gizeh site had been obtained, the first operation involved stripping it of sand and investigating the bare rock. The base of the pyramid was laid out roughly and the site reduced to a level plane, but to save material a mound of rock was left in the center. Most likely the site was then accurately leveled before the final orientation of the structure was determined.

Leveling off the site presented no particular problem to people accustomed to excavating irrigation canals for many centuries. The final accurate leveling was accomplished by building a low bank of mud around the perimeter and filling the enclosure with water, the surface of which served as a plane of reference. It does not greatly matter whether a network of trenches was cut so that the bottom of each trench would be equidistant to the water surface, or whether holes were drilled to the same depth, according to a grid layout. The Egyptian excavators of 2500 B.C. had the means and the know-how of doing it either way.

What is remarkable is the accuracy attained. When checking the site with a modern geodetic level in 1925 the results were as follows:

Side	Elevation, meters above sea level	Difference, millimeters
North	60.405–60.413	8
West	60.413–60.415	2
South	60.415–60.426	11
East	60.419–60.422	3

This is pretty good leveling, much better than the common run of work done by modern engineers for a building site or for roadwork. The west side comes close to satisfying the precision of work laid down by the U.S. Geological Survey which specifies a maximum error of 0.05 foot per mile.

With the site level, there followed the most critical phase in the construction of the pyramid from a theological point of view—the orientation of the structure. It had to be aligned so that all four sides faced the cardinal points. To do so one of the sides—the east or west one—had to be fixed by astronomical observations to true *north*.

The sketch illustrates a method of obtaining astronomic north for the purpose of orientating a pyramid or other important official structure. After the builders had chosen a corner (either the southwest or the southeast one) they erected a semicircular brick wall with a radius of about 10 feet (using the corner as a center) to a height of about 6 feet. The top of the wall was made level and finished with a polished stone slab. On the sketch, a priest is seen holding a plumbline over the corner (a firm support for the plumb bob would have been better) and observing where a star, such as Vega, rises above the wall. This position was marked on the slab, and when the same star descended in the west, a similar mark was made. The chord between the two markings was evenly divided, and the point obtained projected to the ground by means of a plumbline and was marked. The line from the corner to the set point on the ground was true north. Since the resolving power of the human eye is one minute of arc, that was, theoretically, the best accuracy that could be attained, but by making several observations and taking their mean, it may have been possible to establish true north to within a minute or so. When, after repeated observations, using different stars and observers, the head priest was satisfied with the result the wall was torn down and the short line secured on the ground was extended to the full length of the pyramid side.

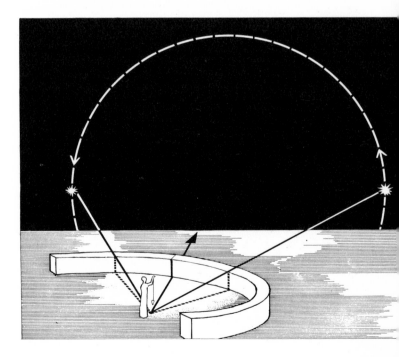

Compared to the Sumerians and their successors, the Egyptian priests were poor astronomers and appear to have contributed little or nothing to the science. But they were able to apply in practice the little they did know, and that included the accurate determination of true north.

Duties of the Astronomer Royal

From the azimuths given on page 51 the west side is seen to deviate 2′30″ to the west of north, while the east side deviates 5′30″, also to the west of north. The original observations were made with the naked eye, and since the resolving power of the human eye is one minute of arc, the best theoretical accuracy that can be attained is one minute, and five minutes is regarded as good for naked-eye sighting. Hence, an error of only 2′30″ must be accepted as a measuring performance that no modern engineer is capable of attaining without instrument aid.

Now, how was it done? Had there been a star in the celestial pole, the operation would have been easy, or simply a matter of aligning two stakes on the ground in the direction of the star. But in 2600 B.C. the nearest star to the pole was Tuban in Drago, and this star revolved around the pole in the manner of Polaris in Ursa Minor today. Without an angle-measuring instrument and access to accurate siderial time, it would not have been possible to use Tuban, and a more complex method would have to be employed.

The Egyptian priests of the Old Kingdom were thoroughly familiar with the regular behavior of the celestial bodies, and the method about to be suggested, although not depicted or described in writing, would occur to them. Briefly, what they would do would be to bisect the angular distance between the rising and setting positions of one or more stars. On a completely flat ground extending to the horizon such observations would present no difficulty. They would mark a point on the ground and hold a plumbline suspended over it. When a star arose on the eastern horizon, they would have somebody place a point some distance away marking that direction. When hours later the same star set in the west, that direction would be marked on the ground. The distance from the observer and the two points set would have been made about equal. Then, by dividing the distance separating the two points in equal parts, the line from the observer to the midpoint would be true north. It is as simple as that, in principle.

In practice it was a great deal more complicated, because nowhere in the Nile valley and definitely not on the Gizeh plateau was there such a plane surface extending toward the horizon. The surveying priests would have to build an *artificial horizon,* most likely consisting of a semicircular wall placed toward the north and somewhat higher than the eye of the observer. The top of the wall would have to be level.

The observer would place himself at the center of the circle, which would be either the southwest or southeast corner of the future pyramid, and with the aid of a plumbline or sighting rod placed over the mark, note where a star rose over the wall. The exact position would be scratched on the top of the wall which probably consisted of a slab of polished limestone. Similarly, when the star vanished below the wall toward the west, that position would be marked.

As star after star rose and set, their positions on the wall would be marked, whereupon the chord between each pair of scratches would be bisected. The mean position of a large number of bisections obtained in this manner would be accepted as true north, and the point projected by means of a plumbline to the ground and marked.

The Cheops pyramid is located at Latitude 30°N, and with the wall facing north the number of suitable stars would be limited to those with declinations between 30° and 60°. A star like Andromeda would transit in zenith and could not be used, and Ursa Major with a 60° declination is a circumpolar star that always remains above the horizon. The number of first-order stars would be limited to Cygnus, Perseus, Capella, and Vega.

If the wall were turned toward the south, the number of good observation stars would increase immensely, since the observer would then get access to all the bright zodiacal stars of the first and second order, such as Arcturus, Regulus, Spica, and many others. The meridian would be marked toward the southern horizon and the line projected northward.

With the point established, the wall was torn down, and the line from the corner to the midpoint was projected to the full length of the side of the pyramid. The length was laid out in Royal cubits and was apparently made 440 cubits long. (A Royal cubit = 524 ± 5 millimeters.) For measuring the length, cords of flax fibers (as shown on page 57) or wooden rods were used, most likely the latter.

Once one side had been established, perhaps the western one, right angles were laid out at both ends. Just how this was done has not been adequately explained. Most authorities seem to rest content accepting the use of an ordinary mason's square, basing their assumption on the near-perfect right angles of many Old Kingdom buildings.

To assume, however, that the leg of a right angle laid out with a set square—no matter how accurate—can be extended a distance of 230 meters (760 feet) is to underestimate the difficulty. It simply cannot be done this way. Since Egyptian cadastral surveyors operated with the so-called *double remen,* i.e., the length of the diagonal of a square with sides equal to a Royal cubit, or the theorem of Pythagoras in an early form, it is reasonable to assume that they applied the theorem to lay out the right angles of the pyramid, using either a long measuring cord or, still better, a long copper or gold wire permitting them to lay out a sufficiently long leg that could be protracted without risk of cumulative errors. (For another alternative see the note on Egyptian survey practice on page 268.)

Once the 440 cubits had been measured along the astronomically determined line, the second corner of the pyramid was established, and a right angle was laid out from this corner and the leg of the angle projected for a distance of 440 cubits. Here the procedure was repeated to obtain the fourth corner. Joining the fourth corner with the first completed the square. Unfortunately, the knoll in the center prevented checking the accuracy of the square by measuring the two diagonals, which differ in length by 169 millimeters. Most likely, on such an important structure, astronomical observations would be taken also at the fourth (southeast) corner to check the orientation of the eastern side, and the lines corrected for any errors detected. The procedure would be repeated, checked and rechecked until the priests were satisfied that the structure was properly oriented and the sides equally long.

After the plan of the pyramid had been laid out in some such fashion, the foundation ceremony followed wherein King Cheops went through the motions of laying out the lines, assisted by a priest, and using a simple instrument known as the *merkhet,* a short calibrated bar with a projection at one end which threw a shadow along the bar and served as a sun clock. It was a sacred instrument deposited in the care of the god Thoth.

When the King, his high priests, and captains had departed after leaving the seal of approval, the construction could begin. A quarry was opened a short distance to the south of the site to obtain a supply of the *nummulitic* limestone, the highly fossiliferous rock of the Gizeh plateau, for use in the core. Across the river in the Mugattam hills work began in the Tura underground quarries breaking out the casing blocks, and far away to the south, at Aswan, quarrymen began pounding away at the weathered granite to reach the sound rock below. Work also commenced on a causeway leading from the edge of the river when in flood, to be used at first as a transport ramp for the Tura and Aswan stone, later as a permanent causeway linking the Valley Temple with the Mortuary Temple close by the pyramid.

Rock quarrying

By the time of the pyramid construction the art of quarrying was thoroughly mastered, and the methods did not differ in principle from those now used. In the open quarries south of the site, the stone was taken out by benching, after the eroded "bark" had been stripped off. To get out a block by blasting, two free surfaces are required. Since no blasting was done, the miners had to excavate two parallel trenches at right angles to the free surface, thus outlining a bench. By cutting a line of wedge slits along the top and

bottom of the block and then driving wedges into the slots, the block was cracked loose and, depending on the size, split into a varying number of $2\frac{1}{2}$-ton stones. Some experts claim that wooden wedges were used which after having been driven into the slots were drenched in water. The resulting swelling produced the force necessary to split a limestone block loose from the "living rock." Whether the investigators actually have tried and succeeded in breaking out such large blocks in this manner does not seem clear. However, whether quarrying in actual fact was performed with hardwood wedges is of minor importance. Being state quarries supplying building stone for the most important monument of the age, they would have been provided with copper for toolmaking from the royal temple stores. Using wedges hardened by hammering and enclosed in feathers of annealed copper strip, the job of splitting rock could be done in the same efficient way as always is done by men who after long practice know automaticaly how a certain rock will behave. (For Egyptian copper metallurgy see the note on page 264.)

In the underground quarries in the Mugattam hills where the close-grained Tura limestone was produced, the method was rather similar, except that to obtain a bench, rock had to be cut to waste also along the top of the bench. The method used can be studied in detail in the Mugattam quarries, as well as in the abandoned chamber under the pyramid. A top heading of the same width as the block was driven at a height permitting the miners to work in a crouched position. At both sides of the heading a trench was sunk to outline the bench. From the bench thus obtained, blocks of desired thickness were wedged out as in the open quarry until the bench was consumed, whereupon the cycle began all over again.

For cutting the wedge slots and trimming the dimensional stone, hammered copper chisels were used. In the Aswan granite quarries the hard rock was worked with dolerite balls, for benching out the block as well as for dressing it after it had been broken out. For the mining necessary to get through the weathered and useless surface "bark" there might have been occasion to use firesetting, i.e. heating the rock with wood fires to make it spall off in slivers an inch or more in thickness.

It has been suggested, indeed reported, that Cheops' granite sarcophagus shows marks that have been recognized as being derived from the use of saws, no doubt in conjunction with some abrasive. As for the abrasive, there have been several suggestions—from sophisticated corundum to quartz sand. Just what would have been accomplished by sawing granite and other hard rock in preference to the straightforward method of trimming the surface with a chisel followed by polishing it with an abrasive, such as quartz sand and water, has never been made clear. Sawing hard rock, a method applied only in recent decades, is done merely to save material and labor: a good piece of granite

The duties of the Egyptian cadastral survey included the restoring and control of the stones marking the borderline of adjoining tax properties after an inundation. The wall painting shown here is from the Menna tomb No. 69 and illustrates a cadastral survey party at work. The painting reveals not only some very interesting technical details of Egyptian land survey, but suggests also the firm establishment of bureaucracy and how little government institutions have really changed since the Eighteenth Dynasty (1400 B.C.). Here is shown how a status-conscious state hierarchy works in the performance of its duties. The nome surveyor (in the center of the picture) is larger in stature and much more impressive than the others in the team; he wears official dress, including what appear to be high boots. Such an official would be loath to be bitten by snakes or have his legs covered with leeches. His assistant, the second man from the left, is also in the career but has as yet not advanced far enough to rate service boots. Both of them hold in their hands a cubit standard to check the length of the measuring rope; the knots or marks on it are clearly indicated. The two men at the ends of the rope are of a much lower order, skillful at their job no

doubt, but their lack of formal education will forever prevent them from wearing the dress of the service. The rest is local labor picked up for the occasion—porters and other nonestablishment characters, perhaps peasants working the fields being surveyed. In the team is another very important person, viz. a blind man carrying a staff and being led by a small boy. He represents the blind abstract government justice.

The annual survey naturally had a more important purpose than settling conflicts between neighbors about moved or missing property corners. This becomes evident from the lower frame of the painting which shows the situation a few months later. The wheat has been threshed and three tax assessors accompanied by two scribes and a tallyman have appeared to collect the State part of the wheat harvest. The State wheat pile (to the right) has already grown much higher than the one the peasant is permitted to keep, but the assessment is still in progress and four taxpayers are kept busy bailing grain over to the State heap. To the left is the tax collectors' chariot which has been driven right into the field to serve as a latent threat and warning against tax evasion. The total social security as determined by the State must be maintained.

57

The three large pyramids on the Gizeh plateau, at the edge of the desert a few kilometers west of the Nile, were constructed in 2600 to 2565 B.C. by pharaohs of the Fourth Dynasty—Cheops, his son Chefren, and his grandson Mycerinus. The Cheops pyramid is the largest with a 755-foot side, followed by Chefren's with 707 feet and Mycerinus' with 356 feet. All three were at one time lined with ground and polished limestone slabs which have been quarried and used as building material. The only slabs still remaining in place are those on top of the Mycerinus pyramid, which also has 16 courses of red granite slabs along the base.

is sawed up into thin slabs to line a wall built of cheap material, such as concrete. Cheops and his overseers in the Aswan quarries were not concerned with saving on rock and the labor used to work it.

The transport of the megaliths, across the river for the Tura stone and a distance of 605 miles downriver for the Aswan granite, presented problems of loading and discharging. The heaviest blocks entering the Great Pyramid weighed only 50 tons, although subsequent pyramids incorporated blocks weighing up to 200 tons. But regardless of the weight, the method of manipulating the blocks in the quarry was the same—they were slid downward on rollers from a higher to a lower level. The Egyptians never achieved a vertical lift, and the only way to move these heavy stone blocks was by building a sloping ramp from the quarry to the river's edge. By means of levers the blocks were tilted on sledges or rollers which slid down the lubricated ramp leading to the barge dock. The barge was ballasted with sand to keep it level or slightly lower than the shore end of the ramp, and when the block had been secured on board, the sand was unloaded to make the barge float. When something went wrong and the block slid off the ramp, nothing could be done to recover it; the block had to be left where it was and a new one quarried to replace it.

When a barge, whether it had crossed the river or made the long passage from Aswan, arrived at Gizeh, the block was unloaded, by lowering the barge by filling it with sand to the level of the quay, and then manhandled on to the ramp leading to the pyramid. When unloading, the sledge was placed on rollers consisting of wooden logs, and the ramp was sprinkled with water or greased to ease the upgrade haulage. It was dragged up the 1:10 grade by hundreds of porters tugging on ropes and lashed along by overseers charged with the job of coordinating, with the aid of a whip, the muscular effort of the men.

The masonry work

In any building operation the transport of material, to the site and on the site, constitutes a major effort. In European building until quite recently masons were supplied with their material by "hod carriers" walking up steep stairs supported by outside scaffolding. All authorites agree that scaffolding was not used in Egyptian construction generally, and definitely not for building the pyramids. The stone was brought up to the working platform on a long ramp that kept increasing in height and length as the work progressed to maintain a gradient of 1:10, or about the same as used in modern underground construction and open-pit mining. For building the Great Pyramid there were probably two such ramps, one leading from the local quarry to the south face, and the other from the river's edge to the east face. During the process of building, therefore, the pyramid—like the temples and other important structures

This picture is taken from the Rekhmara tomb No. 100 from the Eighteenth Dynasty and illustrates the work involved in building a mud-brick wall. The helpers are Negroes, and one of the masons is a Semite, while the Egyptian foreman sitting at the right is taking it easy. The tool used by the Semitic mason (lower left) is a pick. Note that the shape of the handle is almost exactly the same as the American type of ax handle.

—was wholly enclosed by embankments and transport ramps until finished.

The core blocks were delivered from the local quarry with the bedding face trimmed and all other faces left in the rough. The bedding joints of the Great Pyramid are level, which meant that the masons at all times worked on a level platform. A new course was begun at center, and the stones were placed in contact with one another, leaving irregular spaces between the blocks unfilled. No mortar was used. The course was extended uniformly toward the four faces and most likely—no assurance can be had about this important structural detail without tearing down the pyramid—it was interrupted at intervals by internal casings of Tura limestone set with a batter of 1:4, as in the mastabas, to take up lateral movements. Finally, packing blocks with the outside faces cut to conform to the batter of the pyramid were placed and carefully joined. After a full course had been laid, the top surface was trimmed level in preparation for the next course. The chippings were dumped over the west and north sides and represented, according to certain estimates, in total bulk about half the volume of the pyramid, which appears rather excessive.

Before the ramps were raised for the next course, the Tura casing had to be placed. Most likely the casing blocks with the oblique rising joints were worked up on the ground

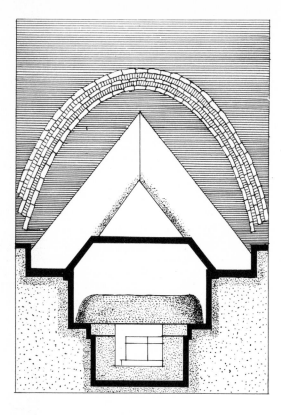

and numbered. It still remains somewhat of a mystery how they actually were placed on the face of the structure. No doubt a block was brought on its sledge to right above the place it was to occupy, whereupon it was slid off the sledge and manipulated into place. But it should be recalled that three sides were ground to produce a joint only one-hundredth of an inch wide, and so the block had to be lowered on to the existing one without damaging the accurate edge. Just how this delicate operation was performed with a block weighing 15 tons or more has never been adequately explained.

One thing is clear: the block had to be eased from its horizontal position on the sledge—of course there is no evidence to the contrary that the block was *not* transported on the sledge lying at a 52° angle, which would greatly simplify the ensuing critical operation—and levered over the edge of the packing blocks. The rest of the operations can be visualized in the following manner. On the embankment a heavy timber construction would be in readiness to receive the block. Assuming that it had been brought to its place standing at an angle corresponding to to the batter of the pyramid face, the first phase would be to get it off the sledge by means of wedges and levers and on to the level timber bed which was propped up by wedged upright timbers. To ease the friction when the slab was slid into place, all contact surfaces would be smeared with a thin coat of gypsum plaster, although in principle this

A problem that always worried Egyptian pyramid builders was how to relieve the tremendous weight of the masonry piled above the burial chamber and other hollow spaces inside the structure. The engineer of the Cheops pyramid believed that he had found a static solution by piling on five compartments atop the burial chamber and placing large granite blocks obliquely to form a V-shaped roof for the top compartment (right). The best and statically most appealing method used by a pyramid builder is shown at the left. It was employed to relieve the burial chamber in the rock under Amenemhet III's pyramid at Hawara. This pyramid has a square plan with a 334-foot side, and was built near the so-called Labyrinth in 1859 B.C. The burial chamber consists of a 22 × 8 × 6-foot block of yellow quartzite which had been hollowed out into a basin and placed in the shallow shaft. The weight of the basin has been estimated to be 110 tons. The sarcophagus was put into the basin which was sealed with three 4-foot-thick quartzite megaliths, each one weighing 45 tons. Above the chamber were constructed two relieving compartments of which the lower one has a flat roof and the other a saddle-shaped roof consisting of limestone blocks weighing 50 tons each. Above these oblique monoliths was built a 3-foot-thick vault spanning the entire substructure.

would not be necessary with the method suggested.

With the block eased in proper position and resting on the top level timber, a set of slightly lower timber uprights would be wedged in place. From now on the ensuing operation would be essentially one of shifting the load, first by knocking away the wedges under the tall timbers, thereby gradually lowering the cap onto the next set of timber uprights. Then a slightly lower set of props would be wedged into place, and the wedges knocked away from the second set, and so on until the casing slab in the end came to be resting only on the cap. The final phase consisted of removing the cap in such a manner that the accurately ground edge would not be damaged when sunk against the top edge of the slab already placed. This without doubt was done with the aid of bosses left on the outside face, when it was ground and polished at the quarry. Oblique timbers were wedged under the bosses to permit the removal of the cap, whereupon the slab, now supported entirely by the boss timbers, was carefully lowered against the lower edge by easing away the wedges under the timbers.

The trimmed casing blocks no doubt overlapped at the corners of the pyramid to escape the difficulty of fitting the last slab between two blocks already fixed in position. By overlapping the edges, the masons would be able to correct for any accumulated errors by trimming the last corner block when in place. However, fitting the very last casing slab of the structure, the one closing the opening of the descending gallery after the interment, must have taxed the ingenuity of the masons to the utmost. This final block had to be fitted between four slabs already in place and, to make the problem more interesting, with oblique rising joints.

Most likely, the ramp was kept in place up to the level of the adit, and the adjacent slabs were left trimmed but not polished. The closing slab with its surface bosses intact would have been supported by timbers, either to the side of or above the adit. Before the closing, the back and four edges would be given a thin coating of mortar and the slab floated into its place, after which it as well as the adjacent slabs would be given their final trim and polish which was extended down to the base as the remainder of the ramp was removed.

The Grand Gallery, the upper reaches of the ascending gallery, the King's Chamber and the relieving compartments above it, as well as the plug blocks, were of course built in the open simultaneously with or before the enveloping core masonry was placed in position. The last job was the placing of the granite capstone, with its boss in the base fitting into a mortise cut into the top course of the core masonry. Lowering the capstone into place was no particular problem. It would be supported on timbers and wedged up so as to permit the sledge carrying it to be removed, and the stone lowered into place by knocking away a series of wedges. Again, it was an operation of transferring loads to successive supports and utilizing bosses left on the four faces for the final drop into place.

The old Egyptian builder's tools do not differ from those used by builders until recent times. Top left shows a plumb rule that still can be found in use. When the plumbline coincides with a mark in the center of the crosspiece, the surface is level. By graduating the crosspiece, different gradients can be obtained. Then there is a mason's square and a device to determine whether a wall is plumb. The mallet is still used by sculptors. This particular one is dated to 2680 B.C. and was found at Saqqara.

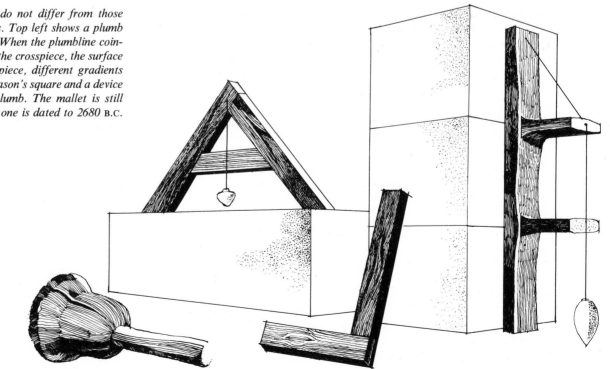

Determining the batter of the pyramid faces

The structure was complete. Now followed the final operation—dressing and polishing the four faces. On this all authorities agree. It was always done from the top downward, the pyramid—or temple—being finished in all details on one level, with the ramp and other embankments in place. As the work progressed, the ramps were lowered, exposing new surfaces for trimming and polishing until, in the end, the embankments were removed and the bottom courses of masonry finished last of all. Once the structure was finished, there was no opportunity for going back to do any retouches on it.

To finish off a conventional temple structure in this fashion is really not a difficult job for experienced masons. The work can be checked all along with a plumbline, the horizontal members are relatively short, and the wall surfaces are broken up in various ways. But trimming down a pyramid starting from the top and continuing over an ever-expanding area down to the base is an altogether different operation. A slight error made at the summit would keep growing as the work proceeded downward. Too little trimming at the top would eventually result in the plane of the polished surface projecting beyond the fabric of the structure. Too much cutting at the top would necessitate the removal of tremendous volumes of rock at the base.

The master masons must have been absolutely sure of the alignment of the structure at all times, and when ordering the final trimming at the top, they must have known beforehand the margin of error permitted. Once the ramps and embankments were removed, it was too late to correct an error.

The question then arises: How did the master masons manage to keep the batter of the face in accurate alignment; indeed, how did they manage to find the accurate batter, the slope of the sides, in the first place?

Knowing the height of the pyramid, and with the four corners firmly established, the surveying priests had two alternatives for determining the batter, one empirical and one theoretical. Naturally, it is now impossible to find out which one was actually used, but this is really of minor importance because the end result would in any case be the same.

Using the practical method, the center of the pyramid would be determined by laying out the two diagonals. Owing to the mound in the middle of the site, running lines between opposite corners, checking and rechecking them, would be a time-consuming job, but in the end the priestly surveyor would accept the mean of a number of diagonal crosses as the true center of the structure.

Around this center would be erected a timber scaffold extending up to the calculated height of the pyramid, and the point on the ground would be projected up to the top of the scaffold by means of a plumbline consisting of a weighted copper wire. The top of the pyramid obtained in this manner would be marked with, say, a polished copper cylinder that could be seen from the ground. Using this as a reference point, set boards giving the batter would be placed at both sides of each corner, and as many as required along each of the four sides.

These set boards extending high enough to establish firm lines of sights to the marker at the top would, as it were, outline the structure, but they could not be used for the actual building construction. As a guidance for the work, a sturdy masonry set would be built outside each set board, carrying a trimmed stone slab set at the correct batter and offset one cubit from the batter board. To facilitate sighting, a line of copper studs with sighting slits could be placed in the batter slab.

With these permanent batter sets in place and the sighting studs adjusted laterally in relation to one another and the top, the scaffold tower could be dismantled, and the work of placing the lower courses with their facing slabs could begin. Since the batter sets were offset one cubit from the finished pyramid face, there would be no risk of their being thrown out of alignment during construction.

The second and theoretical method open to the surveyor priest and permitted by the state of Egyptian mathematics at this time would be to calculate the offsets necessary to establish the batter of the sides. The same kind of masonry structures would be erected at a distance of one cubit outside the finished face of the pyramid, care being taken to keep the outside wall in line and perfectly plumb. Using the outside wall as a reference line, offsets of previously calculated lengths would be laid out toward the face of the pyramid at predetermined elevations from the leveled ground, whereby the batter would be obtained. The stone slab facing the pyramid would be provided with copper studs adjusted for lateral and vertical sighting in the same manner as previously described.

Using these carefully adjusted reference planes to guide the final placing of the facing slabs, either by means of direct sighting or by stretching copper wires through the slits, a large number of courses—perhaps up to a quarter or more of the total height—could be placed with assurance of good alignment. As the transport ramps and embankments grew in height, they would eventually cover the batter sets, and it would be necessary to retain an opening between the embankment and the face of the pyramid to serve as "sighting pits." These would be manned only when placing a few of the facing blocks in each course. The rest of the alignment would be done by the masons working on the platform.

All this is, of course, speculation that does not take into account the uncanny skill of experienced builders in obtaining accurate alignment simply by using their trained eyes. It is not at all impossible that once the original alignment had been given from the batter sets, the rest of the courses were placed by eye alone, lower courses being used as planes of reference as the work progressed.

The heavenly staircase to the stars

So there it was, the mightiest structure ever erected by man. The four battered sides of polished Tura limestone, each triangular area measuring 7.7 acres, were absolutely flat, without a break visible to the human eye when viewed from the side. The precision joining of the casing blocks presented from a distance the impression of one unbroken mirrorlike surface. The granite capstone with sacred hieroglyphics sculptured in low relief was gilded, and lit up like a beacon when struck by the first rays of the rising sun. As the sun rose higher over the eastern horizon, the whole structure glowed in its red reflection. This was indeed a fitting abode for a god-king and a sight that would keep the peasants who had made the structure possible in a permanent state of suppression and in trembling fear before this tangible expression of the cosmic powers endowed in their lord and master. It was not for them to inquire into the cost of this heavenly staircase to be tread by their lord on his way to his eternal abode among the stars.

But now, 4,500 years later, it may be warranted to look into the input of time and labor required to build this mighty structure. Herodotus when visiting the pyramid nearly 2,000 years after it was built, was informed by the local dragomans that 400,000 men worked 20 years on the site. The time seems reasonable, but accepting Flinders Petrie's estimate that the core blocks, weighing on the average 2.5 tons apiece, could be handled by a gang of eight men, the 2.3 million blocks could have been brought to the site by 115,000 men. Since the system of corvée used throughout the long history of the monolithic state was applied usually only during the agricultural off-season, i.e., during the inundation period, Herodotus' figure may well be correct for, say, the three months of the year when the principal transport work was performed, leaving a much smaller porterage force to assist the masons during the rest of the year.

The masons were of course kept busy during the entire construction. They lived, about 4,000 of them, in galleries excavated in the hillside to the west of the pyramid compound. There were 91 such galleries, now buried under the sand, each one 88 feet long, 9.5 feet wide, and 7.5 feet high, and providing sleeping accomodations for about 45 men. A wall ran at right angles to the portals of the subterranean barracks. The corvée peasants called up for porterage slept on the ground in the shelter of flimsy lean-tos, and had to feed themselves with grain and whatever other provisions were provided by the state commissary.

As mentioned repeatedly in passing, there is no doubt but that the monument these wretches sweated and bled and died to erect is from a technical point of view the finest ever made by man. The dry megalithic masonry stood intact for 3,000 years, through earthquakes, floods, wind and sand erosion, scorching heat, and occasional frost. The mighty earth and concrete dams now being constructed may last 300 to 400 years, taking the most optimistic estimate of their service life.

There were to be other pyramids on the Gizeh site. Chefren, son and successor of Cheops, built his tomb there, and so did his son Mycerinus, apparently the best of the lot, who being a "just and pious king" had to have his reign cut down to only six years by the angry gods who had ordained that the poor fellahin should suffer brutish rulers for 150 years. There remained, after Cheops and Chefren had managed to deliver their quota of beastliness, 44 more years of affliction, and when the grandson proved unable to stomach the cruelties of his immediate ancestors, the disappointed gods cut him down. Or so, anyway, says Herodotus. At any rate, the third and last pyramid of the Gizeh group was never finished, and what was accomplished was shoddily built compared to Cheops' standards.

After Cheops, Egyptian building went into a decline, and the structures would have vanished long ago except for the conservation properties of the Egyptian climate and the fact that so many buildings have been protected by a covering of sand. The gouty monstrosities serving as columns of the Hypostyle Hall at Karnak, built during the reign of Ramses II, for example, were erected on a bed of pebbles, and when the temple was flooded in November, 1899, eleven of them collapsed owing to the undermining of their flimsy foundations. The imposing pylons of the New Kingdom temples were built of rubble and always on poor foundations, Compared to the Great Pyramid, all that was subsequently accomplished in the way of building in Egypt was the work of dwarfs.

Building in the West

The long chalcolithic period had run its course and been succeeded by the Bronze Age. Throughout the Fertile Crescent the technology of alloying copper with tin to obtain cutting tools and weapons of a superior quality had been turned into an established production process. The Great Pyramid, the most durable monument to power hubris, was still intact, its glossy eastern face shimmering in the rosy glow of the rising sun as its rays pierced the morning mists layered over the watery lowlands. But it was an empty shell; the red granite sarcophagus had been broken into, the tomb robbed of its treasures, and even the exalted remains of the god-king had been scattered to the winds.

But megalith building had not died with the Old Kingdom. Everywhere, on the northwestern shores of the Mediterranean, even along the coasts of the Atlantic and the North Sea, there followed with the new metal tools an urge to build with mighty stone blocks. It was, without the slightest doubt, an exercise in power. Metalsmiths and prospectors coming from the centers of civilization kept pushing farther and farther beyond the rim of the unknown world, in the manner of their Neolithic forebears, and ran into local chiefs who saw in their shiny hardware the lethal instruments needed to subdue rivals and neighbors. The wandering smiths had also heady tales to tell of the splendors of the their homelands.

And so the old story was repeated again on distant shores, although in a pitiful and even droll fashion. With their new-found power there arose in these homespun rulers a desire to build temples to their local gods and, beyond all, mighty tombs for themselves, and having subdued the neighboring tribes with the aid of the metallurgical magic of the strangers, they could force local peasants to labor on their monuments. Everywhere it was the same. At Mycenae, Tiryns, Argos, the early Bronze Age princes drove their Helladic peasants to labor in distant quarries and drag huge stone blocks up the steep slopes to strongpoints on top of the hills, to pile them into high and insurmountable walls enclosing their modest megaron houses. Later, this ancient construction became known as Cyclopean, the work of giants, because to literate Greeks it was inconceivable that this crude work could have been done by human beings.

Bronze weapons and megaliths

Even in distant Britain this strange exercise in new-found power was going on at about the same time. Close by an ancient cemetery on the Salisbury Plains used by Neolithic settlers, some local chief rounded up as many subdued peasants as his metal power could reach and put them to work on a tomb that appears to have been inspired by the grave circles at Mycenae, namely, Stonehenge.

Just as in Egypt, no local stone would do for the final dwelling place of the Pharaoh but only Aswan granite—so only the finest and largest chunks of rock attainable would do for the anonymous ruler of the Salisbury Plains for his final abode. Large blocks of sandstone, some of them weighing up to 40 tons, were dragged on rollers from Avebury over a distance of 18 miles to the site. Other stones used in the monument, the so-called "bluestones,"

The Grave Circle A at Mycenae has a diameter of 90 feet and is enclosed by a 3- to 4-foot-high wall with a thickness of 4 feet. The wall is constructed as a shell of limestone slabs filled with rubble and topped with stone. Within the ring were six shaft tombs. Owing to Schliemann's scrounging for treasure, the dating has been confounded and vacillates between 1250 and 1500 B.C.

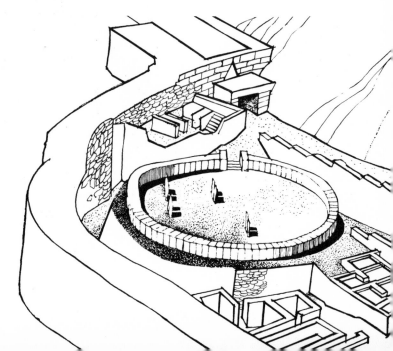

had to be obtained in the Preselly Mountains in Pembrokeshire 150 miles away.

The dressed stones were placed in a wide ring with 26-ton "sarsen" uprights carrying lintels cut to the curve and secured to the uprights in wood-joinery fashion; i.e., each lintel had a mortise hole at each end fitting into a tenon in the upright. Adjacent lintels were joined by tongue and groove. The two 40-ton "trilithons," placed inside the ring in the west and topped by a larger lintel, carried carvings of a bronze dagger and axhead of contemporary Mycenaean design, i.e., about the middle of the Second Millenium B.C.

Everything about the Stonehenge structure suggests foreign influences. Somebody from the Helladic world had inspired the work and possessed enough organization and engineering know-how to raise the megaliths on the site. Raising the stones was accomplished by dragging them up an earth ramp and tilting them, the base end first, down a 60° grade at the end of the ramp, whereupon they were raised to a vertical position by means of a pair of shear legs and ropes. The lintels were likewise dragged up the ramp and placed on the uprights. But there was not sufficient local skill available to produce anything resembling the sophisticated work done in the Aegean. Stonehenge resembles in this respect the early missions built in California by monks working in the wilderness with the help of unskilled Indians.

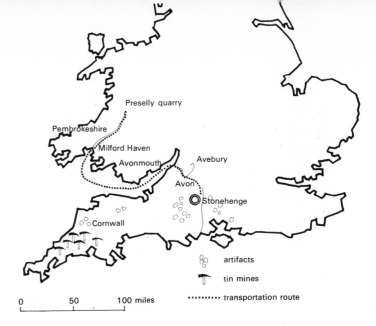

The bluestones used in the Stonehenge monument were quarried in the Preselly Mountains in Pembrokeshire and transported by barge down the river to Milford Haven and thence along the coast to Avonmouth and up the river Avon to its tributary Frome. They were then carried over the divide to the river Wyle and to the site. The total distance was about 150 miles and the transport work can be estimated to 70,000 ton-miles. The megaliths were quarried at Avebury in Wiltshire about 18 miles from the site and must have been hauled overland. The total transport work was on the order of 100,000 ton-miles. The map also shows the concentration of Mycenaean artifacts in the neighborhood of Stonehenge and the tin workings in Cornwall.

Stonehenge on the Salisbury Plain consists of a ring of 29 megaliths, each one weighing 26 tons and rising 13.5 feet aboveground. They are joined in timbering fashion to lintels cut to a curve with a 56.5-foot radius. The two "trilithons" in the west reach 22 feet above ground and weigh 40 tons. Placed in a concentric circle are 6- to 10-foot-high so-called bluestones, and inside the trilithons is a horseshoe of similar stones. The monument is oriented so that the western trilithon is in line with the rising sun at midsummer, or roughly so.

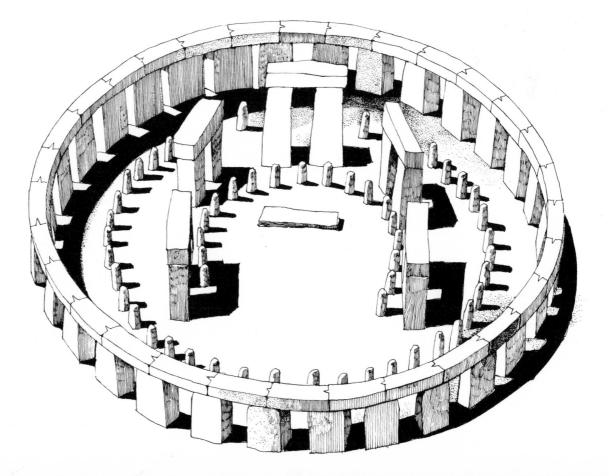

  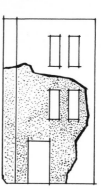 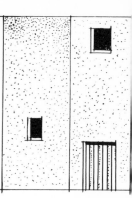

It would be futile to search for a local line of development in such major Bronze Age monuments as Stonehenge and the similar ones in Bretagne, not to speak of the crude dolmen and passage graves along the North Sea coast heaped up with rough and uncut stone slabs with some idea of architectural pretension. Their chief interest lies in the fact that here, too, some ambitious brutes had found the means in the new metal hardware, brought to them by peddlers from distant lands, to extend their personal power so as to be able to force their subdued neighbors to labor on a lasting monument to their lethal prowess. (For further details on Stonehenge see the note on page 264.)

Significance of Minoan building

The bridge from the ancient riverine civilizations to the mainland of Europe crosses the Aegean and it is here, on Crete and the Greek mainland, that the superior metal civilization got its early foothold, accompanied as usual by megalith construction, although coarse and barbaric compared with Egyptian masonry.

An ingenious claim has been made that Crete was settled by refugees from the Lower Kingdom of Egypt who refused to bend to the will of the conquering Followers of Horus. This thesis cannot be proved, but it would help in explaining why this people living "in the midst of the sea," as later Egyptians referred to them, succeeded in developing the most appealing civilization in the Mediterranean world. The sea kings gathering wealth by trade and piracy during the first half of the Second Millenium came in contact with people and ideas in Anatolia, Syria, and Egypt. Indeed, they may have seen the corbeled brick domes of the royal tombs of Ur, or northern copies thereof. But whatever they borrowed, they turned into something new and characteristic of themselves. Like all seafaring people, they had a need for establishing trading posts on distant shores, outposts that grew in size and importance so as to require fortifications and permanent garrisons.

Thus there grew up on the Greek mainland during the second half of the millenium such cities as Mycenae, Tiryns, and Argos, held in vassalage by the Cretan sea kings residing in the palace of Knossos. Eventually, the mainland vassals rose against the Minoan supremacy, and in about 1400 B.C. succeeded in overthrowing the Knossos dynasty, destroying the capital and other cities on the island, which from then on became merely a province of Mycenaean Greece.

But the old Minoan traditions lived on, so that in sketching what became in fact the beginnings of European building history, the Minoan and Mycenaean blend into one. From all accounts the Minoan builders were originally inspired by the traditions in northern Syria and Anatolia, where all the more pretentious buildings were erected on heavy stone foundations carried up to the first floor, above which the walls were built on a framework of timber filled in with brick or stone rubble. The Palace of Knossos —the "Labyrinth" of Greek mythology—derived both in style and construction from Hittite prototypes on the Asiatic mainland. This northern house, as further evolved by the Minoan builders, had its principal rooms not on the ground floor as in Egypt, but on the first floor. The staircase therefore became a permanent feature, and the Grand Staircase in the palace of Minos is the prototype of the innumerable grand staircases on which subsequently the nobility all over Europe ruined themselves; by the time the staircase was built and paid for, the upper floors had to be finished off in slipshod fashion.

Since the glare of the sun was not quite as strong as in the south, the Cretan houses could be provided with windows which almost from the outset were utilized also for artistic effects, to relieve the monotony of the walls and give a feeling of lightness to the elevation. The walls were also made lighter by using mud brick, to a thickness of 3 feet, above the first floor. The walls were plastered over and painted.

The columns used to carry the flat roof over the stair-

The sketches of Cretan house facades have been made on the basis of pottery plaques dated back to 1700 B.C. They suggest that the houses were provided with windows, an unknown feature in Sumerian and Egyptian houses. From the appearance of the windows to the extreme left, it may be concluded that they were fitted with some sort of louvers.

The Greek megaron is the most famous of all house plans since it reached its full development in the Greek temples. The plan is much older; it was used in Anatolia in 3000 B.C., and in Troy II it had already taken on its special function of a public hall for formal receptions and ceremonies. The sketches show a couple of early megaron foundations from Korakou in Greece. The one to the left is from the early Second Millenium and the right one from its end.

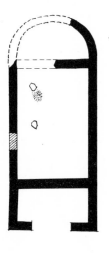
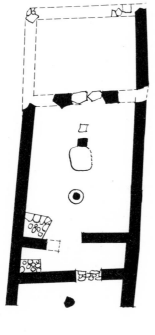

case, the numerous balconies, and the terraces of the Palace of Knossos were of cypress wood and tapered like table legs from the top downward. They rested on tall stone plinths, although later flat disks of limestone were used as bases.

The megaron house

The layout of the Minoan house also derived from Anatolia where as early as in the twenty-second century B.C. —in fact earlier, because this type has been found in the first level of Hissarlik (Troy)—the so-called megaron house had evolved.

The megaron house at the time of Troy consisted of three main parts. It was entered through a columned porch (in the simplest version a trellis framework carried by two posts) into a narrow vestibule leading to the main room— the *megaron*. On the Greek mainland, a small room was added to the back of the megaron for sleeping or storage, as well as a fixed hearth, often circular in form. From this simple megaron plan the elaborate palaces of the Minoan and Mycenaean princes developed, and subsequently also the Greek temples.

There was, however, a fundamental difference between the palaces on Crete and those on the Greek mainland. Mycenae was fortified, the palace was but a part of the citadel situated on a commanding height and enclosed by walls up to 25 feet thick constructed of "Cyclopean" masonry, with blocks weighing several tons apiece. The masonry consisted of boulders which were only partly dressed with hammer; many blocks were not dressed at all. They were used only in the faces of the walls, the core consisting of rubble or earth. Prominent parts of the fortifications, such as the Lion Gate, were built of ashlar—or dressed stone—to lend dignity to the approach. When required, as over the steep stairway leading to the secret well in the Mycenae citadel, or the sally port and galleries at Tiryns, the roof was corbeled by allowing each course of masonry to project beyond the one below. The underside of the corbeled arch was chiseled off to form a false arch.

Tomb of Agamemnon

This method of corbeling found its widest application in Mycenaean beehive tombs, the so-called *tholoi,* which are seen in large numbers in all parts of Greece once controlled by the Mycenaeans. The largest and most well preserved of these typical Mycenaean tombs is the one known as the "Treasury of Atreus" or the "Tomb of Agamemnon" just outside the citadel of Mycenae. It is of tremendous size, measuring 46 feet in diameter, and like all tholoi tombs is constructed at the bottom of a large circular shaft sunk in the hillside. At the bottom of the shaft was built—by corbeling breccia blocks—a pointed dome about 40 feet

67

high, topped with a capping stone. Joined to the main tomb is a smaller beehive-shaped chamber cut in rock. In the side of the main burial chamber is a stately doorway approached from the outside by a wide adit or *dromos* which was also lined with ashlar.

When the tomb was being built, the space between the extrados (the outside of the domed structure) and the shaft was backfilled with tightly packed earth to take up the lateral pressure. When, as frequently happened, the dome extended above the surface, a mound of earth was thrown up to cover it. The entrance with the doorway projected 16 feet out from the base to form a vestibule, the roof of which consisted of a monolith slab weighing 120 tons.

The inside corbels were chiseled smooth and decorated with bronze rosettes and other metal ornaments. The doorway was flanked by two columns of alabaster of the Cretan type, i.e., sloping from the capital to the base, originally

The tholoi tombs on Crete and the mainland once under Mycenaean hegemony are the most interesting and typical structures emanating from the Helladic civilization. They owe their origin to the Royal tombs in Ur and are still used as dwellings in Apulia under the name of trulli. Below is shown the Treasury of Atreus at Mycenae, built with sawed breccia blocks on the bottom of a round shaft. The stones are corbeled to form a beehive-shaped dome, and after each course was laid the shaft was backfilled to take up the lateral pressure. When the dome was finished the inside corbels were chiseled smooth. Such a tomb was used for many centuries, and before each funeral the remains of the body interred were swept aside to leave room for the next corpse. Since the funeral guests were free to partake of the funeral goods, the tombs seldom contain anything of material value, but they have large deposits of bones along the walls.

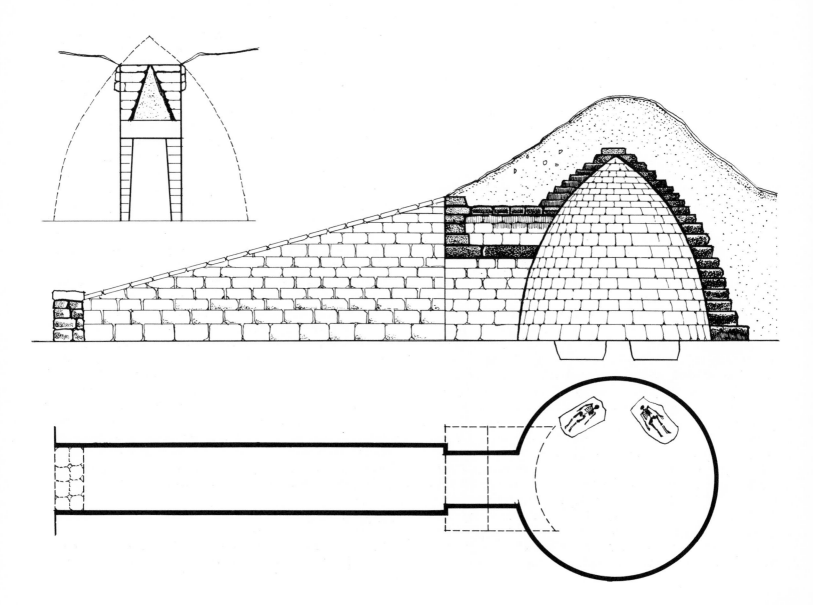

ornamented with chevrons and spirals to suggest metal sheathing over wood. The architrave was also elaborately decorated; it consisted of a triangular panel extending across the entire lintel slab.

Greek building construction

After centuries of conditioning, "Greek architecture" conjures forth a vision to Western man of gleaming marble temples set against a backdrop of an eternally blue sky, an elaborate use of columns and architraves filled with technically perfect sculptures in the half round. This vision of Greek buildings is not altogether wrong, but it applies only to a brief period in the country's history, to the Golden Age of the fifth mid-century B.C. Laymen seldom realize that these magnificent structures not only derived from buildings erected in wood or, more frequently, timber-framed houses with walls of mud brick, but that they reflect old methods of timber construction directly applied to stone.

The earliest Greek temples were modest structures indeed, built entirely of mud brick on stone plinths and roofed with thatch. As their size grew with increasing prosperity, a row of timber posts was added along the main axis in order to carry the increasing weight of the roof. Then timber posts were inserted in the pisé walls, placed in line with the ridgepoles, and in this manner the entire structure came to be carried by the posts while the mud brick took over the subsidiary function of wall fillings. This was the way the Artemis temple in Sparta was built in the ninth century B.C., although here posts were inserted also in the gable ends. The Hera temple at Olympia erected ca. 640 B.C. was entirely surrounded by a timber colonnade with its pisé walls plastered and painted.

The same method of construction was applied to all profane buildings. A timber framework carried the roof, and the space between the uprights was filled in with mud brick. No kiln-fired brick was used at all until the end of the fourth century B.C., and then only sparingly. But eventually the local wealth of stone, limestone, and marble began to be utilized in important buildings. The easily split limestone around Argos was used in ancient Mycenae and Tiryns, but in classical times a limestone known as *póros* found in the west and north of Peloponnesus began to be quarried because its rough texture gave a good base to plaster. The stone temples built of póros were stuccoed and painted—just like the old pisé temples.

But then, with the erection of the Parthenon and other buildings on the Acropolis in Athens, the marble of Mount Pentelicus, a few miles north of Athens, began to be used. The Pentelic marble consists of pure calcium carbonate which when quarried produces blocks with a close-textured and brilliantly white surface. The character of the stone lent itself to fine joints, smooth finish, and all the optical refinements found in the Parthenon and contemporary temples. Since both the Pentelic marble and the póros limestone can be quarried in lengths of up to 15 feet for use as beams or lintels, there was no need for the Greek builders to apply arches, vaults, or domes, and they were content to develop a so-called trabeated architecture employing posts and lintels as construction members, in the manner of their old timber structures.

The Golden Age of Greece

The temples erected during the Golden Age of Pericles (d. 429 B.C.) are noted for their fine masonry jointing, the blocks being ground together to obtain an almost microscopically fine joint. But the master masons did not rely entirely on the dry masonry staying in place by itself, and so the blocks were joined with metal fasteners. The drums of the columns were likewise ground together and provided with sockets in their centers for dowel pins of cypress to hold them intact. According to Pliny, the column bases, as well as the moldings at the top of the capitals, were turned in a lathe.

One interesting feature not noted until recently was the structural use of iron to strengthen the cantilever beams carrying the heavy statuary of the Parthenon. In the Agrigento temple, 5×12-inch iron beams are hidden for a length of 15 feet under the architraves and rest on the capitals of the columns. Iron was also used for such lifting devices as lewis irons and tongs employed for lifting the heavy lintels, and for plugs and clamps to hold the blocks in place. Lead was used in dovetailed joints, and bronze for plugs and dowels.

To facilitate transport and handling of the huge drums and lintels needed, bosses 8 to 10 inches square were left protruding on the finished block. Each drum had four such bosses for attaching ropes to hoist it into position on the site. For transport in the quarry, drums and lintels were provided with wheel rims of wood enabling them to be rolled over short distances. Transport from the quarry to the site was accomplished with low-slung wheeled wagons pulled by 12 to 20 pairs of oxen. These methods of material handling appear to have been used also for the temples in Sicily and on the Italian mainland, where the diameter of the drums approaches 11 feet, the capitals having a spread of 13 feet. However, in the Agrigento temple, where the column drums reach 12 feet in diameter, it appears as though the maximum capacity of the transport facilities had been exceeded, because the column drums were built up in ashlar courses with fine radial joints.

Strange specifications

The roof of the temples, with a pitch of about 30°, was supported by timber construction, judging from the sur-

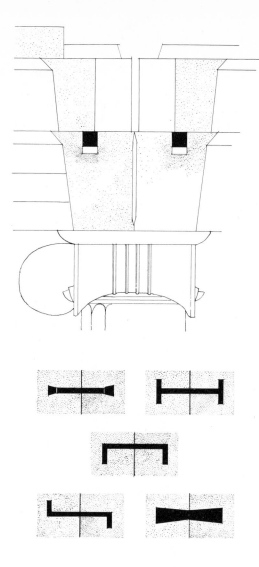

viving holes in the masonry for the footing of joists and rafters. It was covered with tiles or thin marble slabs. Although no roof construction members have survived, the contemporary specifications for constructing the timber roof of the arsenal at Peiraeus have been found on a marble slab and give an interesting insight into the design know-how of the Greek engineers. The arsenal was built in 340 to 330 B.C. and was 434 feet long and 59 feet wide. The walls were of ashlar masonry. Two rows of columns, 32 feet high and 3 feet in diameter, divided the building into three bays, the middle one of which had a span of 21 feet.

The fantastic dimensions specified for the roof members suggest that the Greek engineer responsible for the design had no conception of trussing, i.e., the stiffening of a beam by braces. Along the top of each row of columns —placed on 12-foot centers—were laid heavy longitudinal timbers which were joined with level transverse

A Greek column was built up with a number of drums. In the Parthenon, they varied from 10 to 12. The drums were delivered to the site, each furnished with four bosses for facilitating positioning of the drums without the risk of damaging the brittle marble. But before a drum was hoisted into place, the bedding planes had to be finished in the following manner: In the center of the drum a 4- to 6-inch-square hole was cut to a depth of 2 inches for holding a cypress block (empolia) with a 2-inch hole drilled in its exact center. In this hole was fitted a wooden plug. Around the square was cut and ground a bedding plane, and exactly level with it another bedding plane around the periphery of the drum. Between the two the stone was cut down slightly with a rough chisel. When the drums prepared in this manner were fitted together the joint was hardly visible.

The Greek builders used iron as construction material to a surprisingly large extent. In the Parthenon, for example, the heavy tympanum sculptures are carried on wide iron flats. In the Propyleae (top), the iron beams are 6 feet long and carry a 6.6-ton load exerted by masonry beyond the outside column. The beam has a safety factor of 4, as against a common modern factor of 6. Underneath and along the lintel is a channel permitting a slight deflection of the beam. Since no mortar was used, the ashlars were held together by iron clamps, a few common types of which are shown. They were set in lead. Below are some lifting devices, lewis irons and tongs, used for lifting heavy lintels.

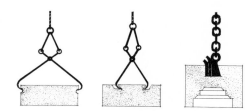

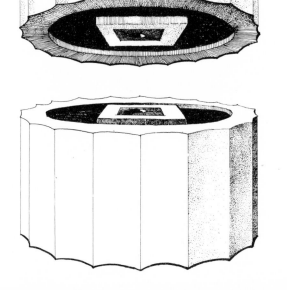

beams at each column. In order to obtain a low-pitched roof, a timber block was placed at the middle of each transverse beam. On these blocks was laid the ridge beam measuring 22×17 inches. The 12×8-inch rafters were spaced 16 inches apart and rested on the masonry wall and the heavy ridge beam, exerting bending stress on the transverse beams. The roof deck carried by the rafters consisted of $6\frac{1}{2} \times 1\frac{5}{8}$-inch boards spaced $3\frac{1}{4}$ inches apart and crossed by $\frac{7}{8}$-inch boarding, on which were laid terracotta tiles bedded in mud. In this roof construction there is no relation between the structural members and the load carried.

Roman building

Formerly, the invention of vaulting, the only method available for bridging long spans before the use of reinforced concrete and steel, was assumed to be Roman. Ancient Greek writers attributed the invention of the vault to Democritus (470–360 B.C.), but corbeled vaulting had been used by the Mycenaeans, and something like 2,000 years before the Treasury of Atreus was built, the city kings of Sumer were laid to rest in beehive tombs. However, it seems

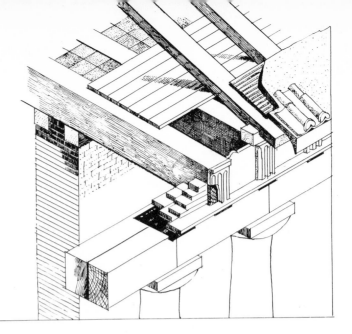

Above is shown a reconstruction of the roof used for the arsenal in Peiraeus. The Greek engineers did not understand or else abstained from trussing. As seen from the section of the arsenal below, the 22×17-inch ridgepole exposes the roof beams to the maximum possible load.

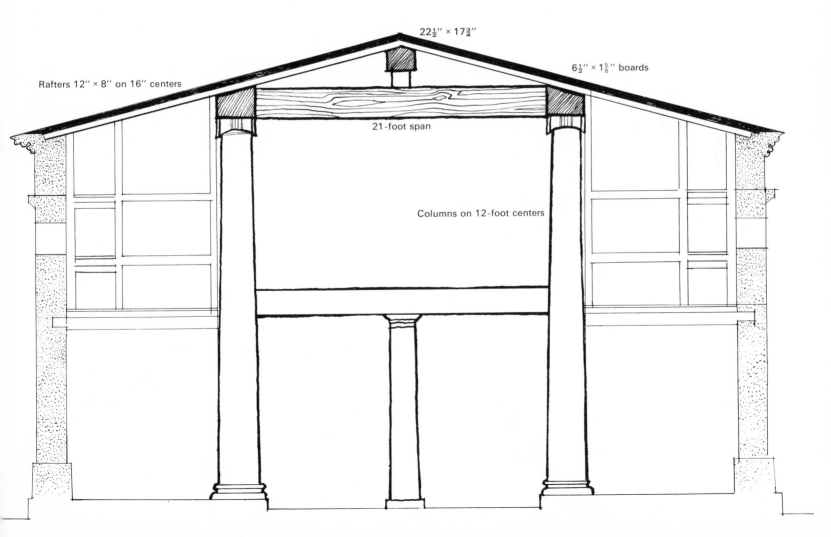

$22\frac{1}{2}'' \times 17\frac{3}{4}''$

Rafters $12'' \times 8''$ on $16''$ centers

$6\frac{1}{2}'' \times 1\frac{5}{8}''$ boards

21-foot span

Columns on 12-foot centers

The Greek engineers in Rome used vaulting exclusively to relieve the weight of the masonry. But they differed from all previous builders, who always relied on corbeling, by employing the "true vault" where the stones, or "voussoirs," are always wedge-shaped. In this copper gravure of the Colosseum made in Rome (1581) the shape of the vaulting stones is clearly shown.

quite clear by now that the Roman builders learned the art of true vaulting from the Etruscans who, judging from recent discoveries, obtained it from the Greeks. The Etruscans used wedge-shaped blocks of stone set without mortar in their bridges and town gates. The bridge at Bieda, north of Rome, built about 500 B.C. and still standing, has a vaulted span of 24 feet.

If, then, the actual invention of the vault cannot be attributed to Roman engineers, they certainly made tremendous contributions to the development of vault and dome building—in brick, concrete, and masonry. From the last century B.C. and onward they constructed numerous vaulted bridges throughout their wide-flung empire. In Rome alone, there still remain a number of ancient bridges capable of carrying heavy modern traffic. Indeed, the Pons Mulvius on Via Flaminia, $1\frac{1}{2}$ miles north of the city, carried the entire military traffic across the Tiber during World War II, first the armored vehicles of the Italian and German armies, later the Allied forces.

But vaulting was also applied in buildings, and the chief difference between Greek and Roman architecture lies in the extensive use of arches by the Roman builders. The vaultings of the vast interiors of the baths and basilicas were superb exercises in civil engineering. The $142\frac{1}{2}$-foot

*Of the many magnificent domed structures erected in Rome by
the emperors, only one—the Pantheon—remains intact, chiefly
because it was turned into a church, now named Santa Maria
Rotonda, by Pope Boniface in 609. The Pantheon (to all the
gods) was originally built by Agrippa in A.D. 27. It burned
down in A.D. 80 and was rebuilt by Emperor Domitian, but the
present structure was erected—and perhaps also designed—by
Emperor Hadrian and dedicated to the seven planetary gods.
The height of the dome is exactly the same as the diameter of
the rotunda, or 142 feet, and the thickness of the walls is 22 feet.
The diameter of the dome is 142½ feet, or 4½ feet longer than
St. Peters'. The opening in the roof is 28 feet in diameter, and
rain is evacuated through a brass drain sunk in the floor. The
dome has five rows of cofferings to reduce the weight. They
were originally lined with gilded brass, but this was plundered
in 663 by the Byzantine Emperor Constans II.*

dome of the Pantheon, the 100-foot vaulted roof over the
hall of the palace of Diocletian, and the 83-foot vaulted
span of Constantine's Basilica were achievements that
could not be duplicated until 1,500 years later, at least in
western Europe.

Invention of ribbed domes

But by the time such giant domes and vaults could be constructed, vaulting methods had been greatly improved by
Roman engineers. The early Roman vaults were invariably circular, or so-called barrel vaults, which possess a
number of serious drawbacks. The barrel vault creates a
dark and depressing room, and it exerts a tremendous
thrust on its supporting walls and piers. Such a vault also

73

requires heavy and costly timber centerings during construction, which moreover are difficult to obtain in countries lacking timber.

It would have been physically impossible to apply the simple barrel vault for roofing the vast buildings mentioned above. Two major improvements had to be made before such work could be undertaken. First, the barrel vault itself had to be lightened, which was achieved by erecting a system of ribs of bricks and filling the spaces between the ribs with light pozzolana cement. By this method the need for timbering was reduced to the inevitable supports for the arched ribs and light shuttering for the concrete. The greatly reduced weight of the vault or dome was taken up by cross-vault buttressing of the walls and piers exposed to the stresses produced by the vaulting. These transverse walls taking up the outward thrust were not visible from the outside, as in medieval construction; they were incorporated in the building itself, forming compartments which in their turn were covered by vaults.

A dome like the 142-foot one of the Pantheon is of course a vault, although built on a circular plan. This hemispherical dome is no doubt erected on a skeleton of brick ribs filled in with concrete, but no one knows for sure. The top of the dome is 4 feet thick; the lower part is considerably thicker but lightened by rows of coffering. It rests on a concrete drum relieved by massive brick arches.

Obviously, such vast and sophisticated structures as mentioned in passing called for advanced technical knowledge, but there is not the slightest evidence that beyond the survey required for the layout the Roman engineers possessed the theoretical and mathematical means of calculating beforehand the stresses produced by such immense domes and vaults. Like the Greeks before them and the cathedral builders a thousand years later, they relied on instinctive know-how, conditioned by experience and the practical wisdom gained from numerous failures.

Vitruvius and materials

But there is one important aspect to be considered. The ancient builders, unlike the majority of modern architects, knew their materials, and they could rely on them. The Roman architect Vitruvius, who took a prominent part in the rebuilding of the city under Emperor Augustus — "who found Rome a city of brick and left it a city of marble" — has a lot to say about building materials in his famous book *De Architectura* written during the reign of Augustus. This is how, according to Vitruvius, building stone should be quarried:

"Let the stone be got out two years before, in summer but not in winter, and let it lie in exposed places. Those stones which in this time are damaged by weathering are to be used in the foundations. Those which are not faulty are tested by Nature and can endure when used in building aboveground."

He goes on to describe in similar detail the properties of different kinds of building stone, travertine, tufa, peperino, and so on, and where and how they should be used in a building. He also describes the making of pozzolana cement, using *pulvis puteolanus,* a volcanic ash found in the Alban Hills, which when mixed with lime made a marvelous cement capable of setting also under water. The pozzolana mortar is in actual fact the reason why the large vaults and domes could be built at all and why so many of them are still standing.

Similarly, the sand used in mortar had to meet stern requirements, and of lime Vitruvius observes "that we must be careful to burn it out of white stone or lava," lime prepared from thick and harder stone being useful for structural purposes, while that made from porous stone is good for plastering. When slaked, lime should be mixed with sand, in the case of pit sand in the proportion of one to three; if shore sand must be used, the proportion should be one to two. In the latter case the mortar will be materially improved if crushed potsherds are used.

Similarly with timber. Each kind of tree has special properties and should be used accordingly. Fir is normally rigid and lends itself to flooring. Alder has the remarkable property of being imperishable underground and will uphold immense weights of walling and preserve them without decaying ... Cypress tends to wrap but lasts long because the oil in it resists dry rot and worms.

When discussing the Etruscan orders Vitruvius points out that the beams carried by columns should be joined by means of dowels and mortises and that a 2-inch space be left between the joints . . . to admit breathing by the passage of air: otherwise "they are heated and decay rapidly." In this context he makes the interesting observation that Roman temples were originally made of wood, and when they were subsequently built of stone, the masons imitated the details of the timber construction. Modern research has established that this observation by the imperial Roman architect is indeed applicable not only to Etruscan and Roman building but to all other — in Greece, Egypt, Mesopotamia, in fact almost everywhere.

Roman masonry and concrete

The Romans inherited the so-called "true masonry," i.e., homogeneous walls of squared stones, from the Etruscans. Vitruvius calls this masonry work *opus quadratum,* which involved laying ashlars with 2-foot-square headers, and stretchers twice as long. When opus quadratum masonry was used in structures exposed to stress, the blocks were joined with iron clamps set in lead. Another common type of walling was *opus incertum,* where small irregular blocks of tufa were laid in thick mortar. In the first century B.C.,

Above is shown a detail of the portico of the court of Amenophis III (1405—1376 B.C.) in the Luxor temple. It serves as a pretty good example of Egyptian trabeatic construction. The columns are formed like a cluster of palm trunks and topped with lotus capitals and a block carrying the lintels. The spacing of the columns was probably determined by the maximum length of lintels that the quarry could deliver. The Old Kingdom did not accept anything but monoliths, but here the columns have been built with drums. To the left is a Greek colonnade from the Parthenon which in its main features recalls the Egyptian. It was built in Pentelic marble in the fifth century B.C. after numerous Greeks had traveled in Egypt.

The most well-preserved Greek temples are not found in Greece but in its former colonies in Sicily and southern Italy. In the sixth century B.C., Greeks from Sybaris founded the colony Poseidonia at present Paestum, 15 miles south of Salerno. During the following three centuries they built three large temples in their colony. The town itself was destroyed by Arabs in A.D. 820, after which the low land rapidly turned into a malaria-infested swamp, and this is the reason why the temples have been so well preserved. The picture shows the end view of the Poseidon temple erected in 460 B.C. It measures 198×80 feet and is carried on 36 husky columns each 28 feet high, having a diameter of $6\frac{1}{2}$ feet at the base, and narrowing to $4\frac{1}{2}$ feet at the neck after an exaggerated entasis. Since the rest of the details are scaled to the columns, the general impression is one of dumpiness. The temple looks early and primitive, perhaps due to the lack of good designers in the colony.

the small squared blocks were laid in diagonal rows suggesting a net, this type of masonry being called *opus recticulatum*. But by that time masonry was yielding to the material characteristic of most subsequent building, namely concrete, which became used for foundations, walls, vaults, and domes. Walls were cast between timber shutterings using a concrete consisting of pozzolana cement and an aggregate of broken peperino, tufa, brick shards, etc., spread in alternate layers of large and small stones. The concrete walls were faced with marble, the first time in Augustus' mausoleum, later in all temples and public buildings. The increasing use of marble led to the opening of the Carrara quarries, but Pentelic marble was also imported at great expense. Later, with the perversion in taste, more gaudy varieties of marble were imported, as well as granite from Naxos and Aswan. Red porphyry was used for a geometrical paving called *opus Alexandrinum* and for monolithic columns.

The Roman engineers working in the homeland succeeded in resisting the temptations of their profession and generally stayed within the human scale in their designs, but their colleagues working abroad, in the colonies, fell victim to the megalithic hubris. In the Roman temple of Jupiter at Baalbek in Syria, the designer used foundation plinths measuring 65 × 13 × 10 feet; indeed, one block left in the quarry measures 70 × 16 × 14 feet and weighs 1,300 tons. In Alexandria, no doubt influenced by local examples, Roman engineers managed to raise a monolithic shaft 68 feet high with a diameter of 8 feet 10 inches.

Another Roman invention was the method of trussing timber roofs whereby such large spans as the one measuring 78 feet in the Basilica of St. Paul could be constructed. The great vaults over the baths were also roofed in with timber, but here the rafters rested on the vaults, which is not a recommended practice.

The houses of the upper classes in Rome were in all ways more liveable and comfortable than those subsequently built up to the end of the last century. They were provided with piped water and drainage, the pipes being of either earthenware or lead. In Italy, the houses were heated by braziers burning charcoal, but in the northern provinces—in Gaul and Britain—the Roman villas were centrally heated by hypocausts; i.e., the hot combustion gases from a furnace were conducted by means of flues built of hollow tiles placed in the walls. The stone floor of the living room was also heated in this manner, a method not applied until recent decades.

Architects and building brotherhoods

Just who were the people, the architects, engineers, craftsmen, and laborers, actually engaged in the vast Roman construction industry? In Etruscan times, the engineers were apparently priests, or at any rate closely connected

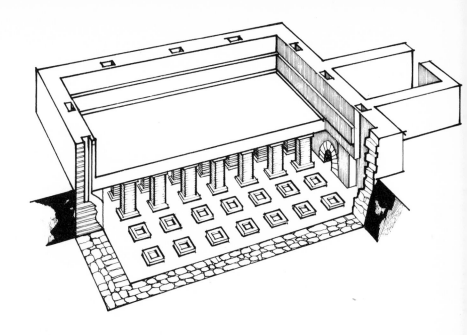

The sketch above shows the principle of a so-called hypocaust, as excavated from a Roman villa at Chichester. From a furnace on the right combustion gases were conducted under the bottom floor which was carried on 2½-foot piers. From this space the gases were evacuated through flues constructed of hollow tiles placed in the walls and plastered over. The hollow tiles with the keys for plaster are shown below.

with the priesthood. Although the papal title "Pontifex" (bridge builder) seems to support this view, it can be interpreted differently. Subsequently, the highest authority concerned with the public building activity rested with the censors who gave their names to the buildings finished under their incumbency. But those actually engaged in the design and construction of the buildings remain nameless shadows. All of them, or nearly all, were slaves. The better ones were Greek; it appears that all architects and

77

engineers were Greek slaves. Hermodorus from Salamis was the architect responsible for the temples of Jupiter Stator and Mars built in the second century B.C. Appolodorus of Damascus, another Greek slave, built the Forum of Trajan and numerous other public buildings in Rome, as well as the great bridge crossing the Danube at Turnu Severin. For some reason he fell into disgrace under Emperor Hadrian and was executed.

But the supply of clever Greeks was not inexhaustible, and Crassus, member of the First Triumvirate (together with Caesar and Pompey), already wealthy, found it good business to run a school for training gifted young slaves as architects and engineers and then hiring out the graduates to contractors.

The procedure followed preliminary to construction was in the main similar to modern practice. The Greek architect, whether imported or a local graduate of Crassus' engineering academy, was requested to present preliminary sketches giving the dimensions of the building. If these were approved, detailed scale drawings were made up, drawn on parchment smoothed with a pumice stone. A detailed schedule was drawn up and the costs were estimated. It was apparently also the architect's job to draw up the documents required for the contractors in order to permit them to bid either on the whole or on part of the job. The contracts contained detailed specifications of the materials to be used, the forfeit money posted as a guarantee for prompt execution of the work, and the period of payment. Building of the Colosseum, for example, was divided among four contractors.

The building trades workers were either slaves or freedmen. The capital resources of a building contractor, who was always a free man, consisted mainly in his supply of slaves, but on public works he could generally rely on being offered additional labor from the pool of prisoners of war. The freedmen building operatives were organized in guilds or brotherhoods, the earliest ones being those formed by masons and timbermen. The vast output of human labor was supplemented by a number of mechanical aids, such as hoists, cranes and pulleys, windlasses, etc. The latter were often driven by treadmill—or squirrel cage—a large wheel in which two or more men kept walking and climbing to hold the wheel in motion.

The beginning of the end

September 10, in the year A.D. 9 will forever remain the darkest day in the history of Europe, particularly for the Germanic peoples. That was the day when a rough coalition of Germanic tribes succeeded in ambushing a Roman column consisting of the XVII, XVIII, and XIX legions commanded by Quintilus Varus and slaughtering it to the last man in the Teutoburger Wald, a few miles from Aliso. The loss of the three legions was by itself not an irreparable blow. It was an annoying incident, and at headquarters the commander was no doubt ticked off as being unforgivably stupid. What was far more serious to the welfare of the northern people was Augustus' reaction to the news. The Emperor ordered the withdrawal of the Roman forces from the Weser to beyond the Rhine and Danube where a line of fortifications was built guarding the marches of the Roman Empire against the barbarians living to the north and east thereof. That decision robbed northern Europe of something like 1,900 years of civilization.

The numerous tribes—Goths, Vandals, Heruli, Burgundians, Lombards, Franks, et al.—continued to live in seminomadic squalor, illiteracy, mental and technical backwardness, rustic ignorance, and savage confusion. "Their personal appearance excited the contempt and abhorrence of any Romanized person who had the misfortune to come in close contact with them. Their rustic manners, voracious appetites, and horrid appearance disgusted the civilized Gauls." (Gibbon.)

"It is for your sake," said one of Emperor Vespasian's governors of Gaul with prophetic insight, "that we guard the barrier of the Rhine against the ferocious Germans who have so often attempted and who will always desire to exchange the solitude of their woods and moraines for the wealth and fertility of Gaul. The fall of Rome would be fatal to the province; and you would be buried in the ruins of that mighty fabric which has been raised by the valor and wisdom of eight hundred years. Your imaginary freedom would be insulted and oppressed by a savage master; and the expulsion of the Romans would be succeeded by the eternal hostilities of the barbarian conquerors." (Tacitus IV: 73, 74.)

Seldom has a political prophecy proved to be so gruesomely true. (For a note on European prehistory see page 265.)

Water Conduction and Drainage in Greece and Rome

"With such an array of indispensable structures carrying so many waters, compare if you will the idle pyramids or the useless, though famous works of the Greeks."

FRONTINUS

The first European civil engineer known by name is Eupalinus of Megara who made engineering history by driving a water tunnel on the island of Samos. This tunnel is another odd piece of engineering which shows that at about 550 B.C. the Greeks did not know much about hydraulic engineering and had failed to learn from the sophisticated work that had been going on for such a long time to the east and south of them.

In any case, Eupalinus was commissioned by the tyrant Polycrates, ruler of Samos, to bring drinking water from a well in the mountains about 1 mile northwest of the city. Between the well and the place of consumption lies the 886-foot Mount Castro that had to be pierced by a tunnel, since it was not possible to bring a pipe over the summit. Moreover, for strategic reasons it was desirable to have the conduit protected against enemy action, and the tyrant ordered the aqueduct to be put underground.

The manner in which Eupalinus carried out his commission shows that he had learned from Persian qanaat builders, but not enough. To give him credit, he was confronted with greater topographical difficulties than the Persian engineers usually had to overcome. The country between the terminals of the conduit was and is pretty rough, and Mount Castro would have necessitated the sinking of shafts to a depth of 300 feet or more, which, had it been technically possible, would have made the project forbiddingly expensive. Eupalinus therefore took the chance of driving a tunnel through the mountain without any intervening shafts, a method used 2,400 years later in advancing the long railroad tunnels through the Alps.

Eupalinus' project had some admirable features. Close to the well, situated in the present village of Agiades, he excavated a triangular reservoir, and since it had to be hidden from view, he built 14 columns on its bottom. They carried stone lintels on which were laid slabs covered with earth. When the aqueduct was completed, he cut through one of the walls to the aquifer in the rock feeding the surface spring.

From this subterranean water reservoir he drove a 2 × 6-foot qanaat along a couple of small streams for a length of 2,790 feet. The heading was advanced in conventional fashion from a number of adjacent shafts. A fireclay pipe was laid on the floor.

The map shows the city of Samos on the Aegean island of the same name and how it was supplied with water through a tunnel. From the source at the village named Agiades, a qanaat was driven from 22 shafts along the ravine of a small winding stream. Owing to the rough terrain on Mt. Castro, it was not possible to sink any shafts to grade, and the 6 × 6-foot tunnel had to be advanced from the portals in a straight line through the mountain. From inside the southern portal, the conduit continues as a qanaat driven from 11 shafts. The tunnel is 3,291 feet long and was constructed about 550 B.C. by Eupalinus of Megara, the first European engineer known by name.

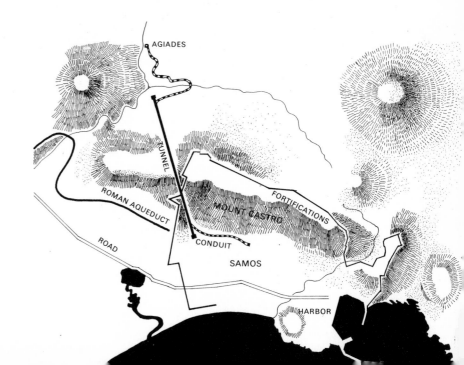

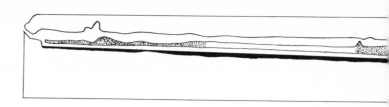

To the right is a longitudinal profile of the Samos tunnel, with the northern portal at the left. The vertical scale is five times the longitudinal, whereby the vertical irregularities have become exaggerated. The actual conduit was laid in a deep trench along the eastern side of the tunnel and is represented by the thick black line at the bottom. The shaded parts indicate how the excavated muck was backfilled after the pipe had been laid in the trench.

Two sections of the Samos tunnel are shown to the right. The left one is the section of tunnel and trench near the upstream northern portal, and the right one the section at the downstream end. The great difference in depth is due to grave errors in leveling. In the qanaats the conduit consisted of jointed fireclay pipes, whereas the tunnel trench was laid with open channels of closely fitted fireclay sections.

The qanaat joined the 3,291-foot-long tunnel through the Castro, and judging from the tool scars, it was excavated from both ends with chisel and hammer. The rock was a rather hard and nonfissured limestone pitching slightly to the west, and the originally flat roof now slopes in the direction of the pitch. About 1,310 feet from the southern portal, the gallery makes a 90° turn to the right and then meanders before joining the northern heading. This is obviously the place of breakthrough, but the strange thing is that the two headings would have met on line but for the angling. It seems as though the miners were led astray by the sound of their tools and abandoned their backsights. Be that as it may, 16 feet of meandering was required to join up the two headings.

This failure to meet on line can perhaps be tolerated, though it shows sloppy supervision at a critical stage, but the more serious errors committed in leveling are unforgivable, and for these the engineer himself must be held accountable. Because when the two headings met, the floor of the northern one was 3 feet above the roof of the south heading, or a total error in grade of 9 feet. But this was not all. Then came the really nasty jolt; when the miners driving the qanaat left the ravine and continued without the guidance of running water through the northern slope of Castro, they lost all sense of grade. The pipeline on the bottom of the qanaat should of course have been brought to the level of the tunnel floor; instead it landed 12 feet below the floor. There can be only one reason for this error: the tunnel drive had begun before the qanaat had reached the northern portal, and Eupalinus had miscalculated the grade.

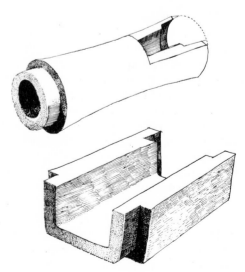

To correct the error he had to excavate a 12-foot trench along the west wall of the tunnel. Had he used a grade of 1:600, as in good Roman practice, the trench would have fallen 18 feet below the floor at the southern portal. Instead, it lies at 31 feet, producing a grade of 1:180, which is much too steep for water conduction. On the other hand, after the error in grade at the breakthrough had been corrected, the bottom of the tunnel itself obtained a grade of 1:570, which would have been much more suitable had the pipeline been laid on the floor as originally intended.

Herodotus regarded Eupalinus' tunnel as "the greatest engineering work in any Greek land." (Of the other great works catalogued by him two more were also on Samos, namely, the long harbor jetty and the Hera Temple.) He is entitled to his opinion and so are modern archeologists who have become interested in the tunnel and think highly of it. Nonetheless, this first European attempt to conduct water underground shows so many elementary technical defects that it is difficult to share their views. Persian engineers would have had a good laugh had they had occasion to inspect the tunnel.

Greek siphon lines

Greek hydraulic engineering was devoted mainly to bringing drinking water to the cities, in the homeland as well as in the colonies. The earliest aqueducts, such as the one supplying Athens with water from springs in Mount Pentelicus, was of the qanaat type with shafts sunk every 50 feet. Then followed a gravity line of fired earthenware pipes laid on stone plinths; when hills interrupted, the pipe was taken through tunnels. The experience gained from laying and maintaining such pipelines would inevitably suggest the development of so-called siphon lines, i.e., high-pressure lines conducting water from an elevated source and utilizing the head to force the water upward across intervening hills between the source and the place of consumption. The benefits obtained from a siphon line are as obvious as they are appealing: instead of the tortuous loops and costly tunneling necessary for a gravity-flow duct, a siphon line could be laid on the ground in a straight line, up hill and down valley, from the source to the city. The solid water column contained by the pipe could be forced over topographical obstacles and could also overcome the friction against the pipes, provided the head, i.e., the elevation of the source, was high enough.

In principle, a siphon line is excellent, but the trouble with such a line is the tremendous pressures to which its lower parts become exposed in crossing a valley. The pressure at the low points might burst any earthenware pipe, but even if the pipe sections could be made strong enough to take the pressure, the valley part of the line would be lifted and fractured if it were not firmly anchored.

The Greek engineers succeeded in overcoming the pressure problems by enclosing the pipe in stone at the critical points and by ballasting the valley crosses. With the modest capacity lines required to supply the relatively small populations of the Greek cities at home and abroad, the engineering problems could be mastered with the conventional materials available, but later, with the large volumes of water required by the Romans, it was no longer possible to apply the technically more sophisticated siphon lines, and Roman engineers reverted to the gravity-flow method of water conduction.

Of the many Greek siphon lines, the one built in 180 B.C. to supply the citadel of Pergamon with water is best known. The citadel was situated on a high knoll with an elevation of 1,089 feet. To supply it with drinking water a spring was tapped in the mountains at an elevation of 3,850 feet, and water conducted by gravity flow to two settling tanks at Hagios Georgius to the east of the city at an elevation of 1,230 feet. From the tanks a siphon line, probably made of bronze or lead and carried on stone plinths set 4 feet apart, brought water down into a valley with a low point of 564 feet, and up over a ridge having an elevation of 764 feet, down into a second valley at 639 feet, and then up the steep slope to the citadel at an elevation of 1,089 feet. The pressure on the pipe crossing the lower valley has been calculated at 20 atmospheres (285 psi).

When Pergamon, like the rest of the Greek cities in Asia Minor, came under Roman rule, the old siphon line was replaced by a gravity-flow aqueduct which was unable to supply the citadel with water, only the city nestling lower down the knoll. From this it should not be concluded that Roman engineers were unfamiliar with siphon lines or

81

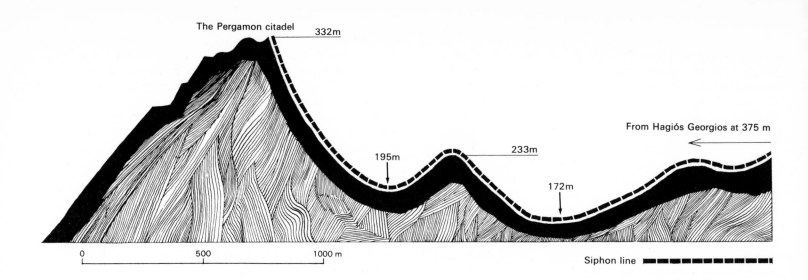

The Pergamon citadel   332m

195m   233m   172m

From Hagiós Georgios at 375 m

0   500   1000 m

Siphon line ▬▬▬▬▬▬▬▬

The Pergamon citadel situated at an elevation of 1089 feet above sea level was supplied with water by one of the most reputable siphon lines in the ancient world. The water pipe, most likely of fireclay (although lead or bronze have also been suggested), was carried on stone plinths and was naturally exposed to tremendous pressure at the low points. Here the pipes were reinforced and ballasted by being enclosed in perforated blocks of stone fitted together with recessed joints. The detailed construction of such a ballast pipe is shown below, although it derives from Pontus Pilate's siphon line to Jerusalem.

abstained from using them when conditions favored such a solution. One of the better-known Roman aqueducts on which siphons were used is the Pont-Siphon de Garon built A.D. 50 during the reign of Emperor Claudius to supply Lugdunum (Lyon) with water. The aqueduct is 32 miles long and consists of three tunnels joined by runs of surface structures that included three siphons. One siphon was constructed of 18 lead pipes that crossed a 217-foot-deep ravine on 52-foot-tall arches. Here the pipes were exposed to a head of a 164-foot water column (or 62 psi). Another Roman siphon was included in the Hebron-Jerusalem aqueduct built on the initiative of a man whose name has become familiar throughout the world—Pontius Pilate.

Beginnings of European land reclamation

Whereas in the ancient riverine civilizations, as previously noted, hydraulic engineering had been largely devoted to irrigation schemes and the canals, dikes, and dams required for large-scale river control schemes, early efforts in Europe were not directed at getting water but rather at getting rid of it. To gain new farmland for the growing population it became necessary at an early date to drain shallow lakes, bogs, and marshes which almost everywhere hemmed in the patches of arable land, turning them into islands of high ground in a watery wasteland.

The Greeks began this job of land reclamation which was continued by the Romans on a huge scale, and then, after an interval of a thousand years, was resumed in the Netherlands, England, France, and Germany. Today, after continuous drainage efforts, there remain in all of Europe only a few major waterlogged areas, in the east and north, as a reminder of the conditions prevailing at the dawn of European history.

In ancient Greece, the major feat of reclamation was carried out by the Ninyae, contemporaries of the Mycenaeans, with the drainage of the Copais lake in Boeotia. The lake was and is situated at the bottom of a mountain-girded basin watered by numerous streams from Parnassus and the Helicon Mountains which flooded the basin in winter and turned it into a reedy marsh for the rest of the year.

The Copais lake was drained in two ways. At first, a system of trenches emptying into canals leading to natural crevices and tunnels cut through the hills was used to drain the area. The second stage employed barrages built of Cyclopean masonry for catching the water before it reached the lake and deviating it to the discharge tunnels. However, such a vast drainage system requires continuous maintenance to keep it in working order, and when for political or other reasons the work ceased, the canals and tunnels silted up. By the time of Alexander, the formerly reclaimed land had returned to its natural state of lake and reedy marsh. Alexander put his engineer Crates to the work of draining the basin anew, and a number of shafts were sunk to grade before tunneling. But the work was not completed, and Copais had to wait more than 2,000 years until, in the last century, a large drainage tunnel was driven through the ridge and turned the Copais basin into the richest farming area in Greece.

Roman reclamation projects

In Rome, water was originally obtained in the same simple manner used everywhere: from the nearest spring or shallow well dug to fetch surface water. When these primary sources did not suffice, the Tiber was a convenient source that could be utilized without much expense and trouble.

It is interesting to note that the growing metropolis was able to satisfy its increasing needs of water in this manner for 441 years, or until 312 B.C., which agrees reasonably well with the experience of the large non-Roman cities north of the Alps. On the whole, it was not until the first half of the nineteenth century that it became necessary for them, to bring in water from distant and unadulterated sources.

But it is also interesting to note that long before the need arose to bring water into Rome, it was necessary to remove (i.e., drain) water from the city. Tarquinus Superbus (500 B.C.), seventh king of Rome, straightened and widened a brook emptying into the Tiber in order to turn it into a more efficient drainage canal. Cloaca Maximus became also a navigable canal, at least for small boats, until inevitably it was turned into a convenient depository for the refuse of the growing city. From being a drainage canal it developed into a sewer—open, evil smelling, and unhealthy—until in imperial days it was roofed over and made into a subterranean sewer of exceptionally large capacity. With dimensions of 11×14 feet it now conducts something like one million cubic meters of waste water per day, or 11 cubic meters per second.

The Roman march from the goat pastures on the Palatine to world power was accompanied by hydraulic engineering on an increasingly large scale. As previously suggested, the primary concern was not to bring water into the territories conquered but to drain them of excess water, to reclaim land for farming. It was work that did not attract Roman poets and historians and has therefore been left unrecorded, except when accompanied by scandal or something else that caught the fleeting interest of the literati.

After subduing the Sabines, the consul Manius Curius Dentalus (ca. 280 B.C.) ordered the marshes at Rieti, their former capital, drained by means of a 2,624-foot-long canal partly cut in rock. This canal still remains, but what has caught the poetic fancy is the fact that the waters were led deliberately to a sheer precipice to form a dramatic waterfall into the river Nera.

At the end of the third century B.C., the turn came to regulating the Tiber to contain its torrential spring floods. This was accomplished by enclosing the river in stone embankments laid out in stepped stages which, incidentally, proved superior to the present scheme of using sheer quays. The Arno was regulated in a similar fashion. The Pontine marshes south of Rome were also drained, although little is known of by whom, when, and with what success. Obviously, a large navigable drainage canal had been dug through the malaria-infested marshland connecting the seaside town of Tarracina with Apii Forum. The poet Horace spent a night on the canal and complained bitterly about the insects, croaking frogs, drunken boatmen, and other inconveniences of canal travel, but beyond that, little is known about this amazing reclamation scheme which proved to be beyond the technical and economic means of the last 500 years until in the 1930s Mussolini's government made a success of it. But the fact that Horace was able to travel through the marshes on scheduled service suggests that this was indeed a major drainage development that could only have been carried through by the Roman state.

The Fucinus Emissarium development

The Po Valley, still exposed to inundations by flash floods, was reclaimed by a scheme ordered by the consul Marcus Aemilius Scaurus in 109 B.C. for the purpose of settling veteran soldiers as farmers on the new land. The drainage canals were also used for navigation, whereby the Po became connected with Parma. The same consul joined this fertile area—Cisalpine Gaul—to a highway *Via Aemelia* leading to Rome. In this previously inhospitable area, which was destined ultimately to become the cockpit of creativity in the western world, reclamation work went on for many centuries. The marshes around Bologna, Cremona, Piacenza, and other cities were drained; the Adige was canalized between Ferrara and Padua. Also the marshes along the Adriatic coast, around Ravenna, were drained and made suitable for human habitation.

Owing to Tacitus and Suetonius, both of whom took great delight in heaping abuse on Emperor Claudius, we are informed about the details of a major Roman reclamation project, the *Fucinus Emissarium,* which involved lowering the level of Lacus Fucinus in order to gain new farmland, estimated at 38,000 acres. Suetonius viewed the Fucinus scheme as a purely speculative venture unworthy of an emperor, while Tacitus treated it as an attempt by Claudius to prove himself equal or even superior to Emperor Augustus who had turned down the project as impractical.

In any case, work on the 3.5-mile-long tunnel got started in A.D. 41 by sinking to grade a large number of vertical or sloping shafts—the so-called *putei*—spaced one actus (120 feet) apart. Some of the shafts had to be sunk 400 feet. The tunnel was driven through rock most of the way and was un-lined, though in soft ground the roof was lined with ashlar.

Some 30,000 men labored for 11 years in the tunnel, ex-cavating about 78,000 cubic yards of rock which was hoisted to the surface in copper buckets with the aid of windlasses placed across the shafts. The size of the tunnel is 9 × 19 feet according to Livy, and this agrees rather well with an eye-witness report from the middle of the last century which gives a height of 20 feet and a width permitting two horse-drawn carts to meet.

The reason why we know about the scheme and the long tunnel emptying into the river Liri is the scandalous opening, as told by Tacitus. Claudius decided to celebrate the comple-tion of the scheme in a truly imperial manner. He ordered the prisons emptied and put 19,000 convicts in arms aboard a fleet of galleys built for the occasion. On the shores were posted detachments of the Praetorian Guard to prevent the escape of any of the convict gladiators. The wretches aboard the galleys did what was expected of them, and a tremendous sea battle went off in a wholly satisfactory fashion. A passable number of gladiators managed to get themselves cut up—to the delight of the imperial establishment—whereupon the Emperor gave a signal for the opening of the sluice gates. But then something went wrong. No water entered the tunnel, probably owing to faulty grade, and the Emperor returned to Rome furious and frustrated by the unexpected outcome.

Another year had to be spent correcting grade, and then another opening party was laid on. This time the imperial banquet was spread on tables set on the muck heaps below the downstream end of the tunnel. Again the opening was preceded by gladiatorial combat, after which the Emperor raised his hand as a signal to open the gates at the upstream end. The time since the previous fiasco had not been wasted. The Fucinus waters came roaring down the tunnel, and backed by a head of 50 feet, a wall of water emerged suddenly through the downstream portal with such force that the tailrace canal failed to contain it. The flood swept away the banquet tables and turned the ceremony into a soggy mess. Empress Agrippina screamed abuse at Narcissus, the equestrian aristocrat in charge of the works, and he replied with a few ill-chosen comments on her wayward life in general and sex life in particular.

The water subsided, but not the human furies; the Empress had her husband poisoned so that her son Nero could succeed him. She also settled her score with Narcissus who was thrown into prison where he died some months later.

The drainage operation appeared to have failed because Lago Fucino continued to exist until 1875, when it was successfully drained by Swiss and French engineers working on behalf of Prince Alessandro Torlonia who was able to add 38,600 acres to his estates. The drainage tunnel driven be-tween 1852 and 1875 followed the same route as the ancient one but is 530 yards longer. In 1951, the reclaimed Conca del Fucino was expropriated by the Italian government, and the following year 8,000 families were settled on 34,000 acres.

The aqueducts of Rome

Each great engineering work in history, particularly the costly innovations, has its origin in the fanaticism and megalomania of one man, and one man only. The ancient rulers, god-kings or not, could order by personal whim the execution of vast hydraulic works satisfying their inflated egos. Thus we have the standard formulas of the stele in-scriptions—"I, Cyrus, king of all worlds, great king, mighty king" or "I Sennacherib dug the canal to the meadows of Nineveh" or "I Darius, the great king, king of lands of many races, king of this wide earth . . . ordered the canal dug . . .," and so on ad nauseum. That such hydraulic developments in the end also benefited the multitudinous wretches that slaved on them was purely incidental.

The Roman emperors were unable to escape the hubris of totalitarian power and could order vast works under such formal titles as "Nero, Emperor and Victor, Crown of the Roman People and Possessor of the World." He commanded convicts and prisoners from all parts of the empire to be brought to the Isthmus of Corinth to dig a 50-yard-wide canal joining the two seas. He sent convicts to dig an inland canal joining Lake Avernus near Naples with Ostia, the seaport of Rome, a 160-mile waterway wide enough to permit two *quinqueremes*—galleys with five banks of oars—to meet. The inland canal was to provide a safe route for the grain cargoes from Africa to the capital. Nothing came of these two major projects, and when Nero died in A.D. 68, they were permitted to lapse.

It is appropriate to recall these questionable features of large-scale engineering works, ancient as well as modern. In societies where the people, directly or indirectly, have some-thing to say in common affairs and realize that they are the ones who will have to pay for them, there will be no large-scale works of any kind. To get anything at all accomplished in the face of compact resistance by the people, in formal assembly or otherwise, who refuse to see any need for a proposed development or, recognizing its merits, shudder at

One of the most well preserved, i.e., well repaired and rebuilt, viaducts on a Roman aqueduct is Pont du Gard crossing the river Gard 12 miles northeast of Nîmes in southern France. It is the major feature in a 31-mile-long aqueduct that brought water from sources at d'Eure and d'Airan to Nîmes, or Nemausus as the city was known in Roman times. The aqueduct and its impressive bridge were constructed in 63 to 13 B.C. in connection with the settling of some 80,000 colonials in the area. The length of the bridge is 903 feet, and its total height is 161 feet. The specus is carried atop the third line of arches and measures 4 × 6 feet. A road was built along the bottom arches in 1743.

The picture overleaf shows the present condition of Aqua Claudia, Rome's largest aqueduct, at Casale di Roma, to the southeast of the city. The arches carry Aqua Claudia, whose specus is still relatively well preserved, and Anio Novus' partly destroyed specus. Together the two aqueducts brought 162 cubic feet of water per second to a castellum a few kilometers north of Roma Vecchia. Aqua Claudia had a length of 42.8 miles, and one of its sources was situated at the existing bridge across the Anio on the road to Anticotti. Anio Novus was 54 miles long, the longest of Rome's aqueducts. Both entered the city through Porta Maggiore built A.D. 52.

Piranesi's map of Rome's aqueducts gives a reasonably correct orientation of the aqueduct system inside the city. Later, the routes were established at numerous points in connection with foundation work for buildings. Aqua Appia's run through the southern parts of the city, as shown on the map, has not been verified except at two places on the Aventine where a century ago its specus was found in a quarry. As an aid to orientation, the figures outlined in black, to distinguish them from Piranesi's own notations, show some of the best known antique remains: (1) Castel Sant'Angelo, (2) Augustus' Mausoleum, (3) Therms of Diocletian, (4) Therms of Titus (of which now only a few traces remain in Parco Appia), (5) Colosseum, (6) Palatine, and (7) Therms of Caracalla.

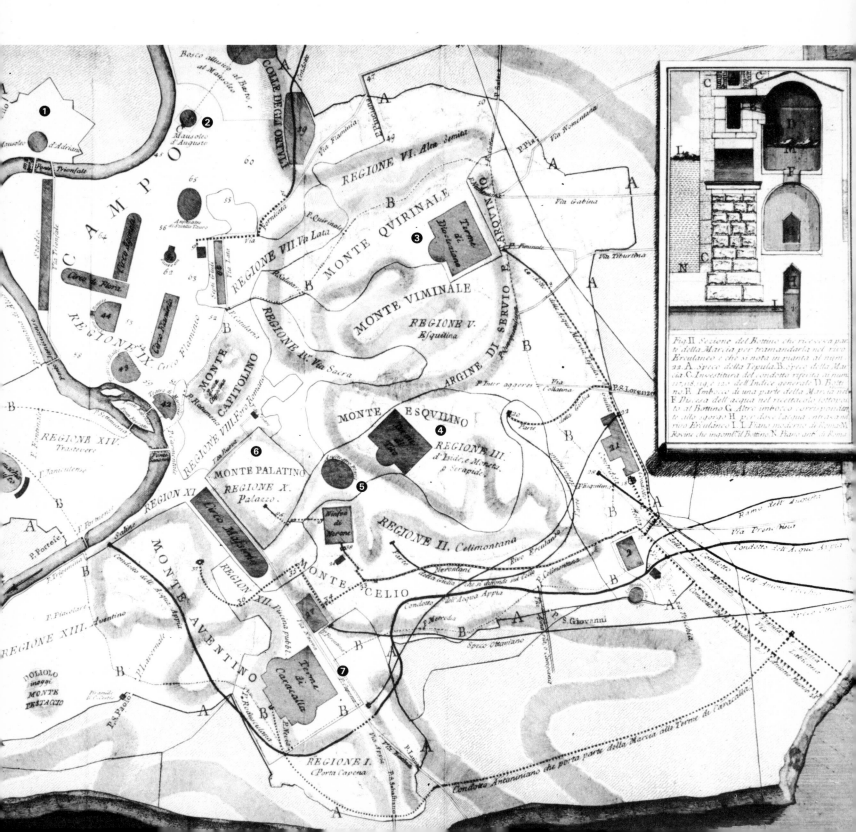

the cost, requires a promoter with a fanatical will and the willingness and ability to play dishonest or unscrupulous politics. That, to a major extent, is the reason for the inability of the modern welfare state to solve problems calling for major engineering developments. The best they can manage is too little, too late.

To connoisseurs of public-works politics it comes as no surprise that the first large-scale schemes pushed through in republican Rome were accompanied by political machinations and outright swindles in order to break down or circumvent the refusal of the people to have anything to do with them. Both the first highway ever built in Europe—Via Appia—and the first aqueduct in Rome—Aqua Appia—were the outcome of dirty politics played by the man who gave his name to both—Appius Claudius Crassus, who served as censor together with Gaius Plautus for an 18-month term ending 312 B.C.

During his term of office, Appius Claudius, by playing plebeians against the aristocratic senate and other trickery, managed to get the Senate's authorization and a vote of money for building the highway and the aqueduct, against a majority that disliked both developments as being too costly. The work started during Appius' incumbency and was far from completed when his term of office expired. Fearing that the Senate would stop the work in progress upon the lapse of his power, he made his colleague believe that both should abide by the rule and abandon office at the end of the term. Gaius did so, only to find to his chagrin that Appius had no thought of doing the same; indeed, Appius managed to get his own term extended until the highway and the aqueduct were completed. Having outmaneuvered his colleague in office, Appius acquired an enduring niche in history by naming the two major works after himself.

Nothing remains of Aqua Appia, and what is known about Rome's first aqueduct is derived from Frontinus' *De Aquis Urbis Romae* written during his term as Curator Aquarium, from A.D. 97 to 104. The Appia drew its water from springs "between the seventh and eight milestones" on the Via Praenestina (modern archeologists place the intake in the stone quarries somewhat to the north of the road). It ran underground for 10.6 miles until it crossed the east wall slightly to the northwest of Porta Maggiore from where it was carried on arches for some hundred feet, after which it dropped underground once more and crossed the city in a southwesterly line until reaching the neighborhood of the present church of S. Saba, just inside the southern wall to the west of the Baths of Caracalla. From here it turned sharply to the northwest under the Aventine and emerged in the open, carried on a short run of arches to a terminal reservoir on the south bank of the Tiber near Clivio dei Publicii. The mean gradient appears to have been 5:1,000.

Having been brought to the city at such a low elevation, the gravity-flow aqueduct and its clearing basin could supply only the lowest parts of southwestern Rome. From the terminal basin, water was piped to 20 distribution reservoirs which received 1,825 quinariae of water per day. (See table on page 95.) The dimensions of the water duct (or specus) were 5×1.75 feet, while the size of the tunnel was 5×5 feet in rock. (Specus is the Roman technical name for the conduit. In good rock, it has the same size as the tunnel; in friable rock and arches, specus refers to the water-conducting area within the lining.)

Anio Vetus and Marcia

Frontinus attributed the underground construction of the Appia to two causes: (1) inability to do accurate leveling, and (2) desire to keep the conduit secure from damage by a besieging army. Doubt can be cast on both: first, because a 10.6-mile-long tunnel could not have been driven without some knowledge of leveling, and second, strategic considerations aside, the best place for a water conduit is below ground. Indeed, of the subsequent aqueducts constructed in Rome, amounting in total length to some 350 miles, only 50 miles or about 13 per cent consisted of surface structures. Appia's engineers knew what they were doing when they placed the conduits in underground tunnels.

The next aqueduct, completed in 269 B.C., was given the name Anio Vetus because it took its water from the Anio, a tributary of the Tiber which has its springs in the mountains to the east of Rome. The actual tap was at Varia (now Vicovaro) a distance of 22 miles from the city. However, the length of Anio Vetus was almost twice as long, or 43,000 paces (41 miles), owing to the necessity of keeping grade in the mountainous country. In picking grade the engineers followed the course of the Anio River, placing the tunnel in the hillside to the left of the river gorge down to Tivoli. There the river cascades down into the plains, extending westward to the city to join the Tiber north of Rome. At Tivoli, therefore, there was a parting of the ways, the aqueduct making a loop in a southeasterly direction, using the high ground as a means of maintaining grade. To get down from the hills the aqueduct, all the time underground, made a series of five long loops until ultimately the contour lines merged into a narrow volcanic ridge to the southeast of the city. The ridge eventually came to carry five aqueducts on two lines of arches one of which had three ducts (Marcia, Tepula, and Julia), and the other two (Claudia and Anio Novus) stacked one on top of the other.

Anio Vetus entered Rome by tunnel and crossed the Aurelian wall in the neighborhood of Porta Maggiore. It made a sharp loop under the present railway station and terminated in a settling tank slightly northwest of the Julia Fountain. According to Frontinus, only 201 paces consisted of arches. The capacity was 4,398 quinariae which supplied the 10 lower wards (vici) in the eastern part of the city. The mean gradient was 3.4:1,000. The cost of the construction was financed by booty captured from Pyrrhus.

These two aqueducts were adequate for 127 years, during which time the residents higher up on the hills had to get their water from local wells, since obviously the water could not be brought above the service level of the terminal settling basins.

By 144 B.C., when the Senate in the face of strong opposition passed the water bill introduced by the two consuls Galba and Cotta, the population of Rome had more than doubled and had outgrown the existing water supply. The two old aqueducts were badly in need of repair, and considerable water was lost owing to illegal taps. The money voted by the Senate for the repair of Appia and Anio Vetus and the building of a new aqueduct—if a suitable source could be found—amounted to 180 million sesterces (equivalent to $8 million in gold). Responsibility for carrying out the work was given to the praetor Quintus Marius Rex who completed the work in four years. When the new aqueduct was opened for service, it was given the name of Marcia in his honor.

Marcia was a much more ambitious undertaking than the two previous aqueducts. It tapped a pool of green water in the Sabine hills, somewhere in the neighborhood of the present hamlet of Marano, to the northwest of Subiaco, at an elevation of 1,030 feet. The straight-line distance to Rome was about 23 miles, but the total length of Marcia has been variously reported as 57 to 58.4 miles, owing to the necessity of maintaining suitable grade. Seven-eighths of the distance consisted of tunnels of varying cross section, the upstream section being 8 × 6 feet.

Marcia also followed the Anio River down to Tivoli, but from here the engineers were more courageous in selecting the route. Instead of the long tortuous loops of the old Anio, they crossed the mountain ranges in a nearly straight line and accepted the consequences of three major river crossings where the conduit had to be carried by huge bridges spanning the gorges. One of them, Ponte Lupos, is still standing and has a length of 370 feet and a span height of 97 feet above the bottom of the gorge. Whether the art of bridge building had advanced since Anio Vetus or whether more money was available cannot be stated with assurance now; whatever the reason, Marcia's route through the foothills south of Tivoli is much more elegant engineering than that of the old Anio.

Coming down from the hills, the underground conduit made a wide swing toward the south, following the volcanic spur used by the old Anio in order not to lose grade too rapidly. Marcia emerged in the open near Roma Vecchia and was at first carried on an embankment, then by low arches, after which the arches kept growing in height, reaching a maximum elevation at Porta Furba where the tall structures continued with a slight curve until the aqueduct crossed the Aurelian wall near present Porta Maggiore.

Inside the city the aqueduct, riding high and handsome on towering arches, followed the town wall and ended in a *castellum aqua* inside the Republican wall. From here a branch crossed the Quirinal to Capitoline. Marcia, then, not only brought the best-quality water to the city; it also brought water under gravity to all the high elevation districts of Rome, which thereby could be served with piped water. The capacity of Marcia was 4,690 quinariae, or about 48 million gallons per day. From the settling tank on top of Capitoline, water was piped to distribution reservoirs serving 10 vici to the west of the Palatine. But there were also numerous branches. The main aqueduct was tapped at a point south of the present Porta Tiburtina where a castellum, or tower tank, was built to receive some of the water. From the castellum an underground conduit was run through the Caelian Hill to serve the low-lying vici along Via Appia.

To serve the houses on top of the Caelian Hill and also the Aventine to the southwest thereof, a specus carried on tall towers crossed the hill. There, under the existing Church of St. John and St. Paul, are still the remains of the ancient distribution reservoir. Except for the arch of Dolabella and Silanus on top of the ridge nothing remains of the two original branches of the Marcia since Nero abandoned them in favor of the great branch of the Aqua Claudia which followed about the same route.

Although of a much later date, another major branch of the Marcia deserves mention, namely the one that supplied water to the Baths of Caracalla. This branch was built in A.D. 212 from a tap in the main conduit at Porta Furba, about seven kilometers to the southeast of the baths. When crossing Via Appia—near present Porta S. Sebastiano—the arch spanning the road was given impressive architectural treatment in the usual manner of road crossings. The crude masonry atop the "Arch of Drusus" marks the elevation of the specus at this point, from which it continued westward a few hundred meters to the terminal reservoir situated southwest of the baths.

Imperial construction

With Marcia ended the Republican concern with aqueducts. It was also the end of solid stone construction, because the surface structures of the three early conduits were built throughout of ashlar, sperone, peperino, and tufa. The road-crossing arches were encased in travertine as were also parts of the structures exposed to heavy loads. They were soundly constructed compared with later structures which consisted of a core of coarse concrete faced with ashlar or brick and usually shoddily built. Owing to the fine stone in the Republican work, nothing or very little remains of these early surface structures since they were used for many centuries as convenient quarries for building stone.

Between the completion of Marcia in 140 B.C. and the reign of Emperor Augustus, four small aqueducts—Tepula, Julia, Virgo, and Alsientina— were added to the system. Tepula was built in 125 B.C. and tapped some tepid springs near the twelfth milestone of Via Latina, at the foot of the

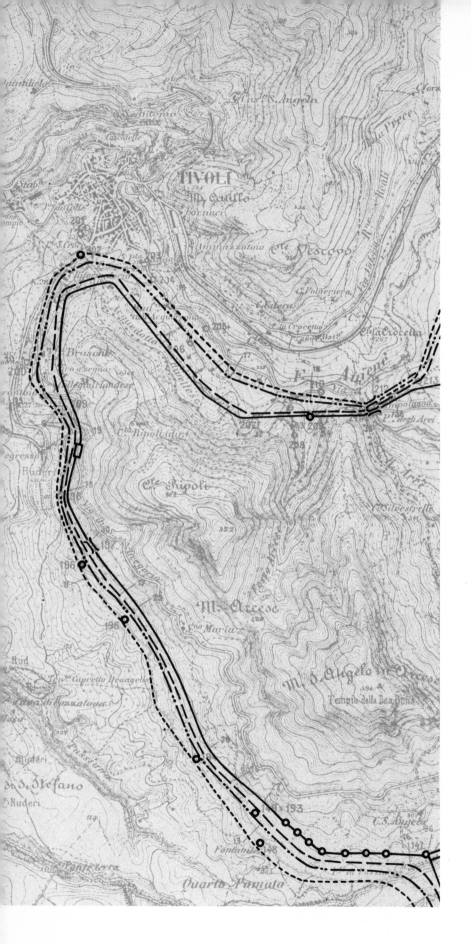

The detailed maps on this spread show the traverses of the four major aqueducts in the difficult Tivoli country. Owing to the reduction in scale, the traverse lines have been accentuated, for otherwise they would have been indistinguishable among the details of the Italian staff maps. The two maps, in themselves rather unique, are presented principally because they show the development in traversing from Anio Vetus (272 B.C.) to Anio Novus (A.D. 52). In the Tivoli terrain proper (left), all four aqueducts are bunched together; whereas farther south in the rough and precipitous Gallicano hills (right), they depart from each other in a significant manner. At Colle Aqua Raminga, Anio Vetus leaves the other and takes off in a series of loops along the sides of the steep ravines. It makes five major loops and several small ones before it begins to approach the others at Colle Farina (bottom left). The engineers succeeded in finding a route that required only one bridge crossing (at Ponte Taulella). Anio Novus, on the other hand, follows a much more elegant but costly traverse, going through the mountainous country with long tangents joined by smooth curves. But to achieve this reduction in length Novus had to be brought across eight viaducts.

Alban Hills. The spring has been identified as Sorgente Preziosa. When Tepula was brought out from the ground, it was carried piggyback on Marcia's arches for the last 6.5 kilometers into Rome. In 33 B.C., Marcus Agrippa, aedile and son-in-law of Augustus and the most progressive man in Rome, built another short aqueduct from a source not far from that of Tepula, whereby Marcia's surface structures had to carry three separate specuses. Owing to their stacked position, Tepula and Julia were able to supply higher elevations in the city than Marcia itself. But their waters were of poor quality. All that remains of Julia is the "Julia Fountain" in Piazzo Vittorio Emanuele on the Esquiline, the only one of the ancient fountains to have survived destruction.

Virgo's link with the present

Of the four minor aqueducts briefly mentioned, Aqua Virgo is the most interesting because roughly speaking it has been in use from June 9, 19 B.C. until the present. It was originally built by Agrippa to supply water to the baths on Campus Martius, near the Pantheon, at that time outside the walls and not yet a built-up area. According to Frontinus, the aqueduct obtained its name from a young girl who pointed out a spring to some soldiers looking for water. The springs were located near the eighth milestone of Via Collatina, i.e., to the east of Rome near the farm Salone. But instead of bringing water into the city via Porta Maggiore in the manner of all previous aqueducts, Virgo made a circuitous route toward the north and west in order to enter the city from the north.

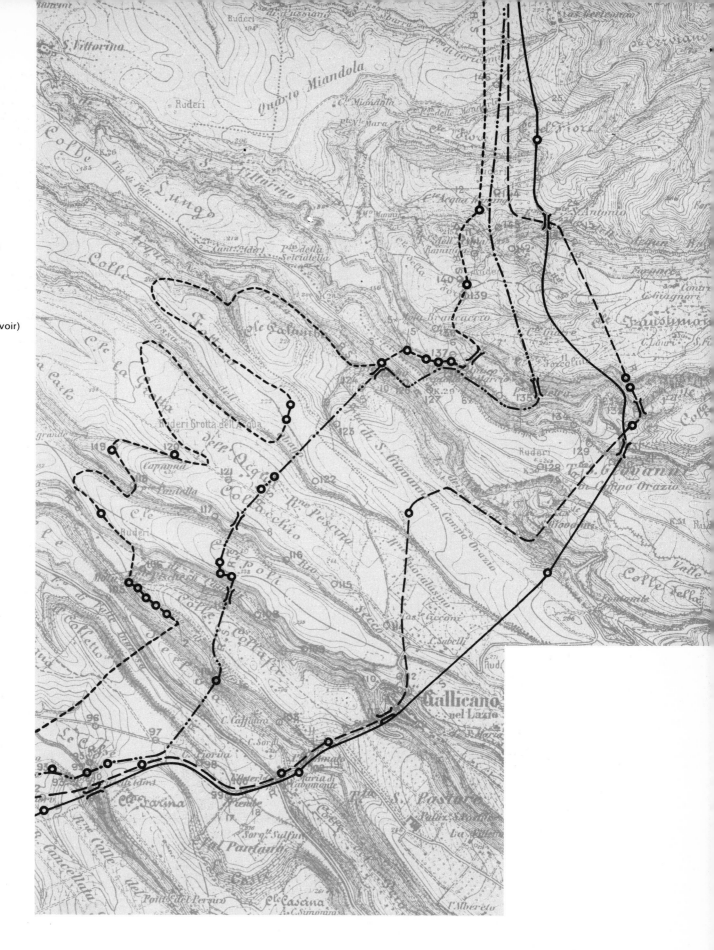

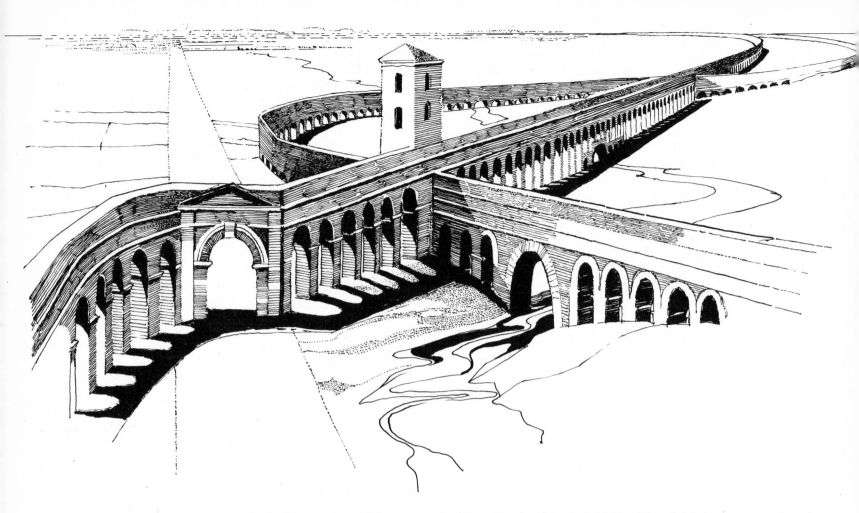

South of Rome, about 3 kilometers north of Roma Vecchia, Marcia (with Tepula's and Julia's specus on top) made a wide loop and intersected Claudia—Anio Vetus twice. The area enclosed by the loop was used by the Goths in A.D. 537 during their siege of Rome. They filled in the arches and put in 7,000 troops in the fortified camp. From this event the place came to be known as Campus Barbaricus. The tower in the foreground pretends to be Tor Fiscale built in 1363 by a papal tax collector. The tower still has remains of the specus of the five aqueducts. However, the placing of the tower is incorrect; it should be at the second intersection, but then, of course, the composition of the picture would have suffered. The sketch is made on the basis of a painting by the German eighteenth century artist Zeno Diemer, and is not much better than fantasy, because when the picture was painted there were no longer any surface remains left. Documented descriptions and archeological investigations suggest that the crossing might have looked something like this ca. A.D. 100.

Virgo's water was comparable to that of Marcia in purity and coldness, and Caligula made himself unpopular when he razed nearly a mile of the arches across Campus Martius and deprived Rome of one of its best sources of drinking water. The conduit was restored in A.D. 46 by Claudius, and of this work there now remains a half-buried monumental arch near the Trevi Fountain. The conduit was repaired again in A.D. 338 by Constantine, destroyed by the Goths in 537, and put back in order once more at the end of the eighth century by Pope Adrian I, and it seems to have been functioning until about the middle of the tenth century. Then it fell into disuse for 500 years until Pope Nicholas V got it working once more in 1453. A hundred years later Sixtus IV and Pius V had the ancient conduit repaired, and in the early years of the last century it was lined with lead. In the present century, the leaded conduit was replaced by cast iron pipes. The existing cast iron conduit of Aqua Vergine follows in the main the underground tunnels originally excavated for Aqua Virgo. (Aqua Virginae as it is now called, was shut down for repairs in 1961.)

To complete the record, Augustus built Aqua Alsientina in 2 B.C., a 20.4-mile underground conduit which drew its waters from Lake Alsiente (now Lake Martigano) to the northwest of Rome. It entered the city from the west and was originally intended to supply water to a *naumachia,* an artificial lake on which sham naval battles were arranged. The water was foul and was not distributed for human consumption.

Another aqueduct entering the city from the west was Aqua Trajana, completed A.D. 109. It was 35.4 miles long and ran underground from Lake Bracciano to a terminal reservoir and settling tank in Janiculum, in the neighborhood of Villa Spada. Finally, in A.D. 226, Alexander Severus built Aqua Alexandrina. It was only 13.7 miles long and drew its waters from springs to the east of the city. It supplied the Therma Alexandrinae, the rebuilt Nero baths near the Pantheon. This was the last aqueduct built in Rome.

Rome, the first western metropolis, at the height of its power and influence, was supplied with water by 11 aqueducts, 9 of which have been identified above. We shall now turn our attention to Aqua Claudia and Aqua Anio, which together had a capacity of 9,345 quinariae and served all parts of Rome.

Pork-barrel public works

Like all major and costly engineering projects, the proposed Claudia system had been gathering dust for many years. The need for additional supplies was laid before Tiberius by Cocceius Nerva, the incumbent Curator Aquarum, who succeeded in getting imperial approval for conducting a study and preparing estimates for a new supply of water. Engineers spent some time surveying possible sources in the neighborhood of Rome in order to find one satisfying their requirements in respect to quality, volume, and elevation. They found two springs meeting their specifications, the Blue Spring and the Curtian Spring, both situated near the 38th milestone on the Sublacentian Road. At the 42d milestone south of Subiaco they decided to tap the River Anio and carry the river water by means of a second aqueduct.

Tiberius paid for the engineering and had detailed plans drawn up. He even called for bids, but in the end he decided that he could not afford to go through with the costly project. His successor Caligula, eager to demonstrate his public spirit upon his accession, began work on Tiberius' scheme, which involved building a 43⅓-mile-long aqueduct—Aqua Claudia—of which 34 miles, or 78 per cent, consisted of tunnels, and a second one 54 miles long—the Aqua Anio Novus—of which 45⅓ miles, or 85 per cent, were tunnels. In order to show progress, Caligula concentrated on the tunnels and left off work on the bridges and other masonry structures. When he ran into political and financial trouble, he used that as an excuse for abandoning the whole project. On his death in A.D. 38 most of the work was unfinished.

It thus fell to his successor Claudius to complete the gigantic project—which he did in A.D. 52 at the cost of 350 million sesterces, according to Pliny, all paid out of Fiscus, the Imperial Treasury.

Both aqueducts ran roughly parallel to Marcia along the Anio River down to Tivoli, but to the south thereof they swung further toward the southwest in long sweeping curves that brought them to the spur south of Rome in the manner of the previous conduits coming from the east. At the place now called Capanella, the subterranean ducts came out into the open, and for the six remaining miles to the city they rode together on a line of tall arches, the specus of Claudia below and that of Anio Novus on top. They were brought into the city at Porta Maggiore which was erected to accommodate the crossing. About a dozen major bridges were constructed in the mountains to bring the underground conduits across the gorges.

The two aqueducts terminated in a large castellum to the west of Porta Maggiore from where the waters were distributed to 92 castelli in the city. Nero built a high-arch extension to the Caelian, using the same route as the Marcia in order to increase the supply of water to his palace, "The Golden House," built to the west of the hill.

As a fitting conclusion to the job, Claudius ordered to be erected the magnificent 105 × 79 × 20-foot Porta Maggiore, built in travertine and containing two large arches and three small ones. The attic was—and still is—divided into three panels, the top one of which contains the commemorative description stating that Claudia was 41.4 miles long and Anio Novus 54 miles.

But these original inscriptions are followed by less flattering ones suggesting that these two engineering works, beyond comparison the largest undertaken by Roman engineers, in actual fact remained in unobstructed service for only 10 years. After being out of order for 9 years, they were restored by Emperor Vespasian in A.D. 71 and again in 81 by his son and successor Titus.

But even these two major restorations were inadequate to keep them in service for any length of time, because Hadrian (117–138), Alexander Severus (222–235), and Diocletian (284–305) had to invest heavily in order to keep the two giants functioning. Times, indeed, had changed. Whatever else the Emperors could accomplish they were no longer able to buy sound work. The seemingly impressive Roman public works had become pork barrels and served chiefly as a means for contractors and their aristocratic associates to grow wealthy at the expense of the Imperial Treasury.

In their heyday when they were all in operation, Rome's 11 aqueducts measured altogether 312 miles in length, of which 262 miles constituted tunnels and the rest arched surface structures. The total cross section of the ducts has been estimated at about 7.5 square meters in area (80.7 square feet) which brought somewhat more than 1 million cubic meters (35 million cubic feet) of water to the city every 24 hours. If, as estimated, the population at the time amounted

to 1 million, the consumption per head works out to 264 gallons per day.

No modern city is capable of supplying even 200 gallons per day per head of population. The District of Columbia, to take a modern example of an administrative center with about the same population as Rome in A.D. 100, consumes only half as much. The per capita consumption in the United States is only 65 gallons per 24 hours, and in Britain 55 gallons, or 20 percent of the Roman.

This Roman figure obviously does not mean that each resident consumed, directly or indirectly, such a volume of water per day. Rome had 11 public and 856 private baths which required considerable water, and the numerous fountains (1,352 in the fourth century) also needed huge volumes. The two aqueducts entering Rome from the west brought undrinkable water that after being used for operating water wheels was fit only for sewer flushing. According to Frontinus, the 35 million cubic feet of water arriving at Rome every day was consumed as follows:

For the Emperor's use	17 percent
Private use	39 percent
Public use	44 percent
Total	100 percent

A further breakdown of the public use gives 3 percent for military barracks, 24 percent for public buildings—including the baths—4 percent for theaters, and 13 percent for fountains.

Since, as described below, the distribution system worked as continuous flow and only a minute fraction was actually consumed, the large volumes of water flowing through the fountains and along the streets served as a refrigerant during the hot summer months.

Some summary data relating to the Roman aqueducts are given in the table on the following page.

Piranesi's engraving of Porta Maggiore shows clearly the greatest problem always confronting the Roman water administration, namely, the tendency to build up against the surface structures. Porta Maggiore was erected in A.D. 52 for the express purpose of bringing Aqua Claudia and Anio Novus across Via Praenestina, but in the 270s it was incorporated in the Aurelian wall. In 1836, the structure was cleared of slums, and it is now free of encroaching buildings, albeit surrounded by an unholy mess of traffic.

The original inscriptions on the three panels on the tympanum can no longer be distinguished, but have been copied by Piranesi and others. They read in somewhat free translation as follows: Top panel: Ti(berius) Claudius Drusus, son of Emperor Augustus Germanicus, Pontif (ex), first tribune, twelfth consul, fifth emperor, 37th pater patriae, (let build) Aqua Claudia at his own expense from the springs called the blue and the curtian at 45th milestone, also Anio Novus from 62d milestone to the city. (Dated A.D. 52.)

Second panel: Imp(erator) Caesar Vespasianus August(us), Pontif(ex), Tribu(nicia) Pot(estate) II, Imp(erator) VI, Co(n)s(ul) III, Desig(natus) p(ater) p(atriae)—i.e., the honorary titles of the emperor—restored with his own means the curtian and blue aqueducts built by sacred Claudius and thereafter falling into disuse. (Vespasian became Trib. Pot. July, A.D. 71.)

Third panel: Im(perator Ti(tus) Caesar Divi, F(ilius) Vespasianus Augustus, Pontifex Maximus (etc., etc.) rebuilt, completely at his own expense the curtian and blue aqueducts that· sacred Claudius had brought to the city, and were repaired by sacred Vespasianus, but had become destroyed by age from the springs. (Dated some time between July 1, A.D. 81 and July 1, A.D. 82.

Status of Roman water supply about A.D. 300

Aqueduct	Length,* kilometers	Head, meters	Gradient, per mill	Quinaria	Liters per second	Cubic meters per day
					Capacity**	
Appia	16.6	8.0	0.5	1,825	876	75,737
Anio Vetus	63.9	219.7	3.4	4,398	2,111	182,517
Marcia	91.4	270.3	3.4	4,690	2,251	194,365
Julia	23.1	279.2	12.0	1,206	579	50,043
Virgo	21.2	5.3	0.3	2,504	1,202	103,916
Alsietina	33.3	136.0	4.1	392	188	16,228
Claudia	68.8	248.2	3.6	4,607	2,211	191,190
Anio Novus	86.9	428.2	4.9	4,738	2,274	196,627
Tepula	18.0	136.0	7.6	445	214	18,467
Trajana	57.0	72.0	1.3	2.846	1,367	118,127
Alexandrina	22.0	9.0	0.4	521	250	21,633
Total	502.2			28,172	13,523	1,168,850

Construction of an aqueduct

The layout of a gravity-flow aqueduct is basically a matter of leveling. To the Roman engineers responsible for the planning of, say, Anio Novus, it became an exercise in bringing water from a source—near Sublaquem—situated at an elevation of about 1,100 feet above sea level to the castellum at Porta Praenestina. The bottom of the settling tank had to possess adequate elevation to permit gravity distribution to the summits of the hills in Rome. There exists no accurate information or even estimate of the actual elevation of the castellum, but assuming for the purpose of discussion that it was 230 feet above sea level, the problem resolved itself into laying out a conduit from the source at 1,100 feet to the terminal tank at 230 feet. The distance as the crow flies between these two points is about 23 miles. Over this distance of approximately 121,440 feet the aqueduct would have to bridge a difference in elevation of 870 feet. In other words, it would have to be given a slope—or gradient—of 1:140; i.e., it would drop one foot in 140 feet.

The Roman engineers would have immediately rejected such a crude simplification of the problem. Any neophyte engineer would have turned it down for a number of good reasons: (1) such a steep gradient would invite rapid water erosion of the channels; (2) the underground conduit when leaving the foothills would have had to be carried on impossibly high arches across the plains east of the city; (3) choosing a short linear route would necessitate bridging such a large number of gorges in order to bring the conduit across them that the development would have become prohibitively expensive.

Referring to the previous table, we know that Anio Novus had a length of 86.9 kilometers, and since the difference in elevation is 428.2 meters, the mean gradient is 4.9:1000 or 1:203. This raises the question of how Roman engineers went about running a traverse with a total length of 86,900 meters and a gradient of 1:203.

To being with, they went about it as any civil engineer before or since, until aerial mapping became sufficiently advanced to eliminate the scouting work. They walked the hills and staked out a number of trial lines until they found a compromise that appeared to the experienced eye to provide suitable grade at minimum cost. From the springs to Tivoli the river Arno had chosen grade for them, and it was merely

*From the source to castellum aqua.
**The estimates were made by Di Fenzío (*Sulla portala degli antichi acquedotti romani e determinazione della quinaria,* Rome, 1916.) A quinarium is given as equivalent to a flow of 0.48 liter per second, or 41.5 cubic meters per 24 hours. There exists no contemporary information on Trajana and Alexandrina, but since the Papal aqueducts Aqua Felice and Aqua Paola used the same sources, the likelihood is that the original Roman aqueducts carried about the same volume of water per day.

a matter of selecting good tunneling ground along the left gorge. They also knew from the experience of their predecessors that in order to escape impossibly high arches they would have to hit the volcanic spur leading from the foothills in a wide arc to the plains southeast of the city.

It was in the hilly country between Tivoli and Roma Vecchia that they had to find their gradient without help of the river. This is a job that a hunter or herder, indeed any man spending his life in the open, does more or less instinctively, but the engineer is also involved in the economic aspects of his lines. The point is that this preliminary but decisive work can be accomplished without any technical aids; it is simply a matter of a man and his ability to master topography.

Having come down from the hills on a meandering course marked on the ground and having it approved by the water administration for shortcuts involving the construction of some costly bridges to span the gorges, the engineers went over the route again, to make detailed investigations of the ground and to determine final grade. To do this the traverse

had to be leveled. For this work Roman engineers used a *chorobates,* a 20-foot-long board that could be leveled up with a plumbline or by filling a long groove in its top surface with water. With a chorobates an experienced man could establish a gradient of 1:2000. This implies, of course, in practice that the engineers established their levels and determined the length of the tangents by means of measuring rods or measuring ropes. Once a level reference plane had been obtained, the vertical distances to the ground could be determined and a profile plotted of the ground, on parchment or by simple reckoning marked on the ground.

From a survey like this it would be possible to establish (1) where the attacks for the tunnel would be made on the hillsides, (2) the direction of the tunnel through the hills, (3) where and to what depth the *putei* (or shafts) leading down to tunnel grade should be placed—if possible one actus apart, and (4) the exact point on the opposite side of the hill where the tunnel would emerge. During this phase the ground would be investigated by sinking trial shafts to determine the substructure.

In sum, the engineering preliminaries to spending 350 million sesterces could be made in the field and the gradient tangents marked permanently on the ground for the construction gangs to follow. It seems quite likely that this was the procedure followed on the early aqueducts. But by the time Claudia and Anio Vetus were being engineered, this simple direct approach would no longer do. A central administration headed by the Emperor had to be provided with documentary evidence of the fieldwork, maps and drawings, and cost estimates, to enable the deciding authority, i.e., in the end the Emperor himself, to be briefed on the overall and detailed features of the project. There were also legal aspects to be considered: the underground and surface structures crossed land held by private owners and right-of-way had to be purchased and deeded to the city of Rome. All this obviously necessitated a tremendous lot of parchment work.

This account of the preliminary engineering work for the construction of an aqueduct refers to Rome, but in the provinces, too, the engineering must have been adequate because each legion had an engineer centurion whose duties included surveying the aqueducts constructed within the administrative area in which the legion was garrisoned. But the local contractors do not always seem to have possessed the know-how required to follow the plan. This becomes evident from an inspection report submitted by the engineer centurion Nonius Datus attached to the III Legion in Mauretania. His report is dated A.D. 152 and reads in part as follows:

"Nonius Datus, engineer centurion of III Legion to the magistrates of Soldae (Algier). I made a survey and took levels of the mountainous ground. I staked most carefully the axis of the tunnel over the ridge. I made up plans and sections of the whole work and submitted them to Petronius Coles, governor of Mauretania. And as an extra precaution I

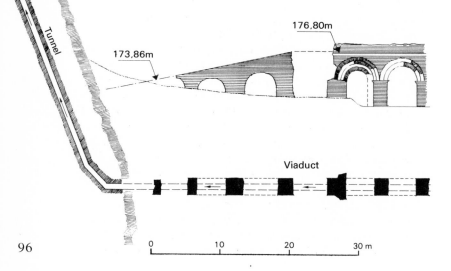

The third viaduct on Anio Vetus was built during the reign of Hadrian, and crossed Mola di San Gregorio. Its purpose was to shorten the line by the elimination of a long loop. It was 510 feet long, and consisted of 41 arches spaced 13 feet apart. The total height was 84 feet. Owing to the difference in elevation at the beginning and end of the eliminated loop, the bridge displays some strange features. Over the first 468 feet, the specus has a grade of 7.66:1000, but over the remaining part of the bridge and to the first bend in the tunnel, a distance of 82 feet, the gradient is increased to 163.5:1000, one of the sharpest on the entire aqueduct system. It seems as though the engineer desired to brake the velocity of the water at the outset to prevent building up excessive hydraulic forces in the straight tunnel following the bend.

96

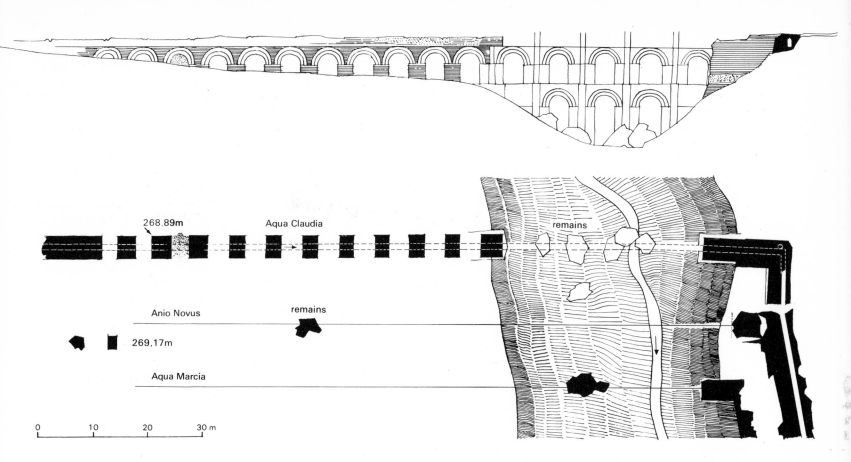

At Fosso della Noce, about half a mile northwest of Castelmadama, the three aqueducts Claudia, Novus, and Marcia are crowded together over a lateral distance of 98 feet. Here are still found remains of arches belonging to a 69-foot-high viaduct carrying Claudia's specus over the ravine. It was built during the reign of Severus. Only traces remain of the double-arched structure that once spanned the ravine. Some 30 feet to the north are the sketchy remains of a second bridge that took Anio Novus' specus across the ravine, and still further north are the traces of what was once a bridge carrying Aqua Marcia. Note the 90° bend in the tunnels after the bridge crossings, a typical feature in mountainous country.

summoned the contractor and his overseers and had a gang of experienced veterans show them how the tunnel was to be excavated. After four years' absence, expecting every day to hear that water had reached Soldae, I returned and found that the contractor and his overseers had made one mistake after the other. Had I waited longer Soldae would have gotten two tunnels instead of one." Nonius Datus had to resurvey the tunnel and supervise the driving of a cross-heading between the two trunks.

The construction work

When at long last, in the case of Claudia and Anio Vetus something like 10 years, the project was approved, the money appropriated, and the contractors selected on the basis of their bids or by insidious influence within the administration, the actual work of building the aqueducts could begin. Hordes of men were mobilized—skilled and unskilled—but all of them slaves. Prisoners of war were also hired out to the contractors. Their names speak of their origin: Spartacus, Allemanis, Paulus, Cyrus, Eras, Copadox, Hermann—men from all corners of the Roman world were put to work in the puteis and on the surface structures. What was admirable about these Roman operations was that the men were maintained in an intelligent fashion. The Roman equestrians who administered the works were commonsense, practical men who saw to it that they got maximum work out of this assorted labor force. Accordingly, the slaves were well fed and sheltered, on the same standard as the legionnaires. If among them there were men of ability—scribes, physicians, skilled tradesmen—they were placed in a capacity suited to their training and ability. There was apparently little if any of the northern corporal mentality of taking delight in inflicting iniquities and hazing better men put to labor under an ignorant plebeian overseer. The capable ones became foremen and eventually freedmen and Roman citizens. Indeed,

97

it appears as though many of these and other operations were engineered and supervised by freedmen originally brought to Rome as slaves.

It would also be wrong to regard this host of slaves as corvée labor of the same character as was called up for canal digging in the East. An aqueduct operation required a large number of specialized trades: tunnelers, shaft sinkers, teamsters, quarrymen, timbermen, masons, plumbers. After the work was completed, the maintenance of the structure needed a corps of skilled men — reservoir keepers, inspectors, pavers, plasterers, and, beyond all, plumbers.

To supply an aqueduct operation with materials was obviously a major business. The early aqueducts were built entirely of dimensional stone, usually volcanic tufa, while travertine was reserved for the decorative facing of road arches. In imperial days the contracts did not call for such solid construction, for then the core of the superstructures was cast with rough concrete with pozzolana cement used as a binder. This volcanic ash was mixed with lime together with an aggregate consisting mostly of pottery shards and turned into *opus signimum* and used to waterproof the specus. Tunnels were left unlined when driven through rock; in soft ground they were lined with masonry.

At the source, the raw water was led into settling tanks provided with sloping floors to facilitate cleaning. But this method did not work at the Anio intake, because when the river was in flood the water became contaminated by silt and organic matter, which made it undrinkable and caused trouble by clogging the ducts. Conditions improved materially when Nero dammed the river and obtained three lakes for his pleasure, from which the existing town Subiaco (Sublaqeum) gets its name. Frontinus shifted the water intake to the middle lake and succeeded thereby in improving the quality of the water to make it comparable to Claudia and Marcia.

Inside the town wall each aqueduct terminated in a large settling tank from which it was led to numerous distribution castelli situated at a lower elevation. At the bottom of each castellum were three outlets serving the fountains, baths, and public buildings in the neighborhood; and higher up were 10 mains for serving private consumers. The overflow of the reservoirs was used for flushing drains or for driving waterwheels. Some private houses took advantage of this free water and, by installing a waterwheel, lifted it into a storage tank inside the house.

Roman water administration

Nearly everything that is known about the Roman aqueducts derives from a book entitled *De Aquis Urbis Romae* written by Sextus Julius Frontinus during his term as seventh Curator Aquarium (A.D. 97–104). Frontinus had served with distinction as a soldier and was provincial governor of Britain (A.D. 74–75), proconsul in Asia, and commanded military expeditions into Germany and Hungary. Returning from his service abroad, he held the office of consul three times during the reigns of Trajan and Nerva. It was during his second term as consul that Nerva persuaded him to take over the office of Curator Aquarium in order to restore the corrupt water administration and the long-neglected hydraulic plant to an acceptable state of efficiency.

Frontinus went about his new administrative assignment in an admirable manner. Since the records dealing with the vast system of water supply were incomplete, faulty, or altogether missing, he began his term of office by making a complete survey of the existing plant and its history. On this material he later based his book, which goes into the technical minutiae of distribution as well as the legal and administrative aspects of water supply.

In Republican times, there was no separate administration of the Roman waterworks, which were supervised in a vague fashion by the annually elected aediles, or magistrates. In 33 B.C. during Octavian's second term as consul, his son-in-law Agrippa was made aedile and put in charge of the Roman water supply, an as yet unnamed office which he held until his death in 12 B.C. After his death, Augustus established a permanent administrative tribunal consisting of a Curator Aquarium with consular rank and two technical members. The tribunal was given a budget enabling it to employ a permanent staff of engineers and inspectors. In addition, it inherited a skilled labor force of 240 slaves which Agrippa deeded to the state. Later, during the reign of Claudius, a further 460 men were added to the original nucleus of "watermen." This was the municipal force. Since 17 percent of the water supply was reserved for the Emperor, there was also a separate imperial organization paid out of the privy purse. During Frontinus' term of office the tribunal payroll was about 250,000 sesterces per year.

Fraudulent practices

From his own account, Frontinus' principal concern was the suppression of the numerous forms of graft and malpractices that had developed during a succession of lax administrations. He was convinced that of the water entering the system vast volumes never reached the city since they were being diverted by means of secret taps to estates along the way. He also suspected that many or most of these illegal taps had been made with the connivance of the watermen. The same type of graft was true of the distribution network within the city: householders bribed the plumbers to put in larger pipes than were recorded, or to connect them to the "imperial level," i.e., the lower row of outlets of the castellum, where owing to the higher head, they would get more water than they were billed for.

In addition to these widespread fraudulent practices in which watermen and consumers connived to rob the city of revenue, Frontinus also had to deal with numerous infringe-

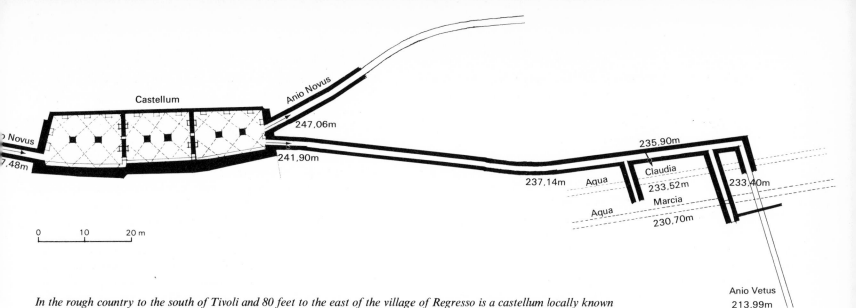

In the rough country to the south of Tivoli and 80 feet to the east of the village of Regresso is a castellum locally known as Grotte Sconce. It consists of three 49 × 43-foot piscinas that were supplied by Anio Novus. Downstream the castellum, Anio Novus takes off in a southwesterly direction, whereas another 3-foot-wide specus runs southward, ending in lateral outlets to the other three aqueducts. The downstream end of the castellum is situated at 241.9 meters above sea level, the Claudia intake is at 233.52 meters, Marcia's at 230.7 meters, and Anio Vetus' at 213.99 meters. Hence the difference in elevation to the intake of Anio Vetus is 27.9 meters over a distance of 150 meters, producing a gradient of 186:1000, the steepest on record. The castellum apparently served as a distribution reservoir from which Novus' water could be sluiced into any one of the other three aqueducts.

ments of the right-of-way, the 15-foot strip of land on each side of the duct that had been purchased and deeded to the city before construction. What inevitably happened was that property owners of the land bordering the strip utilized it for farming, roads, building, and, much more serious, tree planting. The surface structures were damaged by the buildings, and tree roots pierced the lining of the underground ducts and frequently caused the collapse of the tunnel, thus blocking the flow. According to Frontinus, the municipal and imperial water crews were kept busy repairing the ducts for the following reasons: (1) damage to the surface structures due to storms, (2) age, (3) illegal building close to the structures, and (4) secret taps. The bulk of the work consisted of keeping the ducts free from deposits and repairing leaks. No work on the ducts was done in the summer since it necessitated taking them out of service when there was a maximum need for water, and leaks were temporarily stopped by placing a lead lining over the leaky section. Masonry work on the arches was done in summer except on the hottest days.

Before Frontinus' term of office it had been customary for the watermen to take on private work during their paid time. A stop was put to this, and daily work orders were issued to each team. At the end of the working day the foreman had to report on what had been accomplished.

Having stamped out ancient malpractices, Frontinus busied himself with technical improvements. Before his time the aqueducts had been interconnected within the city, and this had resulted in the fine waters of Marcia and Claudia becoming contaminated by the inferior waters of the other aqueducts. He saw to it that the good supplies were kept intact, and extended the distribution network to parts of the city that had not been served by Marcia and Claudia. Then he turned his attention to controls. The reservoir outlets were made of bronze and fitted with adjutages, to regulate the discharge of water. The lead delivery pipes were standardized according to a modular system devised by Vitruvius based on the *quinarium,* a strip of lead 5 digits (3.5 inches) wide and rolled into a pipe. By the application of this module the water service obtained 25 pipe sizes, ranging from 1 to 120 quinaria.

A quinarium was a measure of capacity. It indicated how much water flowed through a pipe of a certain size under constant pressure. But since the pressure varied with the head, it was not possible to determine the actual volume of water delivered through such a pipe per unit of time. Some experts have estimated that under average conditions of pressure a one-quinarium pipe would deliver about 10,810 gallons per 24 hours.

To top off his administration, which in honesty and efficiency can serve as a model for all time, Frontinus wrote a number of bills that were enacted by the Senate to protect the vital structures against ancient abuses and to facilitate their maintenance. When he died in office in A.D. 103, he left his successor Pliny the Younger a water administration that had been brought to the highest pitch of efficiency and a vast hydraulic plant in excellent condition. Succeeding tribunals kept extending the system by the building of Aqua Trajana and Aqua Alexandrina and devoted considerable time and effort to maintaining the shoddily built Claudia and Anio Novus, but from all accounts they never surpassed the standards of administration set by Frontinus.

The end of the aqueducts

As had been the case all along, the physical state of the Roman hydraulic plant reflected the state of the Roman society and its politics. It is in the nature of such plant that it is exposed to attrition from natural causes: the channels are eroded by flowing water, they become clogged up by silt deposited by the water, the surface structures suffer damage from the buffeting of gales and by being struck by lightning. In sum, they have to be maintained; the work on them can never stop. But such permanent maintenance work presupposes reasonable stable political and economic conditions, and when Rome went into decline, with social unrest and uprisings, the repairs were neglected and ultimately ceased altogether. By the time the barbarian tribes beyond the Alps gained sufficient strength to attempt raids which grew into invasions of the heartland of the Empire, the water system had suffered long neglect.

The final destruction of the surface structures occurred in 537 and 538 during the siege of Rome by the Gothic king Vitiges. The Goths camped 50 stades to the southeast of the city where the two lines of arches crossed each other in wide curves. In the 440-foot enclosure they established their fortified camp, and from there they scourged the countryside and tore open the aqueducts in order to deprive the city of water. They even attempted to send storm troops into the city through the dry channels.

This date can be regarded as the end of the Roman system of aqueducts. True, after the siege was lifted some of the channels were restored to service, but they never recovered from the destruction. One after one they fell into disrepair and were abandoned. Aqua Virgo was kept working, on and off, as the Popes kept repairing it, until it finally became incorporated in the present water system, as previously described.

The remains of Marcia and Claudia were used and despoiled by the engineers of Pope Sixtus V when in 1580 a new, greatly needed aqueduct was built by order of the Pope. The rest vanished gradually owing to being used as quarries. The Claudian arches crossing the grounds of the San Salvatore Hospital were auctioned off one by one whenever the Brothers found themselves in need of ready cash. (For a note on aqueduct research see page 269.)

Water supply after the fall

Although the aqueducts of Rome fell into disuse and vanished one by one as described above, it would be wrong to conclude that aqueduct construction ceased altogether. King Theodoric built a new aqueduct at Spoleto in A.D. 541, which included a viaduct with a length of 831 feet and a crown height of 438 feet, i.e., considerably larger than any viaduct in the Roman system.

The Arabs built one at Elvas in Spain, while in Constantinople, Suleiman the Magnificent (1494–1566) repaired the two aqueducts built during the reigns of Emperors Valens and Justinian, in addition to building an entirely new one. The reason for the Turkish interest in water was basically the same as the Roman—their high regard for personal hygiene. The Turkish baths needed plenty of water.

As a contrast to the magnificent ancient hydraulic civilization it may be warranted to conclude with a note on conditions north of the Alps, particularly in Paris which in the Middle Ages had grown to be the center for the upstart northern rulers. Paris actually got an aqueduct in A.D. 360, during the reign of Emperor Julian, but the Aqueduct de Rungis, as it was called, was destroyed by the Normans, and what remained of the ruins was used by Marie de Medici in her d'Arcueil aqueduct which appears to have had a capacity of 572,000 gallons per day. An earthenware pipeline from the springs at Auteuil brought an additional 117,000 gallons to the baths near the Palais Royal.

Then there was a short pipeline originally laid in the sixth century that brought water from Pré-St.-Gervais to the Abbaye de St. Laurent. When in 1082 King Philip Augustus acquired from the Abbey a piece of property used for annual fairs, he also obtained a share of the water, which later was named Fontaines des Halles. This water was highly regarded since it could be used for cooking vegetables, but then of course a bag of potassium carbonate had to be added. The Belleville aqueduct must have been built before A.D. 1300, and subsequently a few more lines were added, but after the monasteries and the nobility had taken what they needed, there was not much left for the rest of the population, which in 1553 had to make do with about one quart per person per day. Henry IV ordered the construction of a water-raising plant at the Seine, but the additional supply soon vanished within the walls of monasteries and residences of the nobility. As late as 1623 the water consumption was not quite one gallon per person per day.

Paris did not get an adequate supply until Napoleon ordered the construction of Canal de l'Ourcq. The 60-mile-long canal collected water from several streams and ended in a $3,660 \times 366 \times 7$-foot basin at La Vilette. But since the basin was used as a terminal for some 500 barges and served as a convenient sewer for the boatmen and their families, the water distributed to the population in Paris must have been a rather lethal concoction.

Conditions were, of course, similar in London where up to the last cholera epidemic in 1853 most parts of the city were supplied with sewer water from the Thames.

The Decline and Rise of Western Europe

"Yet it is not granted to nations as a whole, but only to a few individuals, to have such genius owing to their natural endowment."

VITRUVIUS

The prophecy of Vespasian's governor of Gaul came true, but it was beyond the capacity of human wisdom, Roman or otherwise, to visualize the extent and duration of the damage consequent upon the erosion of Roman military power in Europe. It had taken "eight hundred years of valor and wisdom" to erect the mighty fabric of Roman civilization. It took the Barbarians less than two hundred years to dismantle it, whereupon it required nearly fifteen centuries before the descendants of the savages managed to erect upon the wreckage a civilization bearing their own image.

End of Roman political power

The Barbarian flood broke through the denuded frontier defenses along the Rhine in one violent forward thrust in A.D. 407, and after some tempestuous decades the Franks, Visigoths, and Burgundians managed to form separate kingdoms of sorts in the areas raped by them. But a vestige of Roman influence still remained, and when threatened by the invasion of the Huns, the trashy Germanic kingdoms united their fighting men under the command of the Roman general Aëtius who succeeded in defeating Attila at Chalons in 451. This victory marked the end of the Roman power in Gaul; what remained was the Roman Church.

However, the influence and political acumen of the Church was by no means negligible. The Archbishop of Reims, in a master stroke of *realpolitik,* decided to back the pagan king of the Salian Franks in his power struggle with the others. After his conversion to Christianity and with the blessings of the Church, King Clovis succeeded in subduing the Burgundians and the Visigoths, and by a long series of murders liquidate his rivals among the Frankish tribes who still remained pagan. Before his death in 511, Clovis divided his domains, in the usual Germanic fashion, among his four sons, whereupon followed the murderous sequence of family strife among the "long-haired Merovingians" catalogued in the chronicles of Gregory, Bishop of Tours. The Merovingian state ended in savagery and chaos as complete darkness descended on the seventh century.

While the Merovingians and their kin kept murdering one another in France, Britain became isolated from Rome. No more governors and high officials were posted by the central administration, and the Romanized British were left to fend for themselves. Their military power rapidly weakened, and in order to defend themselves against the incursions from the north, they resorted to the suicidal policy of hiring Saxon mercenaries. It began modestly enough. King Vortigen hired two Saxon pirate captains—Hengist and Horsa—and their ships' companies to help him fight the Scots. In payment they were supplied with provisions and clothing and permitted to settle on the island of Thanet. The news of their new-found wealth and material comforts spread among their kin living in squalor in the swampy river estuaries of the Rhine, Elbe, and Weser, and 5,000 of them with their women and children quickly followed to join the Hengist colony in Kent. It has been estimated that between 455 and 482 something like 300,000 Jutes and Angles and Saxons left their dour homelands and settled violently along the eastern and southern coasts. Before the usual Teutonic pall of darkness descended also over the British Isles, they had formed seven independent kingdoms along the coast, from Humber in the north to Wessex in the south. The Romanized Britains had been driven into the hills of Wales to be absorbed by the Celts.

The northern wastelands

The fabulous "island of the ocean" which for centuries had basked in the admittedly reflected glory of a superior civilization was submerged in a sea of blood. In yet another former Roman province the Teuton invaders proved totally incapable of even maintaining what they had taken. In the desert created by their lethal prowess, the forests closed in on the cultivated land, the Roman villas with their hypocausts turned into rubble, as did largely the cities. Even farming and husbandry declined: one acre of the best land could under the new order support only three sheep. The natives who had managed to stay alive were turned into thralls, and when Wilfred, the Apostle of Sussex, converted his royal master and received as a gift the peninsula of Selsey, near Chichester, he also acquired the ownership of 87 families. In the kingdom of Sussex, 7,000 families slaved on the slaphappy estates of their pagan masters.

Small wonder then that there was no building or construction in the immense wasteland stretching from the foothills of the Alps to the northern ocean. To begin with, there was no need for it. The Barbarian chiefs were content to live in their timbered houses and were incapable of organizing the skills needed to erect a stone building. Indeed, they were incompetent to do anything requiring some measure of economic and technical ability, such as mining and trade. The chief commodities of the times, axes and spears, were fashioned out of lumps of bog ore; getting virgin iron out of the rock was beyond the capability of the new masters. Wealth, metallic or otherwise, was something to be stolen by swinging an ax with better luck than the temporary possessor of the wealth which had been obtained in the same manner.

Nobody could travel. That was a sure way of inviting death. The glories of the Eastern empire and its reflection in Theodoric's Ravenna, the salvage of the ancient civilizations, indeed also of Greek learning, by the empire-building followers of Mohammed emerging out of their desert homeland, remained totally unknown in a continent laid waste by the unbound energy of the northern people.

The triumphs of Byzantium

But during these empty centuries in the West, magnificent buildings were being erected in Byzantium and in the train of the Arab conquests. In Constantinople there was built in the short span of only five years (532–537) the largest domed structure—Hagia Sophia—ever raised outside Rome. The brick dome has a diameter of 102 feet and is carried by giant pendentives which in the east and west are supported by circular semidomes measuring 51 feet in diameter. Haga Sophia is not only an engineering accomplishment of the highest order; it is sublime architecture, religious reverence frozen in stone.

In Italy also some domed structures were erected at this time. In Ravenna the church of San Vitale (526–547) has a large dome constructed with earthenware pots to reduce its weight. The domes of St. Marks in Venice are, like those of Hagia Sophia, built with bricks laid in standing courses. There was also a brief revival of modest megalith construction, as exemplified in the mausoleum of Theodoric in Ravenna where the flat dome consists of a single block of Istrian limestone 35 feet in diameter.

Byzantine construction followed Roman practice. Indeed, in numerous churches, including Hagia Sophia, the columns used were stripped from Roman temples. But the Byzantine builders were more concerned about earthquakes than the Roman, and their buildings give evidence of the numerous precautions taken to save them from damage. For example, the base of the columns was joined to the shafts with a sheet of lead, metal bands were used to strengthen the base and neck of the shafts, and a block of stone was placed atop the capital to receive the vaulting springer. The walls were bonded with timber, horizontal stringers measuring $5\frac{1}{2} \times 3\frac{1}{4}$ inches joined to transverse beams measuring 7×4 inches and placed on vertical 6-foot centers. The iron ties stretching across the domes of Hagia Sophia and St. Marks have not been put in by modern restorers. They were placed there at the outset, to keep the vaults intact against earth tremors.

Attempts at building in the West

In the West, as the dark centuries rolled by and the rough edges of the Franks and Anglo-Saxons were gradually worn off by the incessant efforts of the bishops and their clergy, attempts at church building, mostly in timber, were made.

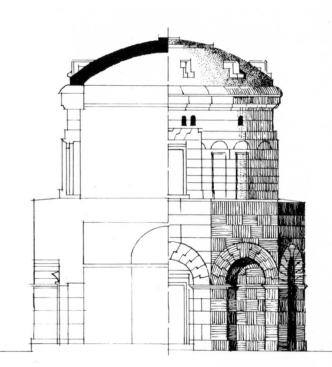

The Roman building traditions continued during the reign of the Gothic kings in Italy, although greatly modified by Byzantine influences. A couple of impressive Byzantine structures still remaining were built at Ravenna (526–547), namely, the Church of San Vitale and the mausoleum of Theodoric. The latter has a flat dome consisting of a 35-ton limestone megalith. When the block was cut into a domed shape, 12 strong angular bosses were left to be used when hoisting the dome into place. The mausoleum has a number of features characterizing Byzantine construction, such as the indented voussoirs clearly seen in the lower arches, probably invented as a safety measure against earth tremors.

102

The interior of Hagia Sophia in present Istanbul is built on such a vast scale that no camera lens can resolve it without distortion. For this reason only the 102-foot dome is shown. Hagia Sophia, built in the amazingly short time of only five years (532—537), remained for a thousand years the largest church in Christendom.

Carcassonne at the river Aude in Languedoc, in southern France, is the best-preserved medieval fortified city in Europe. The strategic site near the Pyrenees and athwart the trade route from Rome to Bordeaux inevitably made it into a strong point. The hill was fortified by the Celts and Romans, and later by the Visigoths who took it in 466 and held it for a couple of hundred years. The existing works were partly built at that time. Of the inside curtain with its 29 towers, one-third were constructed by the Goths. The rest, including the entire outside wall with 17 towers, was built 1270–1285 by Louis IX. Carcassonne lost its military and political importance with the establishment of the Franco-Spanish border in 1659.

A view of Rhodes from the sea is the only reasonably realistic one in the Nürnberg Chronicles, because it is based on a sketch made on the spot by Erhard Reuwich on July 18, 1483. The details agree with the original with two exceptions: the seaside jetty is too close to the town, whereby the harbor basin has been turned into a narrow canal with no space for turning around the ship shown in the basin. The two windmills on the seawall have been added because the artist thought it nice to have them there.

Some small pitiful stone churches were also erected. In order to get started in regions totally lacking in masons and other skilled building labor, the Brothers of the religious building orders chose the simplest structure they knew, i.e., the *basilica,* a simple business building that had become standardized in Rome. According to Vitruvius, a basilica should have a length $1\frac{1}{2}$ times its width, with two tiers of columns separating the nave from the aisles which should be one-third the width of the nave. The basilica was covered with a simple low-pitched roof over the nave and lean-roofs over the aisles.

In the early Christian basilicas built in the wastelands of Europe, a vestibule was added to the west end and an apse to the east one. The walls forming the nave were pierced with windows above the aisle roofs to provide light for the interior. However, the early basilicas were crudely constructed, and practically all of them collapsed after some years owing to faulty building and poor materials.

The first building of any importance erected in western Europe before 1000 is the minster in Aix-la-Chapelle, built by order of Charlemagne on the model of San Vitale in Ravenna. It is an octagonal building with two stories of side aisles around a central domed area. The first story is surrounded by a gallery carried by antique marble and granite columns brought from Rome, Ravenna, and Treves. In the crypt below, Charlemagne was buried, seated on a marble throne. The octagon was begun in 796 and was completed by Pope Leo III in 805. Not unexpectedly, the church was destroyed by Norman raiders in the following century, and the existing building is the one re-erected by Emperor Otto III in 983.

The destruction of Charlemagne's minster by a Norman raiding party was a reminder that the Teuton fury was not yet spent. There still remained a pocket of northern savagery in Scandinavia which after the death of the Emperor began to emit fleets of sea robbers carrying murder and pillage to the coasts of Britain and the Continent. The pitiful attempts to build anew in the ashes of the Roman civilization were frustrated, and along the coasts and estuaries, smoldering fires of monasteries, churches, and towns became the only beacons known to the Viking fleets when, loaded to the gunwales with gold and silver vessels and whatever else had caught the greedy eyes of their crews, they returned to their bleak homelands and the long darkness of winter.

Indeed, the time for the development of an ordered society in Western Europe was not yet at hand. Nothing worthwhile could be accomplished before the last pockets of savage plunderers had been exterminated. So they were eventually, not by sword but through missionary effort. The local chiefs also turned to fighting among themselves in their eagerness to become kings in the fashion of their betters on the Continent, and thus dissipated their lethal energies in their own home valleys. By A.D. 1000, at long last, the time was ripe for western Europe to become civilized.

A painful awakening

The rebirth of building in Europe gave rise to a school of architecture known as Romanesque. This method of building —we are not discussing style in this context—was originally developed in Lombardy where during the reign of Charlemagne legal recognition was given to a builders' guild known as *Maestri Comacurii.* This guild of Lombardian master masons, frustrated by the Roman tradition in building and anxious to escape the fetters of a pagan heritage, devoted themselves to the development of construction better suited to the prevailing conditions in their area. During the next four centuries master masons, most of them Brothers of the Benedictine and Cistercian orders, united in an effort to build an adequately lighted impressive church, as richly decorated as possible, under conditions of inadequate skills and crude materials. They learned by trial and error—from the collapse of numerous buildings—some rudiments of structural engineering, but the walls remained to the end heavily built of brick or small-scale ashlar. The doors and windows were small and round-arched, the columns short and stumpy. In place of columns, use was commonly made of clustered piers to carry the circular vaults.

The door is another typical feature of the Romanesque. The opening was narrow and placed in line with the inside of the thick wall. The doorway widened outward in a number of steps, or orders, each one with a covering arch. The tympanum, or space between the arch and the top of the door, was decorated with reliefs of Biblical stories or the lives of the saints. Each step or order was treated as an entity and richly ornamented with carvings. At some time during the development of the Romanesque, a tower came to be added, first as a campanile, to carry the bells borrowed from Buddhist practice, later incorporated in the church structure.

The sculptures and other ornaments in the doorways of the Romanesque churches are extremely interesting from a psychological point of view. They reflect the mixed-up emotions and terrible frustrations of the men who made them. The builders pitted their ambitions against the inadequacies of material skills. They dreamed of beauty, and achieved ugly perversions of beauty. They dreamed of towering structures soaring into the sky proclaiming the glory of God, and attained earthbound ones crushed under their own weight. This mental turmoil produced sculptures and carvings pregnant with perverse cruelty, terror, and *angst.*

However, of greater interest to subsequent building developments was the continuous experimentation with vaulting that went on. Stone vaults were desirable, owing to the fire hazard of timber roofs. The early Romanesque vaults had barrel form, and were succeeded by groined vaults with their weights and thrusts concentrated to the piers. To escape the need for heavy timber centerings when building large vaults, the vaults were divided into narrow sections by means of ribs carrying the vault paneling. The fully developed rib vault, with cross-ribs and groin ribs, came to be used at about the same time, in churches as far apart as Milan and Durham in northern England.

Many of the small churches still found in the byways of central Europe, in remote Swiss mountain valleys, the Tyrol, and northern Italy, are Romanesque churches erected at this time by peasants with the aid of their lords to whom the land belonged. But the large Romanesque cathedrals, of which many still remain, were built by the religious communities: in France and Spain by the Cluniacs, in Germany by the congregation of Hirsau. They were designed as reliquaries on a giant scale, and as a setting for the great pomp of liturgic ceremonies, with the idea of attracting throngs of pilgrims. Indeed, the major pilgrim routes of western Europe became in the end lined with such Romanesque reliquary churches.

By that time Romanesque building had become divided into two regional types—the Mediterranean and the northern. In the south, in Aquitaine, Anjou, and Saintonge, the church retained its early character of a solid block of masonry, and its barrel vault interior gave the impression of a cavern hewn out of the rock. In the north, there arose a desire to break up the mass of masonry, which led to segmentation of the wall areas into bays covered by rib vaulting. In Germany, Cluny became the pattern for austere and tautly constructed cathedrals, although regional styles flourished in different parts of the country.

However, in central and southern Italy—particularly in Florence and Rome—builders began to invoke against the barbarism of the northern style and preach the virtues of classical simplicity and harmony. This movement found also such expression as a decree in the Roman Senate in 1162 which made injury to Trajan's column a crime punishable by death and put a stop to using the debris of the past as a convenient quarry for building stone for palaces and churches. Cardinal Orsini began to collect Roman antiques; so indeed did Henry of Winchester who visited Rome in 1151 for this purpose.

Thus, while north of the Alps builders struggled with the structural difficulties inherent in the circular arch, in Italy there was already a strong trend, even before the definite breakthrough of Gothic construction, to regard northern buildings as a perversion of esthetic purity. In Florence, the buildings erected at the middle of the twelfth century were a conscious protest against the barbarous Romanesque.

But such sophisticated ideas were a long way off elsewhere in Europe. In Normandy, an area charged with tremendous importance for the future of Britain, an abbot from Pavia by the name of Lefranc was in charge of the monasteries at the middle of the eleventh century, and under his influence Lombardian methods of building were introduced and ingenious experiments in vaulting took place, while Duke William and his knights busied themselves with other plans.

Kings' Houses in England

The Teutonic lords, now seemingly secure in power throughout Europe, never shed their rural instincts. The King and his court, his liegemen in descending pecking order, all lived in and on the country. They had no use for cities and towns except as a source of revenue in the form of customs and other imposts. The bishops, however, chose from the very beginning to reside in the cities founded by the Romans; indeed the word city (cité in medieval French) came to mean the seat of apostolic authority. For all the nobles cared, the cities could turn to debris, and it was only by serving as centers of ecclesiastic influence and power that they were saved from ruin and, in time, began to thrive and gradually expand also as centers of trade.

Britain was no exception to this general rule. An Anglo-Saxon king had no fixed capital; he kept moving with his henchmen—to use the appelation "court" would be pretentious in this context—from one estate to the next. The duration of stay depended on the stores available; when the barns were empty he moved on. The maintenance of a royal township and stocking the "King's House" with provisions was a public duty enforced on peasants and thanes alike.

Medieval chroniclers refer to the rural residence of an Anglo-Saxon king as a "villa regalis." The one recently excavated at Yeavering in Northumberland, where Saint Paulinus preached Christianity in 627, is probably typical of the rest. The King's House proved to have been a simple wooden structure built of squared upright timbers sloppily set in a shallow trench. It contained one large room—the hall—partitioned off at one end to provide sleeping quarters. The hall must obviously have had a shallow hearth, but the digging report does not mention any. The house was divided into a central nave and two aisles by two rows of posts carrying the roof, which was also supported by obliquely set timbers propping up the walls. This grand villa regalis measured 82 × 36 feet (25 × 11 meters) and it says a great deal about the technical incompetence of the builders that the roof of this modest structure had to be propped up by posts and the walls buttressed to keep them standing. Nearby were a few timber hovels measuring 33 × 16 feet or thereabouts, and an irregularly shaped palisaded enclosure which the excavators term "fort."

This, then, was a villa regalis of a Saxon king. It was probably built by Aethelfrith (592–616) and enlarged by Edwin. It was burned down, rebuilt, burned down again, until it was eventually abandoned a hundred years later.

During the following century the "Kings' Houses" improved somewhat. At Cheddar, in Somerset, King Alfred himself ordered a more pretentious residence to be built. The hall measured 76 × 18 feet and tapered toward the ends. It was built of upright timbers set in a double row in a trench, the inner row sloping inward, apparently to support a floor. (In one such two-story hall the upper floor collapsed in 978, injuring King Edward's councillors.) The hall had two opposite entrances at about the middle of the long wall. Close by was a "bower" measuring 28 × 22 feet and built of wattle and daub. The site was undefended except for a shallow ditch to the north, dug for the purpose of carrying off storm water.

Constructing a timber wall by placing logs upright in a shallow trench and then backfilling it strains one's credulity, but it must be accepted in view of the archeological evidence. It reveals beyond doubt a total ignorance of the simplest elements of timber construction, which one would have thought would have been mastered more or less instinctively by people living in timber regions. A timber house, like the early American blockhouses introduced by Scandinavian settlers along the Delaware River, is built by piling logs in level courses and joining the timbers by removing the upper and lower halves of two logs where they meet at a corner. This method of timbering can be learned in a day by a country bumpkin handy with an ax.

The excavation of these two Saxon Kings' Houses is ample evidence of the low constructive sense of the new rulers of Europe. It is not at all surprising that with such beginnings it required a vast span of time before they were able to make some contributions of their own to building and engineering generally.

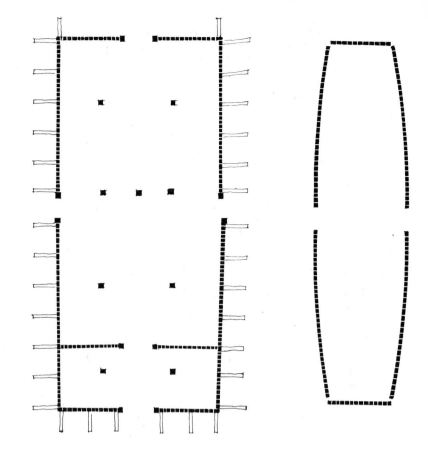

Burh-bot and Burghal hidage

Among the scanty remains from these murky times the dikes and earthworks built for defense required greater mental effort and organizational ability. Offa, King of the Mercians (756–796), had a 40-mile-long ditch with associated earthworks dug from the estuary of the Dee to the estuary of the Severn to defend his realm. This must have necessitated a tremendous input of labor, and the king did not hesitate to put even the clerics to work on his fortification scheme, which brought on him the wrath of the Pope for such brutal encroachments on ecclesiastical privileges.

The incident suggests that at least the King of the Mercians had the means and ability to order any one of his subjects to labor on his works. King Alfred, beset with an influx of raiding pagans, further improved on the crude system applied by the Mercian kings and introduced the "Burghal hidage" whereby the maintenance of the numerous earthworks built by order of the king became the responsibility of the villages protected by them, according to the number of hides they contained. One man had to be produced for each hide, and four men were required to keep up one perch of earthwork. Under the law, the entire perimeter of the earthworks had to be put in a presentable shape within one fortnight after Rogation days each year.

(Rogation days were—and are—the Monday, Tuesday, and Wednesday before Ascension Day. Hidage was a

The typical English "villa regalis," or King's House, proves on archeological scrutiny to have been a modest dwelling built by placing standing timber in a shallow trench and then back-filling it. No stone foundation was used. The left plan is that of the King's House at Yeavering in Northumberland and of the last one on the site used by King Edwin (616–632). It measures 82 × 36 feet. To keep the walls standing, they had to be buttressed by obliquely set timbers and the roof propped up by two lines of posts. The house had an entrance on both sides and on the gables. The King's House in Cheddar built by King Alfred (871–899) had timber walls set in a double row and without buttresses. The inside row sloped inward and appears to have carried the floor to an upper story. Both houses recall contemporary ones along the Elbe estuary and in Denmark, and serve as ample evidence that Roman building know-how had not reached the Germanic world.

measure of land. One hide was equivalent to 80 to 120 acres and regarded as adequate to support a family.)

The early churches built by the Saxon kings, such as by Aethelbert of Kent at Canterbury, Rochester, and London for St. Augustine; by Edwin of Northumbria at York for St. Paulinus; by King Offa of Mercia for St. Peter, were

107

constructed of wood. Just when the first stone churches began to be built has not been definitely established, although tradition points to King Ine of Wessex as having erected one at Glosterbury around 898. It was King Alfred, however, who allotted a fixed part of his income to "craftsmen whom he constantly employed in the erection of new edifices, in a manner surprising and hitherto unknown to the English." *(Gesta Regum.)* From all accounts Alfred sought his inspiration in Carolingian France for his centrally planned churches, all of which have vanished. The new minster at Winchester was founded by him but finished by his son Edward the Elder.

It was not until Edward the Confessor became King in 1042 that major works in stone began to be built in England. The King became the patron of the Benedictine Monastery of Westminster dedicated to St. Peter and apparently founded during the murky seventh century. Out of his veneration for St. Peter, Alfred reserved one-tenth of his revenue—which in any case he should have spent on a pilgrimage to the Holy Land that did not come off—to enlarge the monastery. As a model for the new church he chose Jumièges in Normandy erected by Lefranc in 1040, which he had seen during his exile in that duchy.

Edward's Westminster Church was the first Norman Romanesque structure in England. It was cruciform in shape with a long nave divided into six double bays with alternate cruciform and square piers, and provided with transepts to the north and south and a short presbytery of two bays. The church had three towers, two in the west and one in the center. It was consecrated on December 28, 1065, one week before the King died in his new palace at Westminster. He was buried in the church, of which nothing now remains aboveground.

The Abbey of Westminster was built by imported Norman master masons and with stone imported from Caen. Nonetheless, the only names of the craftsmen that have survived are typically Saxon, such as "Teinfrith my church-wright" and "Leofsi Duddesunu, master mason." From the sparse documentary evidence remaining from the dusk of Anglo-Saxon rule, it appears that out of the gray mass of Saxon peasants there had emerged by the middle of the eleventh century a few skilled men capable of building in stone.

Came the Normans

With the Norman conquest England was brought into the mainstream of European building developments. Duke William immediately set upon erecting castles in all parts of the country in order to secure his conquests by military means, and major halls for securing his usurpation by political means. Like other kings before and after him, William never stayed in one place while in England. At Easter he wore his crown at Gloucester; at Whitsuntide at Westminster; at Christmas he entertained in state the great

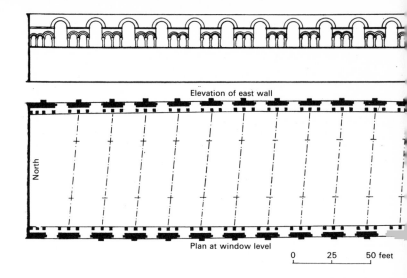

Elevation of east wall

North

Plan at window level

0   25   50 feet

Westminster Hall was rebuilt and enlarged by William II Rufus in 1099. It is the only one remaining of the three Norman halls, or aulae regis as they were called by medieval chroniclers. They were built of stone and in Romanesque style by the Conqueror and his son. Westminster Hall measures 240 × 67½ feet and was the largest aula in medieval Europe. English masons were not too proficient in handling the new building material. The plan shows the hall at window level and the oblique placing of the columns along the walls.

men of England—the archbishops, bishops, abbots, earls, thanes, and knights—at Gloucester. Of these political halls only Westminster survives, owing to the fact that it was reconstructed during the reign of William II in the last decade of the eleventh century. Court was held there for the first time in 1099.

Westminster was an ambitious undertaking, and with dimensions of 240 × 67½ feet, the largest hall built in Europe until then. When court attendants commented on its immense size, the King is reported to have replied "too big for a chamber, not big enough for a hall." The Peterborough Chronicler gives the other side of the story and complains bitterly of how in years of bad harvests multitudes of men were pressed into laboring on the King's hall at Westminster.

The great hall at Westminster still stands, and the walls below the string course are the original. They curve slightly inward, apparently in the tradition of the ancient timber halls. The upper walls are lighted by Romanesque windows set in an arcaded gallery around four sides. On the outside, the building was divided into 12 bays by shallow buttresses decorated with bands of checkered masonry below the parapets of the east and west walls and with a blind arcade on the north gable.

Not unexpectedly, this first major building in England shows by a number of serious errors that the masons were not quite up to the job. There is, for example, a lack of correspondence between the buttresses and arcading on opposite sides of the hall; on the west side the columns are placed four feet to the north of corresponding ones on the east side. Also the stonework is inexpertly put together.

The Pipe Rolls

Before the middle of the twelfth century little is known of the details of building works, in England or anywhere else. However, starting with the second year of the reign of Henry II, i.e., 1155, there began a continuous series of the "Great Rolls of the Pipe," or the records of the Exchequer which listed all expenditures, and much more besides, for the building work ordered and paid for by the English kings. Since the Pipe Rolls record the details of the great splurge on Gothic construction which set in about 1200 in England, much valuable information not available elsewhere can be mined from them, shedding light on the logistics of building.

At the beginning of the twelfth century, a royal building operation was initiated by a writ bearing the Great Seal being sent by the King to the sheriff in the county where the work was to be performed. The earliest writ still extant has the following form:

"John, by the Grace of God King of England, Lord of Ireland, Duke of Normandy and Aquitaine, and Count of Anjou, to the Sheriff of Oxfordshire, greeting. We command you that as soon as you have read these letters you have our houses of Oxford repaired against our coming by the view and testimony of law-worthy men, and the expense shall be allowed to you by the Exchequer. Witness myself at Newbury, the twelfth day of December, in the fifth year of our reign."

Copies of the writ known as *contrabrevia* were made by the Chancery clerks and sent to the Exchequer, where they were filed and retained as a check on the sheriff's claims.

To pay for the work thus ordered, the sheriff drew first on the issues of the royal manors and whatever sources of revenue he may have had. When these were inadequate for a major building, money would be allocated to him in other ways: by diverting profits of escheated honors and bishoprics, drafts on Crown debtors, or direct payments by the Exchequer. On some occasions the men of the county were forced to contribute toward the cost of building a royal house or castle within its boundaries.

When the sheriff attended the Exchequer at Michaelmas to present his accounts, he brought with him his writs as authority for the expenditures. Since the original writ did not stipulate the cost, it was necessary to issue a new writ of *computate* giving the exact expenditures. This Exchequer procedure was laid down in the *Dialogus de Scaccario* which stipulated the writing of half a dozen writs in duplicate in order to account for each building operation.

The phrase "by the view and testimony of law-worthy men" referred to the practice of having men acceptable in a court of law swear before the Exchequer that the sheriff had actually spent the money for which he claimed allowance against his receipts. The viewers had no technical qualifications, and their function was entirely fiscal. A doctor and a parson viewed the work done at Nottingham Castle in 1183, and Edmund Blund, purveyor of wine and food to the king, also served as viewer of the work done on the Tower and at Westminster. The practice of laymen viewers suggests that local craftsmen were employed by the sheriffs and that there was no central technical supervision of the work done.

The King's ingeniatores

However, such works as the great stone keeps and aisled halls required skilled technicians and artificers of an altogether different background and experience from the local help mustered by the sheriffs. They were employed by the Crown and were usually foreigners invited by the kings to take charge of major operations.

The leading members of this small group of specialists were termed *ingeniatores*. The first engineer active in England and known by name was Geoffrey, who was called in by Henry I and rewarded for his services by the keepership of Westminster which carried a stipend of 7 pence a day. He was succeeded by Ailnoth, apparently of English descent, who served for 30 years building and repairing the king's houses and keeps. In the north, Richard the ingeniator was responsible for the design and construction of the keep of Bowes Castle, for which he received a payment of 20 shillings in 1170. Dover Castle was built by Maurice, another ingeniator, between the years 1180 and 1187 for which he received 1 shilling a day and 3 pounds 3 pence in lieu of his robes.

Ralph of Grosmont was employed on the King's works in the Welsh marshes; Wulfrie supervised work at Carlisle Castle (1172–1173); Fortinus repaired Colchester Castle (1203–1204). Master Elyas of Oxford, apparently the first native ingeniator, appears on the Pipe Rolls of 1187 as "Elyas the stone mason of Oxford" who received 1 shilling a day for the custody of the King's house at Oxford. With this town as a base, he was kept busy until 1203 working on the Tower of London, Rochester, Hastings, Marlborough, and numerous other royal keeps and residences. He was also good at constructing siege engines.

The duties of the ingeniatores ranged from supervision of building, civil and military, to the design and construction of siege engines, from which they derived their appellation. They served the King in his field campaigns, as well as in his castles and houses. They did not fit into the building trades of masons, carpenters, smiths, plumbers, glaziers, which had begun to be differentiated by the time they appeared in England, and the Exchequer clerks had difficulties at first fitting them into the simple establishment. With a few exceptions they were all foreigners and familiar with building developments on the Continent. They were widely traveled men; those in the service of Richard I accompanied him on his crusade.

The King's building masters

The King's *ingeniatores* had need of skilled masons and carpenters, at first not available in England. Some of them appear in the early Pipe Rolls, such as Godwin the mason who worked at Windsor in 1167 and 1168. Ralph, "the King's mason at Dover," received 40 shillings for his work in 1170 and 1171. In 1207, Master Nicholas de Andeli, the carpenter, came from Normandy, together with a small group of fellow craftsmen who had previously been employed on Richard I's castle, Château Gaillard, on the rock above the town of Andeli on the Seine. He remained in the King's service for 30 years, building castles and constructing siege engines. In 1215 he is recorded to have been paid 9 pence a day and a robe; that year he also received a gift of land from the King.

Of the other master craftsmen permanently employed in the King's service mention should be made of Osbert the *quareator,* in charge of quarry workers but also engaged in military trenching. Master Pinell, the *minator,* supervised miners in the cellars of Dover Castle in 1205 and accompanied the King on his Irish expedition in 1210.

Thus, from 1150 onward there grew up a nucleus of skilled specialists permanently engaged in the King's services, and most of them were recruited abroad, such as *ingeniatores, petrarii* and *quareatores* (stone cutters and quarrymen), *carpentarii, minatores* and *fossatores, hurdatores* (who prepared the battlements of the castles with wooden galleries). The engineers were the best paid: besides grants of land they received 7 pence a day; master craftsmen were paid 6 pence a day when working and 3 pence when idle, plus money in lieu of their robes; carpenters and stone cutters got 4 pence. a day; quarrymen and miners 3 pence.

These, then, were the King's craftsmen who in addition to their military duties also served to direct and instruct the native labor in times of peace when they were put to designing and constructing castles and halls, churches and monasteries, subsequently also harbors.

Some space has been devoted to the early English Pipe Rolls because in their curt accountant's language they reflect the mechanics whereby the superior skills of the Greek artisan slaves employed on Roman building and construction filtered through into the wastelands of Saxon and Frankish Europe. The ancient skills, originally developed in the East, had never vanished altogether in Italy. The religious orders, such as the Benedictines, later also the Cistercians, had salvaged at any rate the elements of these skills, and when the political structure of the Empire came crashing down, the monasteries in Lombardy became the repositories of these ancient skills and engineering know-how. And, when in the fullness of time the lands north of the Alps became sufficiently mentally advanced to require these construction skills—and many more besides—the Church through its Orders was able to supply them. By the time of the Norman and Angevin kings, there were in such regions as Savoy also ingeniatores, master masons, and carpenters for hire by northern kings who began to feel a need for their services. But what they knew, whether they themselves realized it or not, derived from the ecclesiastical building orders in Italy.

San Marco in Venice began as a wooden church built by the Doge Particiaro in 828 as an abode for the bones of the evangelist Marcus imported from Alexandria. When this church burned down, Byzantine masters began work on the present one in 1063.

The Gothic Revolution in Building

"As if turned by a magic wand the turrets rose to the sky..."

FRENCH CHRONICLER

Europe during the twelfth century was an open society, without internal political borders and with fluid frontiers toward the east and to the south and west. Everywhere in the former wastelands there were stirrings of an intellectual awakening. Young clerks with a thirst for learning flocked in great numbers to the cathedral schools in Germany and France and to the municipal schools in Italy. They arrived with open minds, and absorbed eagerly an exciting blend of knowledge recently acquired from pagan antiquity and the non-Christian world of the east. The Church was equally responsive to the intellectual fervor and accommodated itself to it. The twelfth century Church was buoyant, pliable, and colorful, and was served by bishops and clergy who led boisterous, cheerful lives. It was an exciting century when anything could happen.

Geometry and Crusades

We are not concerned here with the wider aspects of the intellectual awakening of the recent pagans north of the Alps. However, in view of the tremendous development in building that accompanied it, a brief mention must be made of the principal cause of the ferment: the discovery of the classical learning of the Greeks as transmitted by Arabs and Jews. During the lost European centuries, scientific learning had flourished throughout Islam, and the works of Greek philosophers had been translated into Arabic. Among the many Islamic centers of learning, those in Spain—Cordova, Saragossa, Toledo—became the dispersal points for the rest of Europe. Particularly in Toledo, a number of Jewish translation schools did outstanding work from the tenth century onward, turning Greek and Arab works into Latin.

To these Toledo schools came also pioneers from other parts of Europe, to work and study. Adelard of Bath (1090–1150) spent many years in Toledo translating Euclid from Arabic, as well as "Arithmetic" by Al-Kwarizmi (whereby Indian numerals became known in Europe). Robert of Chichester spent six years in Spain translating "Algebra" (1145) and some astronomical tables. Gerald of Cremona (1114–1187) turned Archimedes' "On the measurement of the circle" and Euclid's "Elements" into Latin. Eugenius of Palermo turned the astronomical and mathematical system of Ptolemy (the Almagest) into Latin (1160). Michael the Scot (1175–1235) translated Aristotle's biology and brought Averroës' (1126–1198) attack on traditional astronomy to the attention of Latin scholars.

Then came the Crusades. It would lead us too far afield to try to account for the numerous complex causes of this strange Holy War. Suffice it to say that one of the basic causes was an attempt of the Church to deflect the warlike instincts of the feudal barons from fighting among themselves to the more worthwhile aim of recovering the Holy Sepulcher from the infidel. In the far-off Holy Land, the Christian knights could butcher to their heart's content all day long, and at night drenched in the blood "from the winepress of the Lord" kneel sobbing with joy at the altar of the Sepulcher. Here was also a chance for adventurous princes and younger sons to carve out principalities in the East at a comfortable distance from the ancestral estates. It suited the merchant republics of Venice and Genoa who were eager to open up short and direct routes with the East.

After killing some 10,000 Jews in Hungary and ravaging Greece on their way, the bands and divisions began to arrive in Constantinople in 1096. The following year a host estimated at 150,000 strong crossed the Bosporus, and the First Crusade was on its way. There were to be seven more before an end was put to the macabre affair with the capture of Acre by the Mamelukes in 1291.

The importance of the Crusades from the point of view of building and engineering was that after 500 years of isolation, hundreds of thousands of Europeans of all stations—princes, knights, priests and monks, common soldiers, merchants and camp followers—were brought into contact with a superior technical civilization. The fighting knights learned from their own wounds what could be accomplished with superior steel; the less warlike observed and learned from their encounter with the East. In building, as well as in many other ways, Europe was set on a new track.

The beginnings of Gothic construction

What some of them found in Syria was the pointed arch with a history of development going back to Sumerian times. When the Norman knights began to carve out their king-

doms from the scorched hills of the Holy Land and found a need for buildings of their own, the convenience of the pointed arch became apparent to the builders. In Acre, for example, the refectory in the castle built by the Order of St. John is spanned by pointed arches carried on clumsy piers. Seeing the range of the crusaders' fortresses and palaces along the Palestinian coast, one is tempted to regard their buildings as the first exercises in Gothic construction.

In Europe, the pointed arch was gradually introduced and appeared in the same buildings together with round-headed windows and doorways in the Romanesque tradition. The first Gothic structure, the abbey of St. Denis near Paris, built in 1140 to 1144 also retains some rounded arches, but in the flood of cathedral construction released by this venture, the masons rapidly developed the characteristic features of

Gothic construction, as exemplified in Chartres (1145), Notre Dame (1163), Laon (1160).

The speed with which Gothic buildings evolved during the latter half of the twelfth century was mostly dictated by political reasons. Abbé Suger, the man wholly responsible for the building of St. Denis and thus the acknowledged pioneer of the Gothic—or *Opus Francigenum* as it was called —was of peasant origin and had risen to power as a councillor and friend of Louis VI and his son Louis VII; indeed he acted as regent on some occasions. To him it was of vital importance that Ile de France, the seat of Capetan power but in numerous ways the most backward region of France, should have a monument worthy of a mighty Christian king. He allied himself with the Archbishop of Reims and five other bishops, all of them sharing his hostility to the aristoc-

The Romanesque builders worked entirely with circular arches, which presented them with difficult problems. In a square bay, for example, the diagonal ribs owing to the larger span extended much higher than the transverse ribs across the nave. It therefore become necessary to mainipulate the arches so as to obtain the same crown height. If the transverse arches were kept circular, the diagonal ribs had to be made elliptical and the side of the bay divided into two stilted arches, as shown to the left. By such manipulations with radii, the typical Romanesque sexpartite vault eventually developed, which despite the clever geometrical arrangement obtains an undesirable height at the crossing of the diagonal ribs. With the pointed arch (right) no such problems arise, because the transverse arch can be made twice as wide and more bluntly pointed than the narrow and sharply pointed wall arches, and the diagonal ribs become circular. And yet the crowns of all the arches will be level with one another.

The design principles involved in the construction of a High Gothic church are shown in this section of a bay in the Amiens Cathedral, erected 1230 to 1240. The height from the floor to the keystone of the vault is 140 feet. The vault over the nave is carried by four clustered piers (only two are shown) on which also rest the blunt arcade arch and the wall panels with their window tracery. The wall above the arch was divided into a lower part, the triforium, and a higher one, the clerestory. From the clustered piers a tall and slim shaft with its stones bonded to the wall, extends to support the vault over the nave. But only a part of the weight of the vault is carried to the ground by the shaft and pier; the rest of it exerts an outward pressure or a lateral force which would tear the wall asunder were it not taken up and absorbed by a buttress placed in line with the piers. Hence, the mass of masonry in the buttress must be heavy enough to absorb the lateral pressure exerted by the vault. In the French cathedrals the lateral forces are transferred by means of one or two flying buttresses, but in the majority of other Gothic churches the buttresses are placed flush with the wall. The timber roofs rest on the walls and not on the vaults.

racy challenging the Capets, and they contributed to the cost of developing a church suited to a *Rex Christianissimus*.

Suger himself was not a building master, but he supervised the erection of his abbey to the smallest detail. He selected the timbers, arranged for the transport of the columns, and was the overseer of the numerous craftsmen employed. Upon completion of the work in 1144, he wrote the dedication ("Liber de consecratione ecclesia") and arranged to have 19 archbishops and bishops consecrate the altars. In the ceremonies, King Louis VII played the part of Christ, and 12 knights tried their best to act the apostles. From a public relations point of view the new system of building could not have had a better launching.

St. Denis became, as planned from the outset, the supreme witness to the sanctity of Capetan kingship. Upon its comple-

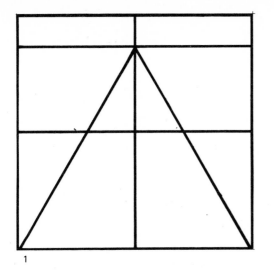
1

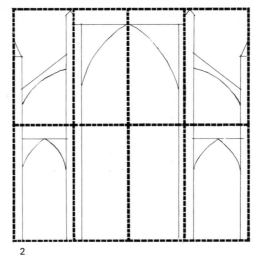
2

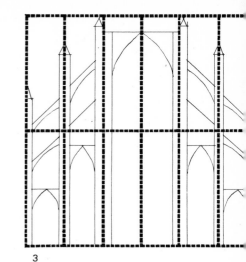
3

tion, this former sterile region of France blossomed with cathedrals erected in rapid succession, in Paris, Chartres, Reims, Amiens, and elsewhere. All these as well as subsequent French cathedrals were the products of cold, calculating intellect. They were exercises in a new building technology yoked to a vision chiefly preoccupied with mathematics and geometry, and like St. Denis serving definite political ends.

Refinements in building techniques

But to the ecclesiastic master masons the pointed arch must have come as a revelation. It relieved them of their long struggle with the difficulties involved in the rounded arch. All the vaulting problems that had occupied the master masons for centuries now vanished overnight. Instead of manipulating with rounded arches of different radii, depressing or stilting them to make them fit, they now could apply the pointed arch which could be varied between wide limits with the height of the vault kept constant. It did not matter whether the transverse rib across the nave became twice as wide and more bluntly pointed than the narrow and sharply pointed wall arches. There was no longer a need to distort any arches.

By the end of the century the French masons had thoroughly mastered the structural problems involved and devoted themselves to refinement of detail, as well as to increasing the height and lightness of the structures. This second phase of French Gothic began in 1194 with the building of the nave and choir of Chartres, and is characterized by the further slimming of the columns and the dissolving of the walls into window tracery. Reims, Amiens, Bourges, Beauvais, and others belong to this second period. The striving for still higher vaults, carried on still slimmer piers, reached its ultimate folly in the Beauvais Cathedral where the vault was brought to a height of 154 feet—whereupon it collapsed. It

was rebuilt in 1274 only to collapse again. In the end, the builders succeeded in keeping the tall vault intact by installing intermediate piers.

What happened during the 65 years, from 1155 to 1220, in the development of Gothic construction can be put briefly as follows:

1. The builders succeeded in covering a maximum space with a minimum of material and in raising the naves to unexcelled heights.

2. A clear distinction came to be made between bearing columns and nonbearing walls, the latter becoming mostly windows.

3. The pointed arch was universally applied.

4. The difference in character between the vault-supporting ribs and lightweight vault was recognized.

5. To absorb the vault thrust, a system of buttresses—in France also "flying buttresses"—was introduced and perfected.

Intuitive engineering

Thus a Gothic building had become a framed structure where all the structural members were able to absorb and transfer thrust. Since the builders could not use tension members and stiff-jointed frames, the structures became more intricate than necessary. By concentrating the solid stone material to those points where it had to fulfill a static function, considerable savings in material could be made elsewhere in the structure, for example, between buttresses.

The principal reason for the successful development of the Gothic structures is of course the static efficiency of the pointed arch, which conforms better to the line of thrust than the semicircular arch. As the rise of the pointed arch is greater relative to the span, the arch thrust is smaller and

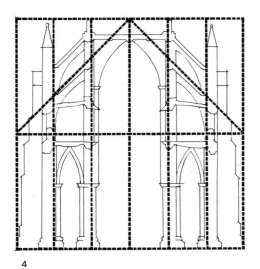

4

The two geometric figures used for the planning of a Gothic church on a particular site and for determining its height were the equilateral triangle, or opus ad triangulum *and the square,* opus ad quadratum. *A structure built ad triangulum was always lower than one built ad quadratum (1). The Gothic masters preferred building ad quadratum, which permitted them to choose between dividing the square into four or six equal parts. In the first case (2) they obtained a wide nave with two broad aisles and the relation 1:2 between the width of the nave and its height. The French masters preferred to divide the square into six parts (3), whereby they got a structure with double aisles and the relation 1:3 between the width and height of the nave. Actually, the real height of the nave is the same in both cases, but with the relation 1:3 it appears taller. Westminster Abbey (4) is a compromise between the two alternatives. Because of the crowded site, the master had to settle for a single-aisle church, but he nevertheless divided the square into six parts and thus obtained the relation 1:3 in the nave. In this way he created an illusion of the same height as in contemporary French churches, although the real height is lower.*

the thrust line steeper. By using bearing ribs, the intermediate panels could be reduced to a thickness of 6 inches, and the weight of the vault—and hence thrust—materially reduced.

Nonetheless, the greatest accomplishment of the Gothic master masons was the development of the buttresses, in order to absorb and transmit the arch thrust acting on the springing of the nave vaults. At these points the thrust is divided into two components, one of which acts in a vertical direction and is transmitted to the ground through the pier. The other component acts obliquely and is transmitted by the flying buttresses to the next column or pier, in some cases to the one beyond. In this transfer the downward resultant is absorbed by the weight of the pinnacles, and the remaining horizontal one is compensated by the large mass of the outermost buttresses.

When at its best, the structural framework of a Gothic cathedral is a perfectly contrived static system which invites the inevitable deduction that it was derived mathematically. But this was not the case at all. The formulas and geometrical rules of construction handed down from master to master were concerned only with form and composition, somewhat in the manner of the Greek temple builders, and did not involve structural design. The Gothic rib vaults and buttresses were based entirely on experience and the intuition of the craftsmen. Each detail of a Gothic structure served a double function, structural and esthetic. Structure and decoration became one, and no member could be disregarded without bringing down the whole.

These statements apply primarily to the French cathedrals of the late twelfth and early thirteenth centuries. When later the original static functions were forgotten or ignored, the forms degenerated to elements of embellishments as expressed in the intricacy of the vault ribbing, such as are found in English and German cathedrals. There, most of the ribs no longer served a structural function but were merely plastered on as a decorative feature.

Itinerant master builders

The masters who built the French and other Gothic cathedrals were not engineers, nor were they architects. They were designers, artists, and craftsmen, all wrapped into one. They were members of an international organization; and a lodge of free masons was set up on every Gothic building site. The great masters moved, alone or with a team of skilled masons, from one site to another. Villard de Homecourt went from northern France to Hungary; and his sketchbook is an important source of information about the structural details of Gothic vaults and the building methods used in the early thirteenth century. The Cologne Cathedral was designed and built by Gerardus Lapicide, *rector fabricae;* Strasburg was built by Master Erwan von Steinbach who left behind his large parchments showing the elevations of both these cathedrals. The major difference between the master masons and carpenters and the skilled men working for them was the masters' knowledge of geometry which enabled them to develop features of a structure from a given plan without the intervention of working drawings.

The two geometrical figures applied by the Gothic master masons when laying out a church were the square—the *quadratum*—and the equilateral triangle—the *triangulum.* From these two figures they obtained the geometrical proportions of the structure. In either case the combined width of the nave and aisles was the same, but the height became greater in a church designed *ad quadratum,* and this was the scheme most favored by the Continental masters. In a single-aisle church, the entire structure was laid out within the confines of the square which by division into quarters determined the width of the nave (two quarters) and the aisles (one quarter each), which gives a width-height ratio of 1:2. In a church provided with double aisles, the square was divided into sixths instead of quarters, producing a relation between width and height of 1:3.

115

Among the Gothic masters—or rector fabricae—*were many dynasties or families whose members worked everywhere in Europe and left their sketchbooks to their descendants. This sketch of a window is culled from Peter Parter's (b. 1325) collection of designs. He worked on the cathedral in Prague, while his brother Michael worked on the Strasburg Cathedral where the* rector fabricae *was Erwan von Steinbach whose "patrons" (designs) for the structure have been found. Another well-known French master was Villard de Honnecourt who also left behind him his sketchbooks.*

The great Gothic cathedrals of France generally have double aisles and were thus based on the ratio 1:3. Westminster, on the other hand, is a single-aisle church built according to the system of a double-aisle one by dividing the square into sixths. This is a unique case of Gothic geometry, and it was apparently chosen to achieve the soaring grandeur of a French cathedral on an English plan. Although the height to the keystone of the vault is 103 feet, as against 140 feet in Amiens, for example, the vertical effect is comparable to that of most French churches.

The masons and other skilled men working on the cathedrals were, like the masters, largely itinerant and at least in the early stages mostly Italian. They worked in Bavaria in the eleventh century and in Saxony a hundred years later. Since Italy was the only country in Europe with an ample supply of skilled masons, carpenters, and smiths, there is no doubt but that the core of the skilled labor on all the early Gothic cathedrals was Italian. After a few generations they had transmitted their know-how to the natives, and the French in particular had become skilled masons and stone carvers by the end of the twelfth century.

The Cistercian Order adopted the Gothic at an early stage, and through its efforts the Gothic system of construction was introduced to the rest of Europe. In Germany, the Gothic emerged with the building of St. Elizabeth's Church in Warburg in 1240, whereupon followed a whole line of red brick churches erected in the cities being founded along the shores of the Baltic—Lübeck, Wismar, Rostock, Stralsund, Copenhagen, Malmö, Danzig, Dorpat, Reval, and Riga. Throughout Europe the Cistercians were responsible for the erection of about 350 Gothic churches. But this was an altogether different Gothic than the autocratic French cathedrals, being simpler and much more naïve since it retained some of the Romanesque features of its early beginnings.

Materials used

In the French cathedrals the material used was limestone dressed on all sides into ashlar. But in England the walls of the churches were frequently made of rubble plastered on both sides, or of rubble core lined with ashlar on one or both sides. Only thin walls such as parapets and choir screens were built of ashlar. Dressed stone was also used in window tracery and doorways of churches where the walls were made of rubble or flint, stone grit or sandstone. In sum, the walling materials used varied depending on the local supplies; in northern Europe, nearly all the Gothic cathedrals were built of red brick, as previously stated.

The mastery of the Gothic stone masons culminated in the "stereotomy," the highly skilled special craft applied to window tracery. A large Gothic window was built up of as many as 90 or more blocks, all elaborately molded. The making of such a window required an intimate knowledge of geometry, since the pieces were so ingeniously jointed and arranged that any single member could be removed for repairs or replacement without endangering the fabric. An iron grille reinforced the window tracery and strengthened the glazing. The system of iron reinforcement included a horizontal stay bar at the springing of the arch and extended from joint to joint through the mullions, carrying thin and small horizontal and vertical bars to strengthen the glazing between the mullions.

Medieval English timbermen spent considerable effort inventing roof trusses that would eliminate the tie beams obstructing the view of the east and west windows. The solution was the invention of the hammer beam truss which served the primary purpose of permitting an unobstructed view of the interior, although it is not a statically perfect truss. The hammer beam made possible the construction of beautiful timber roofs, such as the one in Westminster Hall where the art of timbering reached skillful refinements that have never since been excelled.

In England, most country churches and halls were covered with timber roofs. To strengthen a pitched roof a tie or collar beam is required to prevent the rafters from spreading outward and exerting an outward stress on the walls. The ideal position for a collar beam is at the level of the wall plates (on which the rafters rest). But a beam placed so low will block the view of the east and west windows, and to overcome this difficulty English carpenters devised the *hammer beam* truss.

A hammer beam truss is a tie beam from which the middle post has been removed. The ridge timber is supported by a vertical leg resting on a hammer beam which tends to drive the rafters outward, but this thrust is resisted by the strut below which presses firmly against the jackleg set against the wall. The purpose of the jackleg is to transmit the thrust lower down the wall. The rafters were sometimes tapered from 8×7 inches at the bottom to 7×5 inches at the top. Boarded or paneled wooden ceilings were used with trussed rafters or curved braces.

Churches and large Gothic buildings were covered with lead roofs weighing up to 13 pounds per square foot. Each sheet was cast on a bed of sand. The roof of Chartres consists of lead sheets $\frac{5}{32}$ inch thick, 2 feet wide, and 3 to 4 feet long, fastened to the boarding with large-headed iron nails. The side of each sheet is rolled up with the edge of the adjoining sheet into a $1\frac{1}{2}$-inch roll. The bottom edge is secured by two iron clips to prevent the wind from taking hold of it, the tails of the clips being nailed to the boarding. The roof is still in good order after more than 600 years, and has acquired a beautiful patina due to the silver and arsenic in the lead.

Water from the roof was generally discharged directly to the ground by means of stone gargoyles, thus avoiding the need for downpipes, although lead gutters 2 feet wide with downpipes were used occasionally.

These construction notes apply to the pretentious buildings of the age, to cathedrals and palaces. The houses of the skilled men who erected them were a great deal more modest, consisting of wattle and daub, or timber-framed, or built of rough stone, depending on the nature of the local building materials. The roof was thatched.

After this brief account of the development of Gothic architecture, it may be of some interest to devote the concluding part of this chapter to a few details about the building of the earliest and best known of the English cathedrals—Westminster Abbey.

The building of Westminster Abbey

In July, 1245, Pope Innocent IV issued a bull urging all Englishmen to contribute to the rebuilding of the church of the Abbot and Convent of Westminster "which was very old and decayed, and which they began to repair in so sumptuous a manner that they were unable to complete it with their own resources."

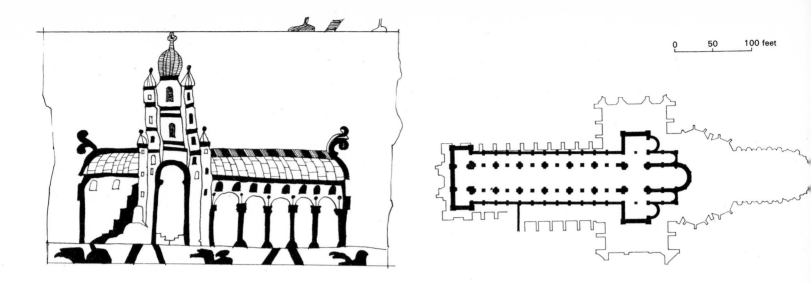

The only remaining illustration of Edward the Confessor's church in Westminster, consecrated December 28, 1065, is found on the Bayeux tapestry (left). In all its naïveté the picture is correct in one detail, the central tower. But it had also two towers in the west. The Abbey was built by Norman masons who used the church at Jumiège in the duchy as their model. The Confessor's church was the first Romanesque building in England and the largest stone structure erected since Roman days. Its scale compared to Westminster Abbey, built on the same site, is shown by the plans (right) revealed by excavations.

The response to the papal bull was hardly enthusiastic. No contributions were forthcoming and the building of the Lady Chapel begun by the monks in 1220 in "so sumptuous a manner" could no longer continue for lack of funds. In this situation Henry III stepped in and offered to rebuild the 200-year-old and decrepit Confessor's Church at his own expense. What St. Denis and Reims were to the House of Capet, Westminster was to become to the House of Plantagenet.

Conditions for undertaking such a major work were at this time altogether different in England than merely a hundred years earlier. There was to begin with a supervisory body—"the keeper of the works"—with paymasters and accountants capable of maintaining both technical and economic control of the building operations. There was also at Westminster a corps of skilled men permanently engaged on the King's works.

The first administrative step taken by the King was to appoint Edward of Westminster as keeper of the works, responsible for the financial side. Master Henry was appointed as master mason in charge of the design and construction of the Abbey. Master Alexander, another trusted craftsman in the King's service, became carpenter in charge of all timbering. These three were the key men who started up the work that was not to be completed until a couple of hundred years after their death. Master Henry de Reyns, the mason,

was either a Frenchman from Reims or an Englishman who had worked on the Reims Cathedral and become familiar with contemporary developments in French Gothic. He had been working on the new chapel and cloisters at Windsor, and in 1243 while there he was given the robe of office—"with good squirrel fur"—as master of the King's masons. Master Alexander, the carpenter, was also given a robe. Both received a stipend of 1 shilling a day.

The Exchequer was ordered to pay 3,000 marks (£2,000) a year to the keeper toward meeting the building costs. This basic allowance was supplemented by writs of *liberate* for specific sums. As the King's financial position grew worse, the Exchequer found difficulty raising ready money, and during the last nine years of the King's reign the keeper received only £2,818, as against expenditures totaling £10,400. The balance had to be made up in numerous ways, in part by issues of the Great Seal which produced £300 to 400 per year, in part from the mints in London and Canterbury. The Wardrobe made contributions, the sacrist of Westminster handed over offerings made at the altar of St. Edward. Much more came from fines, debts, and amercements ordered by the King to be paid direct to the keeper of the works. Persons in all stations of life found their financial obligations to the Crown diverted to the cathedral works: for example, the widowed Countess of Lincoln was fined £4,000 for retaining the custody of her infant son; a Jewess paid 4,000 marks to be allowed to keep her deceased husband's property. The Earl of Gloucester, Isabella Countess of Aumale, and Hawise de Chaworth were also fined substantial sums which they paid into the building fund.

In 1250, the King obtained from the Pope a relaxation of one year's penance for all who "lent a helping hand to the fabric of the church of wonderful beauty being built by him at Westminster." But even this brought little response, and the only voluntary contribution recorded during the early years was a legacy of 100 marks under the will of the late Archbishop of Canterbury, Boniface of Savoy.

Under the circumstances, the office of Keeper of the Works was not an envious one. He had to purchase materials, engage and pay hundreds of craftsmen, hunt down Crown debtors, extract money from recalcitrant Jews, and wrest cash from obstinate government departments which used all means at their disposal to escape paying their dues. A special works exchequer was set up, using the same checkered cloth for the calculations as the one used by the barons of the Great Exchequer. The works accounts show that by Michaelmas 1261 the keeper had spent altogether £29,345 19s. 6d.

(Another "keeper of the works" more familiar to English literature students bears the name of Geoffrey Chaucer. He was responsible for the accounting of work done on the Tower of London from July 12, 1389 to June 17, 1391. Later, he spent £1130 8s. 11½d. on the St. George Chapel at Windsor. He was not concerned with the Westminster Cathedral.)

Building the Cathedral

The building work was directed by the master mason and the master carpenter, each one with unlimited powers in his department. Being the "Master of the Works," Master Henry planned the new church on the French pattern, with a chevet at the east end and a great north portal modeled on the west portal of Reims. Particular emphasis was put on the north transept to provide a state entrance from the adjoining palace. But the site was hemmed in by existing buildings, particularly the cloister quadrangle, and Master Henry was forced to forego the French plan. Instead of laying out the single-aisle church *ad quadratum,* he used the division of six whereby he obtained a ratio between width and height of 1:3.

Demolition of the Confessor's Church began on July 6, 1245, after the bodies of the King and his Queen had been removed and a temporary choir for the monks arranged in the former nave. The foundations of the new church were laid in the spring of 1246, and the Constable of the Tower was ordered to release some supplies of building stone, timber, and other materials. The materials used were Kentish ragstone for the foundations and rough walling, freestone for moldings and carvings, marble for piers and colonnettes, and hard chalk for vault paneling. (A freestone is any stone, but particularly limestone and sandstone, that can be cut freely

without splitting.) The ragstone and Caen freestone reached the site by water. In April, 1246, the sheriff of Kent was ordered by the King to warn all ships trading in his area that no stone could be sold except to the keeper of the works at Westminster. He was expected to arrange for 200 boatloads of stone to be delivered to the site within the next five weeks.

The marble came from the Purbeck quarries near Corfe in Dorset. An invoice for three shiploads of Purbeck marble in 1253 shows that the freight came to £4 per shipload. Caen stone cost from £5 to £12 per shipload, whereas 14 boatloads of ragstone cost £4 12s. 3d. Reigate freestone was sold by the piece: Roger of Reigate, the principal supplier of this building stone, charged 2d. apiece.

However, in order to obtain the materials needed, both the master mason and master carpenter had to travel far afield. In 1250, Master Alexander visited the royal woods in Essex and the Weald of Kent to select structural timber for the roof. He obtained additional supplies of oak timber from Roger de Mortimer's park at Stratfield in Berkshire, as well as from the woods of the Grevecoeur family at Bockingfold in Kent. In 1266, Roger of Leybourne made the King a present of 40 oaks from his woods near Maidstone in Kent.

William of Ingleby, sheriff of Lincolnshire, purchased 63 fothers of lead at Boston in July 1259, and while the lead was being shipped in two vessels, tallies of the weight were sent by special messenger to the keeper of the works so that the cargoes could be checked on arrival.

Labor used

The labor force varied greatly over the years and at different seasons of the same year. The annual mean appears to have been about 300. In 1251, the King ordered the keeper of the works to put on 600 to 800 men and to proceed with the marble work throughout the winter, but lack of money prevented his order from being implemented. In 1253, the labor force varied from a maximum of 428 men in June 23 to 29 to a minimum of 100 in November 23 to 29. The summer force was made up as follows:

White cutters	53	Smiths	17
Marblers	49	Glaziers	14
Layers	28	Plumbers	4
Carpenters	28	Laborers	220
Polishers	15	Total	428

By that time Master Henry had been succeeded by Master John of Gloucester who was aided by two master masons, William de Wauz and Richard of Eltam. Master Alexander, the carpenter, was still in charge of timbering, but work on the site was supervised by Master Odo, his next in command. Walter the clerk kept the accounts.

From Walter's accounts in 1253 the wages paid are known in great detail. Only two men, Masters John and Alexander,

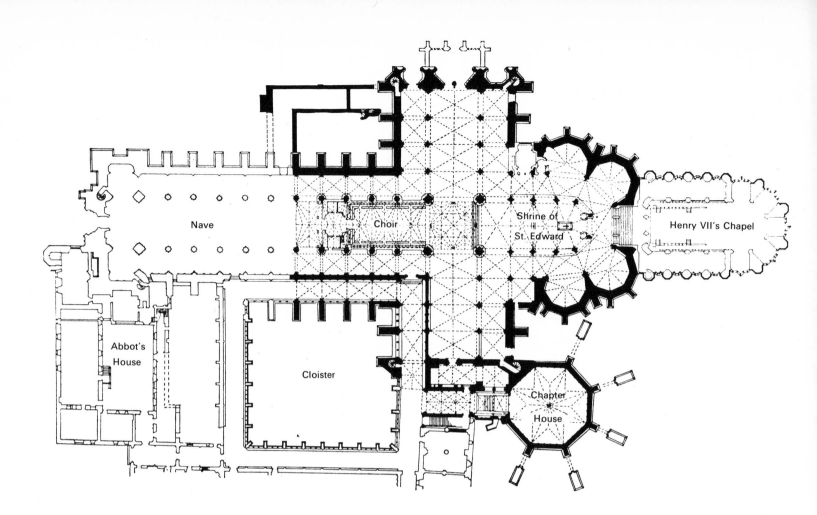

The plan of Westminster Abbey shows the completed church in thin outline. The black portions are those completed before the death of Henry III in 1272 when the building work had been in progress for 27 years. Westminster Abbey was not the first Gothic church in England. The transition to the pointed arch occurred earlier with the rebuilding of the Canterbury Cathedral in 1174. When work on the Abbey began in 1246, the Salisbury Cathedral was half finished, the Benedictines were busy at Worcester and Ely, the Augustinians with Carlisle, and the Premonstrantians with Titchfield. But Westminster Abbey differs from all these other cathedrals by being planned at the outset as a so-called "Königskirche" similar to St. Denis and Reims, which were used as models. As previously mentioned, the geometry of English structure deviates from the that of the French, and the workmanship differs in other ways too. In other words, the English masons had by the middle of the thirteenth century acquired a know-how of their own which permitted them to adapt the French inspiration to their own idiom.

keeper of the works, were paid a regular stipend of 1 shilling a day, which was doubled when they traveled. All the rest were paid by the week or for special tasks at fixed rates.

Master Aubrey, a subcontractor, was paid 4 pence a foot for 53 feet of parapet stones and 3½ pence a foot for 59 feet of voussoirs with fillets, and 5 pence apiece for 50 facing blocks. In the fourth week after Whitsun, Master Aubrey and his three *socii* undertook to make three "form-pieces" of window tracery by task, and after 10 weeks he was paid £6 0s. 10d. Other masons mentioned by name were Bernard of St.

Osyth, Henry of Carshalton, and Adam of Aldwych who received 2s. 6d. to 3s. a week. Carpenters were paid at the same rate, and smiths 2s. 8d. Pickmen *(pickarii)*, barrowmen *(bayard)*, hodmen *(libri hottarii)*, and similar laborers were paid 1½d. a day.

But the clerk was not always able to pay the men in such an orderly manner. In November, 1254, and again in February, 1256, the coffers were empty, and the workmen laid down their tools and did not resume work until, after strenuous efforts, money was raised to pay them off.

Mont-Saint-Michel is a rock rising 164 feet above the sea off the coast of Normandy. The small island, with its circumference of only 3,300 feet, has been crowded with Gothic buildings. The first monastery was built on the summit in 708, but most of the remaining structures date from the middle of the thirteenth century. The cellars of the houses present an unbroken series of vault constructions from the beginning of the eleventh century. The island fortress served as a prison for political offenders until 1863.

121

Rose windows are commonly associated with Gothic churches, and rightly so; but similar to the originals of the Gothic, the round window derives from the East where the Crusaders found many variations of it. With the rapid breakthrough of Gothic construction, the new craft of "stereotomy," or window tracery in stone, soon developed in France. A French rose window generally comprised some 90 or more pieces shaped into a spoked wheel with concentric circles radially divided into narrow sectors ending in foils, the spaces being filled with colored glass. To the right is a rose window in Notre Dame, the finest of them all.

St. Denis, immediately to the north of Paris, is the first Gothic structure in Europe. It was built in the amazingly short time of four years (1140–1144) under the personal supervision of Abbé Suger, chancellor of King Louis VI and his son Louis VII. The western facade presents so many Romanesque features that a layman has difficulty identifying it as Gothic. From the outside only the rose window and the buttresses suggest a departure from the old style. The interior, however, is unmistakably Gothic with its high nave and pointed arches, as well as many other features that have come to be associated with the new style.

Building accomplishments

During the first building stage, from 1246 to 1259, the transept crossing, chapter house, and vestibule were completed in one continuous effort. The octagonal chapter house was a daring piece of construction since the vault was supported by a slender central pillar only, aided by lightly buttressed walls reinforced by iron ties. After a hundred years the structure was threatened with collapse, and in 1350 four flying buttresses were added to save it. The vault was destroyed in the eighteenth century.

The east end of the church, including the transept, was finished in the summer of 1259, at which time the King ordered the old fabric to be demolished as far as the vestry and rebuilt in accordance with the work already done. This implied that work was started on the nave. Master John died in the summer of 1260 and was succeeded by Master Robert of Beverley who had been working at Westminster for some years. Master Robert introduced some variations in detail; piers to the west of the crossing were given eight shafts instead of four, and the vaults were enriched by intermediate ribs. But rebellions and other trouble slowed down the work after 1264, and during the next two years only £2,000 was spent on the Abbey. In the last six years of Henry's reign, an average of only £1,163 a year could be raised, and by the time of his death in 1273 less than five bays of the nave had been completed. When the work came to a standstill the fifth bay had been finished to the top of the triforium, but it lacked clerestory and vault. There was therefore a gap in the roof between the new and the old work which was not filled in until 200 years later. The new king—Edward II—and his successors showed no disposition to spend any money on Henry III's expensive church.

Completion of the work

A legacy by Cardinal Langham in 1376 got the work started again, but it required 150 years before the seven remaining bays could be finished—at a cost of £25,000 of which £6,000 was contributed by Richard II, Henry V, and Edward IV.

The remaining £19,000 had to be raised by the monks themselves. Succeeding master masons followed the original patterns closely—except for the west front—without allowing themselves to be influenced by contemporary fashions.

Thus, although the building history of Westminster Abbey embraces the entire Gothic Age, the abbey remains true to the original conception worked out in the reign of Henry III. As such, it is more French than English, as seen in the radiating chapels at the east end, the system of flying buttresses, the great portal in the north transept, and the use of bar tracery in the fenestration. The north front resembles the front of Amiens, the apsidal chapels were designed by someone who had been at Reims. The windows of the chapter house are copies of those at Amiens and Sainte Chapelle; the rose windows are French and were designed by a master mason conversant with the latest developments in building technique practiced in the French cathedrals in the thirteenth century. The presence of French masters and masons when building the Abbey is also betrayed by numerous technical details such as the extensive use of iron ties to link the piers together, the manner in which the vertical joints of the piers are concealed by the attached shafts, the stilting of the vaulting on either side of the clerestory windows in order to concentrate the thrust on the axis of the buttresses, and, not least, by the use of large blocks of stone bonded with the wall so as to integrate the structure at the springing of the vaults.

For these reasons Westminster Abbey was destined to remain unique in English building history. Although it exerted an influence on subsequent English churches, this was confined only to details, and no other church was ever built according to the system of proportions adopted by Master Henry for the Abbey.

The expenditures on Westminster Abbey by Henry III between 1245 and 1272, as accounted for by the keepers of the work, came to £41,269. To this should be added £4,000 to £5,000 to cover the cost of the shrine. The total of about £50,000 is equivalent to the entire revenue of the King for two years. To this should also be added the £25,000 spent during the 150 years required to complete the nave. The total bill comes to £75,000, equivalent to about £10 million or more in present purchasing power.

English masons' marks from ca. 1200.

The Fortified City

"A city is well fortified when surrounded by brave men and not by a wall of brick."

LYCURGUS

As the sand is shifted and archeologists dig into succeeding layers of ancient towns they inevitably run into the foundations of defense works, walls, ditches, and towers. So was the case in the tells hiding what remained of Jericho and the temple cities in the Euphrates valley. Impressive as these fortified towns appeared when found, they dwindle in comparison with the more recently investigated Catal Hüyük in southern Anatolia which during its brief period of commercial flowering as the center of the obsidian trade (6500–5700 B.C.) occupied an area of 32 acres enclosed by walls.

Such immense defensive works required the input of tens of thousands of man-days that could ill be spared from farming and, in the case of Catal Hüyük, the obsidian industry, and only dire necessity would make the population undertake the sacrifices involved. Clearly, practically from the outset there was a need for the town dwellers in the Fertile Crescent to protect themselves and their accumulated wealth from the pastoral brigands roaming the highlands.

In Neolithic Europe, conditions were altogether different. There, as previously discussed, the early peasant settlements were left unfortified, explainable only by the general poverty of their Neolithic economy which did not leave a surplus that could be converted into wealth. In the Near East, natural conditions—with the aid of human effort—obviously were much more favorable and responsive to primitive methods of cultivation. For one thing, there was no need to abandon land on which much labor had been invested, owing to the exhaustion of the soil. The peasant population could stay put and increase in size, producing more labor to enable more land to be cultivated from which still larger harvests could be reaped, until the land yielded a surplus to feed people not earning their food by their own labor in the fields.

Thus there arose at an early stage not only priests and scribes who as liegemen of the local god insisted on the right to receive the surplus, but also more directly wealth-producing traders and artisans. Owing to their efforts, the early peasant settlements were converted into temple cities and nuclei of wealth and became dangerous places in which to live. There have always been people eager to share what others possess, to the extent of killing them to facilitate the transfer of their chattel.

It is this eternal conflict between the settled *haves* and the roving *have-nots* that made necessary the building of defensive works around the early towns. As the towns grew in size and more wealth was gathered within them, the walls were built higher, more towers were added to protect the walls, and more ingenuity was expanded on the defense of the gate, always the weakest point in the enceinte. For thousands of years, or long enough to turn into a lasting inbred faith defying all reason, the static defense came up to expectation. The ditches and walls protected the lives and wealth of the town dwellers from raids by pastoral tribes roaming the highlands. In the absence of siege engines and mines, the walls could only be overcome by escalade, i.e., climbing over them, and that was too formidable a venture for a raiding tribe.

Rise of imperalism

All this changed with the mastery of metals. Ancient wall-enclosed towns that hitherto had been permitted to grow in relative safety into important religious and commercial centers in political control of large uplands began to attract the interest of much more formidable enemies from afar. Ambitious local rulers found in their new metal weapons the means to subdue their neighbors and, having enlarged their own power base, launch military campaigns into distant lands.

Sargon of Agàde (ca. 2250 B.C.) seems to have been the first local ruler in the Euphrates valley to become aware of the possibility offered by the metal technology of extending his campaign into distant lands. His drive to "the silver mountains and cedar forests"—in other words Syria—was for the purpose not of acquiring new land to settle but of robbing distant lands and cities of their wealth. With Sargon the ancient world experienced the rise of imperialistic power, and after him the Fertile Crescent was turned into a battlefield for imperialistic contenders which has endured into present times.

After Sargon the introduction of the horse-drawn chariot created a revolution in warfare which brought in train a number of clearly distinguished developments that have become durable features of any power bent on imperialistic conquest. The aims and character of the wars required new types of military strongpoints serving altogether different purposes from the passive defense of the old simple town walls. They were fortified military camps founded in far-off lands to keep the subdued population in check, to collect custom and taxes, to safeguard communications with the imperial capital, and as advance bases for further campaigns.

War had become a serious business requiring experts. The chariot had also made it into an expensive business which favored large property owners, since they were the only ones able to acquire such costly vehicles and to breed and feed the stallions for them. Thus there arose everywhere an equestrian estate of professional soldiers sworn to personal allegiance to the ruling dynasty and in return favored by tax exemption and grants of land. Bands of landless mercenary adventurers, many of them recruited in distant barbarian lands, and eager for plunder, also became available to any ruler in need of their services. In the imperial establishment flourished new guilds of artisans: wheelrights, chassis builders, armorers, bow makers, sword and javelin smiths, etc. Under the forced draft of armaments and war, the entire economy forged ahead and became increasingly more complex and wealth-producing.

It was the same everywhere; indeed, it has always been the same. In the following, Egypt is chosen as an illustrative example, simply because of the wealth of documentary and archeological detail available to facilitate tracing the developments. As briefly discussed in the previous chapter, after the conquest of Lower Egypt by the Horus clan, the dynasties of the Old Kingdom remained content with what they had and devoted their energies to public works. When their reign came to an end with Pepi II (2261 B.C.), the nome (provincial) nobles erupted in violent quarrels that turned into a civil war lasting until 2000 B.C. when a military commander of the Thebes nome-king by the name of Amenemhet succeeded in putting an end to the chaotic interregnum and once more united the country under a ruthless central administration, with himself as pharaoh and founder of the Twelfth Dynasty. Led by him and his dynastic successors, Egypt was launched on an imperialistic expansion that brought the country into enduring conflict with its neighbors to the east, west, and south. Amenemhet also set up a permanent military establishment with all the privileges and appurtenances pertaining to an equestrian order, except that it was still lacking horses and chariots.

The initial drive was directed toward the south, to the conquest of the gold-bearing wadis in Nubia, beyond the First Cataract. The campaign was directed from the vice-regal administrative center on Elephantine and was sustained by the building of fortresses along the Nile. In time, there were eight between the First and Second Cataracts and six to the south thereof.

The names of these fortresses (with their modern 'place names in parentheses) were as follows: Semmet (Bigeh), Yebu (Elephantine), Baki (Kubban), Ikkur Malam (Aniba), Hesef-Medju (Serra-East), Ink-tawi (Faras), Buhen (Wadi Halfa), Iken (Kor). South of Buhen there were six more: (Mirgissa), (Shelfak), (Uronarti), plus three forts at (Semna).

All these strongpoints along the Nile were planned in the same fashion as was subsequently used by the Romans, and they can be regarded as the earliest recorded attempts at town planning. The space within the walls was divided into square blocks by a main street crossed at right angles by a number of side streets. The blocks contained administrative buildings and houses for the garrison and its families. The garrisons were rather small: the one in Semna East had 50 fighting men, Semna West had 150, and Uronarti 100. Buhen, on the other hand, was a large city and headquarters of the commandant of the cataract area. One side of the town stretched along the Nile and was provided with docks. From both ends of the river front ran a wide and deep ditch, the far side of which was provided with a brick rampart serving as first line of defense. On the town side of the ditch rose a second line consisting of a brick wall having circular towers at close intervals. The rampart as well as the towers had numerous loopholes arranged in such a way that archers placed at them could fire into the ditch. Behind this second rampart ran a narrow connecting way beyond which rose the 30-foot-tall walls. The thickness of the walls was about 16 to 20 feet. The curtain was interrupted by square bastions and ended in tall square towers at the corners.

The inland access to the fortress town was enclosed by two high walls built at right angles to the curtain and crossing the ditch and outside ramparts. The roadway between the walls led across a bridge which was retracted on rollers. The gateway was closed by double doors.

The masonry consists of mud brick reinforced with timber, built on a base of rock and rubble. Like some of the other fortresses between the two cataracts, Buhen structures are well preserved, having been enclosed in a protective mantle of sand for the last couple of thousand years. With the completion of the High Dam at Aswan, they will be submerged and the brick turned into what it was 4,000 years ago, i.e., mud.

If Buhen et al. were imperialistic strongpoints in the desert marches, Babylon and Nineveh, each in succession, were seats of imperial power. At Nineveh, the development of the ancient methods of fortification reached its ugly perfection. The city had a perimeter of 50 miles outlined by a huge wall 120 feet high and 30 feet wide, including 1,500 towers. The wall was constructed between two shells of brick masonry and filled with rubble. The mighty walls of Nineveh stood

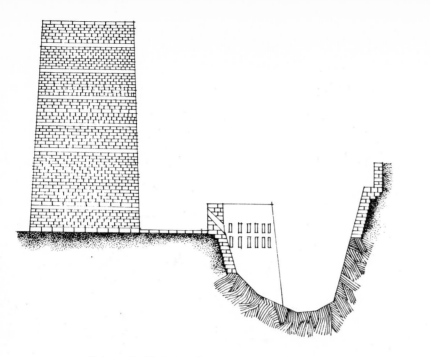

Buhen, the Egyptian fortress town in Nubia, was surrounded by a deep moat, the field side of which had a "covered way," and the town side a low rampart with round towers. The wall and towers had loopholes at regular intervals, arranged so that archers could fire into the ditch in three directions (sketch below). In the rear of the rampart ran a connecting way behind which rose the town wall to a height of 30 feet. The timber-reinforced mud-brick wall tapered from a thickness of 20 feet at the stone plinth to 16 feet at the top and was provided with square towers. A drawbridge mounted on rollers led over the moat. Buhen was built ca. 2000 B.C., and the works were in a state of excellent preservation before being submerged by Lake Nasser. A color photograph taken a year before the inundation is shown on page 131.

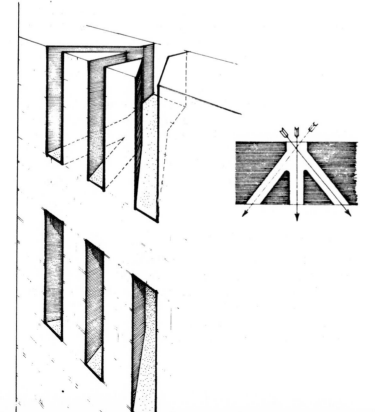

intact for several hundred years but ultimately fell to a Scythian army, not by escalade but by capitulation. With the fall of Nineveh to the Scythians, the Second Assyrian Empire fell into ruin and never recovered.

End of Babylon

In 691 B.C., the founder of the Second Assyrian Empire, King Sennacherib, had shocked the religious conscience of Asia by razing to the ground the holy city of Babylon, capital of Hammurabi. After a successful rebellion, the new king of Babylon Nabopolassar and his son Nebuchadressar II rebuilt the city from the ground and once more made Babylon the mistress of the world. This is the city that has become known from the writings of classical authors, beginning with Herodotus.

The new Babylon was laid out as a rectangle, divided roughly into two parts by the Euphrates which here runs broad and swift. Around the perimeter, variously given as 365, 385, or 480 stades by ancient writers, was dug a broad and deep ditch from which material was obtained for building the walls. The clay was molded into brick and fired in kilns established on the site. The walls were erected inside the moat to a height of 200 cubits and a width of 30 cubits. They were made of fired brick laid in hot bitumen, obtained from a source 8 miles away. At every thirteenth course was inserted a mat of wattled reeds to relieve stresses in the masonry. The walls were brought down to the steep shores and continued in line on the opposite shores. The eastern half of the

This Assyrian stone relief (now in the British Museum) illustrates the siege of an unknown city ca. 880 B.C., and gives valuable information about contemporary siegecraft. To the left two sappers are busy prying loose the facing ashlars from the wall. In the center another sapping team is undermining a tower, while the main assault is mounted from a siege tower at the right. Bowmen in the tower are shooting through arrow loops, the one on top of the tower being protected by a shield bearer. The defenders have succeeded in catching the ram with an iron chain, and two helmeted soldiers using hooks are trying to get the chain off the ram. The principal weapon on both sides is the compound bow, but the defenders are dropping heavy boulders on the besiegers and throwing firebrands at the siege tower.

city was also walled along the river front. The curtain was interrupted by 250 defensive towers, "their height and width corresponding to the massive scale of the wall." Inside the walls the residential areas were laid out in squares, and the streets leading to the river were closed by brass gates at the bank. The houses were three and four stories high.

The building history of the new city of Babylon contains the fascinating but naturally somewhat garbled account of constructing an underwater roadway connecting the two shore palaces. According to Diodorus, a tremendous reservoir measuring 300 stades along each side was excavated to a depth of 25 feet at the lowest point in the city and lined with baked bricks set in bitumen. When the structure was completed, the river was diverted into it, and during the seven days that it was filling up, a vaulted passage was hurriedly built on the river bottom. The vaulted structure was 15 feet wide and 12 feet to the shoulders of the barrel vault. The bricks in the vault were laid in hot bitumen to a thickness of 4 cubits (78 inches) and the walls were made 20 bricks wide.

The rebuilding of Babylon took place during the 43-year reign of Nebuchadressar II during the first half of the seventh century B.C. But its glory was brief; it did not last even a century. In 538 B.C. Cyrus, king of the young and rising power of Persia, rebelled against his Scythian suzerain and invaded Babylonia. He defeated an army commanded by Belshazzar, son of Nabonidos, last king of the ancient kingdom of Babylonia. Nabonidos took refuge behind the sturdy walls of his capital, but with the king in hiding, the garrison became demoralized, and on the 16th of Tammuz (June), Cyrus' forces passed through the opened gates and took

possession of the city. Cyrus made his formal entry into Babylon on the 3d of Marchesvan (October). The power but not the cultural influence of the ancient East had come to an end.

The ultimate fate of Nineveh and Babylon, the most heavily fortified cities on record, illustrates the fallacy of trusting to static defense. They were impregnable, and if they had been stocked with provisions and water, a determined garrison could have held out longer than the besiegers were able to remain in the field. Tyre and Jerusalem withstood successfully sieges lasting up to 10 years. It should be noted in this context that from the time of Jericho, i.e., over a span of some 5,000 years, there were no technical means except escalade and siege to subdue a walled city. The ancient commanders had no ballistas or battering rams at their disposal, nor did they seem to have used sapping to undermine a wall. Although the Assyrian armies possessed siege engines, their effect would have been insignificant against the tremendous walls of Nineveh and Babylon. Yet these ancient seats of power fell owing to the loss of heart of the defenders.

Hippodamus from Miletus

Among the two million workers brought in to rebuild the city of Babylon were "architects and skilled artisans of the world," but needless to say, the names of the architects and engineers responsible for its splendors are forever lost to history, if indeed they were ever recognized. Such was not the case in Greece where, on the contrary, hero worship of

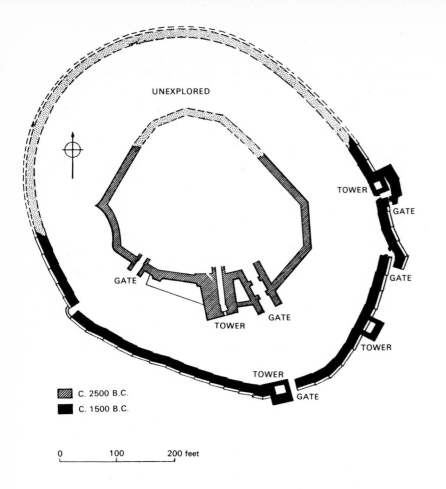

UNEXPLORED

TOWER

GATE

GATE

GATE

TOWER

TOWER

GATE

TOWER

GATE

░░ C. 2500 B.C.
▬▬ C. 1500 B.C.

0   100   200 feet

*The first wall around Troy (the second town) was built ca.
2500 B.C. It had a perimeter of 820 feet and a heavy gatehouse
protected by a tower. Sixth Troy, ca. 1500 B.C., probably
Homer's Ilium, was surrounded by a 900-yard-long and 20-foot-
wide high wall with a prominent taper. The site demanded a
circular wall, but for some reason—perhaps inability to build
it—the engineer chose to construct the wall in 16- to 26-foot-
long straight sections with 6-inch offsets. Only three towers,
each defending a gate, have been excavated. The gate between
the two towers on the right was hidden by a projection of the
wall and was apparently the postern, defensible from three
directions.*

the individual obscured the collective effort. This Greek
concern with individual creative effort has contributed to an
exaggerated appreciation of Greek engineering and building
by posterity at the expense of the much more magnificent,
albeit anonymous, accomplishments in the East.

Greek cities such as Athens, Sparta, Corinth, and others
did not differ much from the oriental ones: Athens was a
jumble of squalid houses built without plan along narrow
winding streets unpaved and unlighted, nestling at the foot

of the Acropolis with its gleaming marble temples and other
status buildings. The *agora,* serving the double purpose of
public assembly place and market, was surrounded by
colonnaded buildings of pleasing design which, however,
were obscured most of the time by a litter of stalls and
shanties. It was not until Pericles employed Hippodamus
from Miletus (in Asia Minor) to build Peiraeus, outport of
Athens, that an attempt was made to break away from the
fortuitous jumble and develop a city along orderly lines.
Hippodamus introduced the principle of straight wide streets
and made provision for the proper groupings of dwelling
houses. (In actual fact, he was not the first to do so; the towns
built in Egypt to house construction workers, such as the
one at Kahun and others in the neighborhood of Thebes,
were laid out in checkerboard fashion.) Hippodamus also
combined the different parts of the city into a harmonious
whole centered around the agora.

In addition to Peiraeus, Hippodamus laid out the town of
Thurii in southern Italy, and he also seems to have had a hand
in the planning of his home town of Miletus. But his influence on subsequent town planning was more important
than his own work. Numerous cities around the Mediterranean, such as Priene, Delos, Ephesus, Pompeii, Olynthos in
Macedonia, Corinth, Antioch, and not the least Alexandria
—planned by Alexander's engineer Dinocrates—all adhere
to the gridiron plan and possess similar features. A wide
parade street, usually colonnaded, led to the agora, and
parallel to this axis ran a number of streets each about 20
feet wide. These streets were crossed at right angles by
narrower streets or even alleys. This layout resulted in blocks,
subsequently called *insulae* by the Romans, which were
subdivided into house lots, the insulae around the agora
being reserved for public buildings. Around the perimeter of
the city ran a wide and deep ditch and a wall.

Roman town planning

"Since we have to seek healthiness in laying out the walls of
cities ... I emphatically vote for the revival of the old method.
For the ancients verified the beasts which were feeding in
these places where a town or fixed camp was being placed,
and they used to inspect the livers ... When they had made
trial of many and tested the livers in accordance with the
water and pasture, they established there the fortifications ...
If they found them faulty, they judged that the supply of
food and water ... would be pestilential also in case of
human bodies. And so they removed elsewhere, seeking
salubrity in every respect."

This is how Vitruvius writing before 27 B.C. begins his
book on town planning. He goes on to explain that a settlement can be laid out in marshy land, so long as the site is on
the coast. "For if dikes are cut, there is made an outflow of

water to the shore, and when the sea is swollen by storm, there is an overflow into the marshes which . . . being mixed with sea salt does not permit the various kinds of marsh creatures to be born there . . . The Gallic marshes around Altinum, Ravenna, and Aquileia are for this reason incredibly salubrious. But the Pontine marshes by standing become foul and send forth heavy and pestilential moisture."

This passage makes it abundantly clear that Roman engineers were aware of the danger of stagnant water to public health, and although they did not actually know the link between the malaria-carrying mosquito and the fever it caused, they knew better than to place a camp or city in the proximity of marshes where "minute and invisible animals grow and cause diseases." It took the medical men north of the Alps nearly 2,000 years to become aware of this. The simplicity of the tests suggested preliminary to the siting of a town or camp is wholly admirable: catch a number of wild animals, and cut them open. If their livers are healthy, they live in a healthy milieu; if they look sickly, it is an augury to move on and find a healthier place.

Vitruvius also goes into the principles guiding the layout of the defenses. The planner should strive for a circular wall because a square and angular shape favors the besiegers. The distance between towers should be within a bow shot, so that an attack against one tower can be fended off by fire from "scorpions" and similar missile engines from neighboring towers. The towers should be round or polygonal in order to better withstand a battering ram. They should be projected beyond the face of the curtain wall so that the enemy becomes exposed to flanking fire. The approach to the gates should not be straight, but winding in such a way that attacking troops will have their right side, unprotected by a shield, exposed to the walls.

The foundation of the wall should preferably be carried down to bedrock and be made wider than the wall extending aboveground. The wall is to be constructed of two masonry faces set 20 feet apart and the space between them filled with rubble. They should be reinforced by timbers extending from face to face. Best are charred olive trunks which possess "everlasting strength." The top of the wall should be made wide enough to allow two armed men to pass without hindrance. Particularly exposed places should have ditches of "amplest possible width and depth."

Thus Vitruvius, who enters into the minute details of the layout and construction of a defensive wall. Roman fortifications throughout the Empire were laid out according to his rules, which had become codified into standard practice during the first century B.C.. The preferred site for a fortified town was high ground sloping toward a river on one side and with steep slopes falling away on the other sides. Elsewhere, the plan was adapted to the topography of the site; where the approach was easy, the walls were made taller, the flanking towers stronger, and the ditch wide and deep. Around large cities necessitating the enceinte wall to be

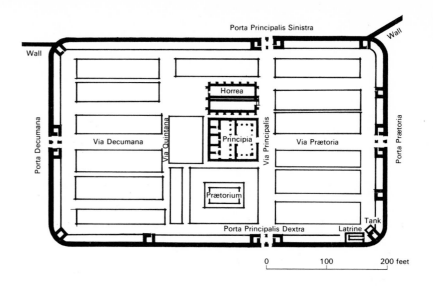

A Roman camp was always planned in the same manner regardless of location or whether permanent or temporary. The rectangular camp was surrounded by earthworks, later replaced by a stone curtain with a gate on each side. The gates on the long sides were connected by the main street, named Via Principalis after the headquarters (Principia) with the personal and administrative quarters of the commanding officer. On one side of the Principia were the officers' quarters (Praetoria) and on the other the storage buildings (horrea). The rest of the insulae *had barracks for the troops. The plan shows the camp at Borcovicum near the Hadrian wall. It had in its southeast corner a 33-foot-long latrine flushed by running water from a stone tank sealed with lead.*

carried over hills and valleys, it would be drawn back somewhat at the low points to permit flanking fire from the hills.

Within the enceinte the town was invariably laid out on a gridiron plan, whether inspired by Hellenistic practice or, more likely, simply along the lines of the traditional Roman military camp. A legion camp was square or rectangular and divided by two main streets, one running north-south, the other east-west. The headquarters buildings were sited around the cross, and barracks and stores in the insulae formed by the gridiron plan. Around the perimeter ran a ditch and a wall with towers at the four corners. Since most Roman towns began life as military camps, they conformed to this standard pattern.

Outside Italy the provincial towns, settled by veterans and

termed coloniae, were also laid out in this fashion, but pre-Roman settlements, granted Roman charters and called *municipia,* retained their ancient planless jumble. In Britain, place names ending with *-chester* (castra) speak of this military origin, although Silchester belonged to a class of *civitates* that had existed before the Roman occupation. Like many other tribal communities, it was retained as an administrative center and rebuilt according to a modern, i.e., gridiron plan, with a forum, basilica, and other public buildings at the center cross. Although the area of Silchester was about 100 acres, it contained only 80 houses since the native population insisted on building their houses as they pleased within the framework laid out by the Roman surveyors. The wall enclosing Silchester was polygonal in plan, no doubt owing to the fact that it was erected on the ancient earthworks of the tribal town.

Londinium was laid out on a rectangular plan on a gravel plateau on the left bank of the Thames, rising to a maximum elevation of 50 feet. The total area of 325 acres was bisected by a stream—the Walbrok—and enclosed by a wall beginning at the Tower and running in a northwesterly direction for 3,500 feet, westerly for 3,000 feet, after which it meandered south toward the river, reaching the bank a couple of hundred feet east of Fleet Ditch. The total length of the land enceinte was about 11,000 feet, to which should be added 6,500 feet of wall along the river. Access to the city was by six gates. The population around A.D. 60 was about 20,000, which suggests a density of 60 persons per acre.

On the Continent, Treves, Cologne, Belgrade, among others, still retain streets that follow those laid out by Roman engineers when they planned the original camps. Aosta and Turin, initially fortified camps guarding the invasion route taken by Hannibal down the Aosta valley, still retain in their centers the Roman chessboard plan, as do Florence and Lucca, although to a lesser extent. But the center of the world, Rome itself, despite its Augustan architectural splendors, was permitted to grow into a crowded jungle of ramshackle multistory tenements. In the second century A.D., the area enclosed by the walls was about 3,000 acres, and since the population has been estimated at $1\frac{1}{4}$ to $1\frac{1}{2}$ million, the density must have been on the order of 400 to 500 persons per acre, not a pleasant figure to contemplate.

Roman siegecraft

Although the Roman city planners continued by force of habit to enclose new towns with walls, towers, and moats incorporating all the refinements of fortification, the Greek and Roman military engineers devoted their efforts to making static defenses obsolete. Under their guidance, the legions became masters in siegecraft. No wall built by man could in the end stand up against their siege engines. The Greeks had devised methods of mining and approach trenches, i.e., the art of sapping, which the Romans adopted. They drove a tunnel under the foundation of a tower and excavated a large subterranean room under the structure, propping it up with timber until the foundation had been completely undermined. Then they put fire to the timbering, thereby causing the whole tower to collapse.

Missile engines—ballistas—originally invented and constructed by Greek engineers, threw darts a yard long and weighing 5 pounds and stones weighing 10 pounds over an effective range of 500 yards and directed at the defenders manning the walls. Several hundred engines throwing darts and stones swept the Jewish defenders off the walls of Jerusalem when it was besieged by Vespasian in A.D. 70, and knocked down the battlements.

But the really most fantastic engine of them all was the "helepolis" (destroyer of cities)—originally devised by the Macedonians but greatly improved by the Romans. It consisted of a huge tower that could reach a height of 150 feet and weighed 100 tons or more. It was mounted on a chassis provided with a large number of wheels 6 to 12 feet in diameter. The tower contained 20 or more floors joined by ladders. It was strongly built of timber and boarded, with the outside lined with green hides to ward off firepots. The ground floor contained one or more battering rams with ironclad heads; the upper floors were provided with missile engines and archers. In the tower were a water tank and force pumps working with hoses made of the intestines of cattle. At different elevations there were drawbridges that could be let down on the top of the wall to permit sallies by storm troops.

The main function of the tower was to dominate the defense and keep down the fire from the walls and towers while the breaching operations were going on. To bring the tower up against the wall in the face of defensive fire, it was of course necessary to fill up the ditch and level the ground. For this purpose an "approach tortoise" was used. This consisted of a much smaller wheeled structure provided with a strong roof under which men could work in reasonably security.

Building such a tremendous engine under field conditions and manipulating it against the wall and keeping the ram operating day and night until the wall was breached was of course a magnificent military accomplishment that has not been mastered by any soldiers other than the Roman, except for French and German in recent centuries. It required a disciplined body of men who besides being soldiers were also highly skilled craftsmen.

After the Decline and Fall

Seen in retrospect, it seems evident that the decline of Roman power set in at the height of Rome's military strength, with the decision by Augustus, after the catastrophe in the Teutoburgerwald, to withdraw his no doubt overextended lines

On the western shore of the Nile, immediately north of the Second Cataract, lies Buhen, Egypt's strongest fortress in Nubia, and probably established chiefly for the purpose of guarding the gold mines in the wadis east of the Nile. Buhen was buried in sand until excavated in 1957 by a team of British archeologists. The defense works consisted of a counter escarp with a glacis (left), a deep moat, a low escarp with round bastions, a rather wide protected way and, finally, the main curtain with its projecting square towers. The round bastions had loopholes (right) from which a bowman had a 60° field over the counter escarp and into the moat. Buhen, built ca. 2000 B.C., was a fully developed system of static defense that was not to be improved upon, except in scale, until the seventeenth century A.D. The building material was mud brick which now, after being submerged by Lake Nasser, is in the process of being converted to its original state.

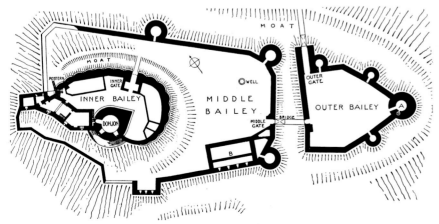

Château Gaillard on the Seine near Les Andeleys was built by Richard Coeur de Lion in 1197 and is commonly regarded as the first castle erected by a returning crusader in accordance with Eastern principles. It soon become apparent that the early builders had misinterpreted the basic lines of the layout of Byzantine fortifications. Château Gaillard fell to King Philip August in 1204 after a siege lasting only seven months. The strong tower defending the outer bailey was mined, and the gate to the second one taken by a party entering through the latrine. The plan of the castle (left) is suggested by the remains. Facing inland and jutting out like a prow was a round strong tower defending the outer bailey, which was joined by a drawbridge to a middle bailey. A ramp led to the gatehouse of the inner bailey with its powerful donjon at the edge of the shore escarp. The main trouble with the layout was the inadequate space allotted to the different works which prevented the garrison from coming to the relief of the point attacked.

To the right is an aerial view of Dover Castle situated at the narrows of the English Channel. The original keep is still surrounded by two concentric curtains, a third one having been demolished in Napoleonic times to provide ramparts for artillery. The mound with the Roman lighthouse and Anglo-Saxon church is also clearly shown. Above is a detail of the tower gate with its barbican. The square keep was already obsolete when completed in 1190 at a cost of £4,000.

Caernarvon Castle is the largest and strongest of the many castles built in Wales by Edward I from 1274 to 1330. It is strategically well sited in the angle of the river Seiant and the Menai Strait and could be supplied from the sea. Caernarvon was originally intended to serve as the residence of a viceroy of Wales but no English royalty has ever lived there. The octagonal towers with their horizontal decorative bands recall the towers of the Justinian wall around Constantinople and the likelihood is that this was the source of inspiration.

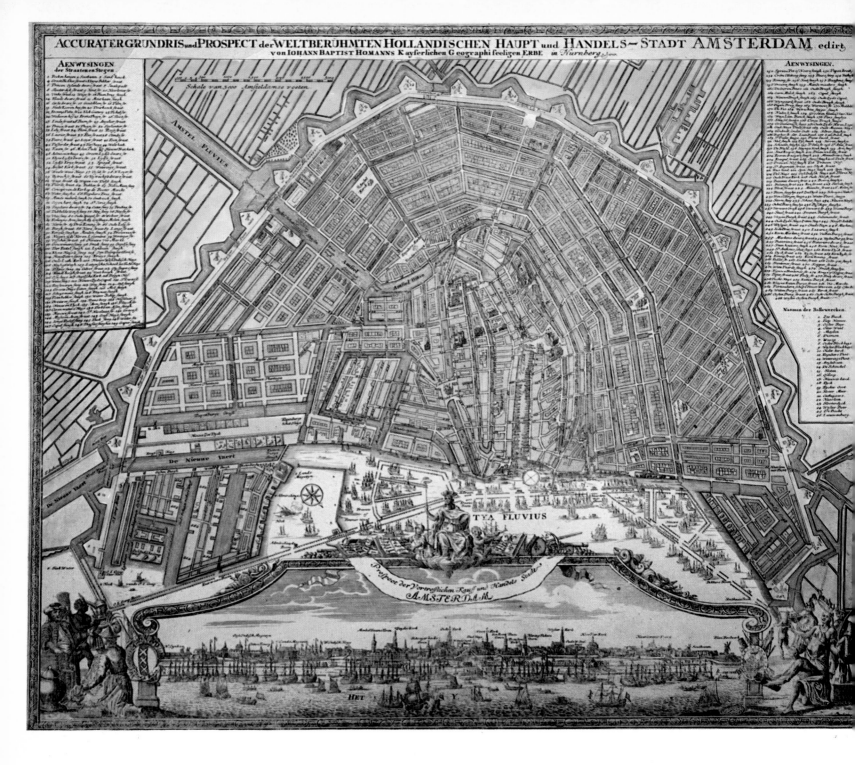

Few cities in Europe have had to struggle with such poor foundations as Amsterdam. The city was built on a 50-foot layer of clay overlaying 10 feet of sand, and rests on oak piling driven through the clay into the sand. Owing to the city's hospitality to the victims of the religious persecutions in the fifteenth and sixteenth centuries, the population increased rapidly and likewise the economic and cultural importance of the city. In order to get additional building space, a new town plan was adopted in 1612, whereby the area was quadrupled. The work of expanding the city went on until 1670 and consisted primarily of con-structing a number of concentric canals. The space between them was drained and turned into housing blocks. In this manner Amsterdam obtained its famous concentric canal system which from the medieval core of the city comprises the Singel (1383), Heerengracht (1593), Keisersgracht and Prinsengracht (1630), and Singelgracht (1658). The latter is a 6.5-mile-long canal around the perimeter of the city protected by a curtain with triangular bastions built by Coehoorn. The concentric canals were crossed by radial ones, thus forming some hundred man-made islands connected by 300 bridges.

beyond the Rhine and the Danube. The rest followed in logical sequence. His successors decided to contain the barbarians by means of gigantic border fortifications beyond which they hoped to live peacefully and civilized forever. Emperor Hadrian built a 73½-mile wall extending from the Atlantic to the North Sea to keep out the troublesome Picts and Scots.

On the Continent, a vast enceinte was constructed extending from the Danube north of Ingolstadt for a distance of 170 miles to the river Main and from Frankfurt via a 120-mile salient into Limes Germaniae to the Rhine. The earth and masonry walls were studded with watchtowers, guard posts, and strongpoints; and in the rear were fortified legion camps. The work began in the second century A.D., by which time the Han emperors at the opposite end of the Euro-Asiatic continent had completed the longest frontier wall ever constructed.

The Chinese wall was begun by Emperor Shih Huang Ti (246–210 B.C.) and continued by the Han dynasty ending A.D. 220. It ran through a barren country for a distance of 2,500 miles without a break except for strongly defended gates.

The idea behind both these vast heaps of masonry and earth was the same—to contain the Barbarians. By the time the Romans put their legionnaires and impressed local labor to work on their wall, the vast Chinese enceinte had proved a dismal failure—the Mongols had broken through and put an end to the Han dynasty and made themselves masters of China.

Masonry walls are no protection for an empire rotting at the core. The garrisons manning the long walls were thinned out, the strongpoints abandoned as the legions, one after the other, were withdrawn to take part in the internecine struggle for power at the center. And so, once again, the inevitable happened. Ill-smelling, brutish, hungry for land and loot, the Barbarians appeared—in bands, in droves, in tribes and nations. They crossed the Rhine and the Danube on reed floats, climbed over the unmanned walls at the watershed of the two rivers, and took possession of the civilized lands beyond. And once more the pall of darkness descended over western Europe.

The stupidity and ignorance of Europe's new masters prevented them from adopting what should have been of the greatest appeal to them, i.e., the superior disciplined and efficient methods of killing one's enemies developed by intelligent commanders during the previous 3,000 years. With the abandonment of the Roman ramparts in central Europe was lost also the art of warfare, including building and subduing fortifications. The warriors of the German chieftains knew nothing but personal killing with poorly made weapons, soft iron hardware made of bog iron inadequately hardened. Franks and Saxons fought on foot with spear and ax; their ancestors had been scared witless by the horse-riding archers of Huns and Magyars who had swept in from

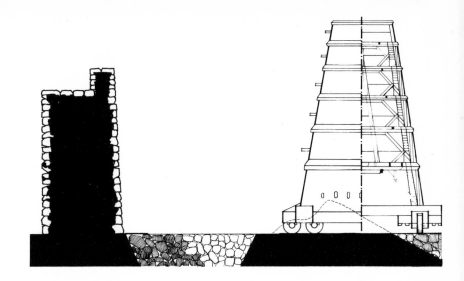

The helepolis, or "destroyer of cities," was a type of siege engine invented by the Macedonians, and greatly favored by Alexander. Demetrius Poliorcetes used one with a battering ram that required a thousand men to work it against the curtain of Rhodes in 305 B.C. The Romans adapted and improved the siege engine, which ultimately obtained a height of 150 feet and a weight of a hundred tons or more, and was propelled on wheels.

the steppes and annihilated their tribal forces. The best the Saxons could think of in the way of a missile was to throw an ax at somebody.

So it might have remained but for the Crusades. The so-called chivalry of England and France learned to their amazement, at Constantinople, in Syria and Palestine, that there was more to warfare than individual combat with a two-handed sword and a "francisca." Five hundred years had passed and some of them at least had become mentally capable of understanding and appreciating what they saw.

Although neither space nor the general line of the narrative permit a more detailed discussion of the post-Roman art of fortification, no matter how fascinating a subject, nonetheless a brief note on the tremendous military construction that went on in the Byzantine world during Europe's lost centuries seems warranted, in view of subsequent developments in Europe.

In Byzantium, military construction on a grand scale began in 413 with the building of Constantinople's enceinte across the neck of land in the west to serve as a defense against the assault of Goths and Huns. This original Theodosian wall was destroyed by an earthquake in 447, and when it was rebuilt it was provided with an outer curtain and a wide moat. The walls were a shell of ashlar with a core of Roman concrete. The towers were decorated with horizontal bands of brick. Despite earthquakes, numerous sieges, and centuries of

135

neglect, the solidly built walls still remain largely intact. The outside wall is 6½ feet thick, and the inner one 15 feet.

Nicea was also fortified at this time and provided with a double enceinte the inner wall of which was constructed exactly like the line at Constantinople. Both walls had round towers spaced zigzag fashion. A powerful keep in the southern works successfully resisted a long siege by the Crusaders and could not be subdued without recourse to mining. It seems quite likely that it was this particular experience that made the keep, or donjon, so popular in the Crusaders' home countries during the following 200 years.

In the early fifth century, Byzantine engineers surrounded Cairo with a wall which included two mighty towers 90 feet in diameter and placed 60 feet apart. Together with the curtain along the Nile, the two towers controlled the river traffic upstream to the delta and the important caravan route from North Africa to the east. But the towers are noteworthy also from a technical point of view. They were constructed with two concentric shells placed 15 feet apart and connected with eight strong ribs, whereby the same strength was obtained as with a 28-foot wall, with considerable savings in material. This was also clever construction from a military point of view, because if one of the eight cells was damaged during a siege, it could easily be repaired; and if the enemy's troops had succeeded in entering the breach, they would be exposed to cross fire from archers shooting through arrow slits in the adjoining cells. When the citadel of Cairo erected by Saladin in 1170 to 1182 was rebuilt in 1520, it was provided with two such towers, Burg Magattam and Burg Al-Wustany, carrying cannon on two levels.

However, it was during the reign of Justinian (483–565) that Byzantine military construction culminated both technically and in scope. This emperor busied himself with building and reconstructing some 700 fortifications throughout the Byzantine empire. That is a record in military construction which has never been beaten, and it was these Justinian defense works that served as models for Arabs and Europeans when at different times and places they came into bloody contact with these fortifications around the Mediterranean.

As protection against the Persians, Justinian rebuilt the curtain around Dara, about 140 miles northwest of Mosul. The walls were raised to 60 feet by building vaults atop the original curtain; the old crenellations were filled in and turned into arrow slits, whereby two vertical lines of defense were obtained. The walls were 30 feet thick at the base and tapered. As in Nicaea, one of the highest towers was made stronger than the others. It was originally called a "watchtower," and subsequently it became still another model for the donjon.

Many of the forts in North Africa, until recently garrisoned by the French Foreign Legion, were built by Justinian, and so were the fortifications at the Marmara Sea, the Bosporus, the Dardanelles, and along the Danube. Rome's defenses were also strengthened and modernized by adding fly walls to the merlons and digging a wide moat along the curtain. The work was managed by Justinian's general Belisarius upon his recovery of the city from the Goths in 536.

It need hardly be added that these magnificent static defenses, whether cities, citadels or castles, fell like a row of kingpins before the Arab assault. After the Arabs in the course of merely a few decades had swept over the entire Near East and placed Syria, Egypt, Persia, Cyprus, and Rhodes under their hegemony, they appeared before the walls of Constantinople in 668. Here the defenses withstood their repeated assaults, and after a siege of seven years they abandoned their thrust into southeastern Europe. But the reason for the Arab failure was not entirely the strength of the defense works; it was rather the "Greek fire"—ancestor to napalm—which could not be extinguished by water and which annihilated ships, siege engines, and storm troops. A final attempt in 716 to take the city by storm was also beaten back by Greek fire.

But elsewhere, along the march routes to Palestine, Justinian fortifications became garrisoned by Arab troops. When in the fullness of time the Crusaders set out on their bloody pilgrimage, it was on this road studded with Justinian strongpoints that the masters of Europe at long last learned how to build castles.

Having learned the twin arts of fortification and siegecraft, the barons on their return put their oriental education to good use for enhancing their economic and political fortunes. They established themselves on the tops of hills or in mountainous country on inaccessible crags. Since a stronghold was intended only for a baron and his small body of followers, the site chosen could be altogether different from that suitable for a fortified town. The principles followed in the building of the robber barons' castles were borrowed from Byzantine engineers, although they incorporated some changes not then apparent to the copiers. Whereas in Europe the *donjon,* or central keep, was regarded as the fortress, and the double walls subsidiary to the defense, in the East each envelope with its curtain walls and projecting towers was a fortress in itself, and the central keep was merely the last refuge of the garrison.

Château Gaillard on the Seine, near Les Andeleys, built by Richard Coeur de Lion in 1197, is commonly regarded by castle connoisseurs as the finest example of an early Crusader castle embodying all the principles of fortification. It stood on high ground and had three concentric walls in addition to the keep. The towers were round, in order to stand up to battering rams that by now had come into use also in western Europe.

Unfortunately, the art of sapping had also been learned. After a siege by Philip Augustus lasting seven months, one of the towers in the first enceinte was collapsed by mines, and the wall abandoned by the defenders. A gate in the second wall was then taken by surprise, and the garrison prevented

from gaining the safety of the donjon. At the time it fell, Château Gaillard was only seven years old.

In sum, Château Gaillard and many lesser seignorial castles built during the twelfth century were not as strong as their builders intended, chiefly because the difficulties placed in the way of the attackers also prevented the defenders from coming to the aid of the points attacked. Gradually, some ingeniatores began to grasp the principles of static defense, as mastered by the builders of Buhen, for example, 3,000 years earlier. The size and height of the works were enlarged, the area of defense was expanded, and the layout allowed for flanking. Attention was given to step-by-step defense. Entrances were provided with traps and numerous gates, the towers had machicoulis galleries for vertical defense, and the gates had flanking towers and barbicans.

But such sound developments in fortress construction took a century or more to mature. The early seignorial castles such as Gaillard, Covey, and Pierrefons, although naïvely designed and ill-suited for their intended purpose, nevertheless required a tremendous input of labor and material. The crude labor and transport presented no particular problem: the baron and his men dominated the local territory and rounded up the peasants by force to work on the castle. The trouble with this simple method was that no one succeeded in turning a peasant into a mason by prodding him with a spear. A great deal of the early failures of castle building was due, in addition to faulty planning, to the lack of competent masons.

Nowhere in Europe is this early development of castle building so easily followed as in England. Space unfortunately does not permit more than a brief sketch of the methods chosen by successive British kings to fortify their power and the unintentional effects of fortress construction on the establishment and growth of altogether new towns nestling close to the walls. Elsewhere in Europe developments were the same, albeit with local variations.

Having successfully invaded England, William the Conqueror immediately set about to secure his power by impressing the Saxon peasants to work on his "motte and bailey" castles. The first was thrown up at Hastings and is shown on the Bayeux tapestry. A motte *was a man-made mound surrounded by a trench, and on it was erected a tower of timber. The* bailey *was a kidney-shaped enclosure at the bottom of the mound and intended for horses and cattle. The works were surrounded by a timber palisade. The mound measured 100 to 300 feet in diameter, and its height 10 to 100 feet. With enough labor, the earthworks could be finished in a week. Below are plans of a couple of typical motte and bailey castles from different places in England built immediately after 1066.*

Motte and bailey castles of the Conquest

To secure his conquest and maintain his authority over the Norman barons, William relied on his mounted knights and moated castles. The early type of castle as constructed by Odo, Bishop of Bayeux working out of Dover, and William fiz Osbern of Norwich consisted of a flat-topped mound or "motte" encircled by a ditch and provided with a kidney-shaped enclosure or "bailey" on one side. Both the motte and the bailey were protected by timber palisades, and on the motte was erected a wooden tower placed on stilts. The purpose of the bailey was to shelter the horses and cattle of the garrison and to protect the stairway leading to the motte. The building of such a castle was done by forced labor, and all the able men in the shires were impressed to work on the

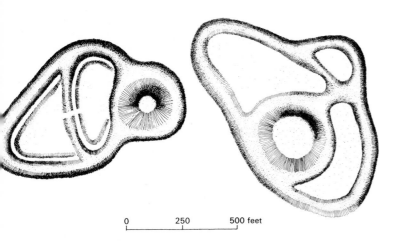

0   250   500 feet

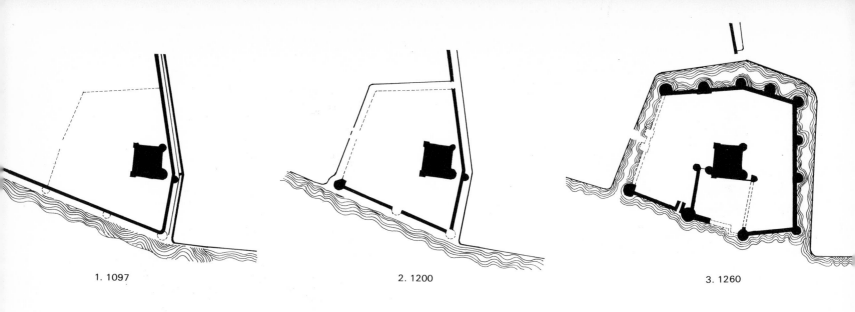

1. 1097   2. 1200   3. 1260

earthworks. With a large input of labor, construction went fast. The mottes at Dover and York were thrown up in eight days, and the Tower of London just about as fast by impressed labor from Surrey, Middlesex, Essex, and Kent.

Motte means height in Medieval French, and *bailey* was the contemporary term for a wall or timber palisade, although it also referred to the enclosure. Hence "motte and bailey" is a man-made mound enclosed by one or two timber palisades. It was a common type of contemporary defense works in Germany, Italy, and Denmark, although particularly favored in Normandy.

Although the motte and bailey work was supervised by archers, there was on each site a man familiar with timbering. The Domesday Book gives the names of six carpenters listed as "the King's servants," among them Stefanus, Durabus, Raynerius. Since timbering skills were rare, the *carpentarii* held important positions in the establishment and were granted one or more manors for their services. All told, they erected 40 motte and bailey castles at strategic points throughout England during the first 20 years after the Conquest.

The most celebrated of the early castle builders was Gondolf, Bishop of Rochester, who accompanied Lefranc to England when the latter was appointed Archbishop of Canterbury in 1070. He was obviously versed in stone building, and in return for a manor he undertook at his own expense to erect a stone keep at Rochester. This very first stone castle in England was erected between September, 1087, and May, 1089, at a cost of £60. However, his chief claim to fame was his supervision of the building of the Tower of London, a 107 × 188-foot stone keep replacing the old motte and bailey. Completed in 1097, it was a three-story-high tower with massive 15-foot walls at the base and containing the necessary accommodations for a royal residence, including a great hall and a chapel. Entrance was by means of a doorway on the first floor level, accessible by a wooden

staircase that could be withdrawn. With the tower completed, King William ordered men from the neighboring shires to erect an enceinte wall around it in 1097.

A somewhat similar stone tower measuring 110 × 151 feet was built at about the same time at Colchester, as a defense against Danish invasions in East Anglia. There are reasons to believe that both towers were designed by the same man, but not by Gondolf who only kept the accounts. Who this master mason was is not known, but he must have been called in from the Continent because these two towers bear no relation to the Norman timber towers or to any contemporary donjons elsewhere. The design is apparently inspired by the fortified houses in France following the breakup of the Carolingian empire. During the ensuing anarchy, the leading nobles plotted and fought against one another, and it became necessary for them to turn their country residences into fortified strongholds. The source of inspiration is also reflected in the early references to the two stone keeps, i.e. *arx palatina* (fortress palace) rather than the usual *castellum* (military fortress). They remained unique not only in England but on the Continent as well, since subsequent construction followed different lines.

During the reign of three descendants of William, the hurriedly built motte and bailey castles were gradually replaced by stone structures—either by the kings or by their castellans. Purely private seigniorial castles were also being built, but they could not be made too strong or too high, or the moats too wide or deep. Nor could the walls be crenellated without risking forfeiture. But such regulations could not be maintained during the reigns of weak kings when royal castles, including even the Tower of London, fell into the hands of ambitious barons. During the reign of King Stephen, last successor of William I, there were in England alone about 1,500 such adulterine, i.e., illegal, seigniorial castles.

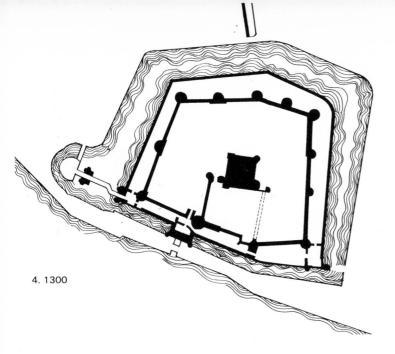

4. 1300

Plan of the Tower

The development of the Tower of London is shown in the four sketches above. (1) The keep sited near the southeast corner of the Roman wall was finished in 1097. The bailey was enclosed by earthworks on two sides. (2) A trench was dug along the Roman wall and the existing earthworks. A defensive tower, called the Bell Tower, was erected in the southwest corner (1200). (3) Henry III converted the Tower into a concentric fortification by building a stone curtain with towers along three sides and widening the moat. The Roman wall was also strengthened (1225–1260). (4) The old moat was filled in, and a new and much wider one was dug; the west curtain built in stone, and the gatehouse provided with a barbican. The shore was filled in, and among the works built on the fill was a strong sea gate called the "Traitors' Gate" (1300). The layout has not been changed since. To the right are a plan and elevation of the Tower.

During the reign of the Angevin kings (1154–1216), some £46,000 was spent on castle building, £21,000 of which was contributed by Henry II for building and maintaining 90 royal castles. Dover alone cost him £6,500. The military theory behind such new castles as Corfe, Harley, and Kenilworth remained the same as previously; they consisted of two parts, a bailey and a keep. At Arundul and Windsor, the old timber stockades were replaced by a ring wall around the summit of the motte to form a so-called shell keep. The towers were rectangular in plan and normally entered at the first- or second-floor level. Compared to the previous keeps, the walls were made so thick that chambers and wall passages could be built in their upper parts. No vaulting was as yet employed; instead, the Angevin castles were provided with a vertical internal partition rising to the full height of the

The White Tower

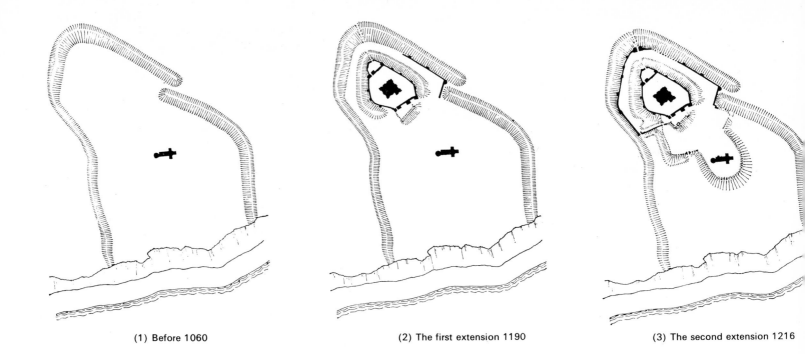

(1) Before 1060   (2) The first extension 1190   (3) The second extension 1216

tower, to aid in supporting the roof in two spans with a central gutter. The castles were intended for passive resistance and were impregnable against missiles of stone, lances, and arrows.

The Dover castle, strongest and costliest of them all, was the last square tower. It was built between 1180 and 1190 at a cost of £4,000 and was obsolete when finished. The 98 × 96-foot donjon rises to a height of 83 feet, and its massive walls taper from 21 feet at the plinth to 17 feet at the crenellations. The walls are buttressed in the center and at the ends, where the buttresses rise above them to form angle turrets 12 feet high. Internally, the tower is divided by a vertical partition supporting the roof. A massive forebuilding against the south and east sides contains three flights by means of which the keep was entered. On the second level was a drawbridge and on the third a guardroom covering the entrance to the main hall. Nearby is an elevated wellhead with a recess for a tank from which lead pipes carried water to different parts of the keep.

The double-saddle roof was, as in the Tower, built of timber and shingled, which of course made it combustible. Therefore, the walls were heightened to mask the roof. A besieger armed with missile engines would have had no difficulty throwing firebrands above the walls and thereby igniting the roof. It may therefore be assumed that such missile weapons as the petraria, ballista, and trebuchet—in common use by the end of the twelfth century—had not yet been developed enough to worry the castle builders.

However, by the time Dover was completed, the military weaknesses of the square tower had become apparent to contemporary engineers. The four salient angles offered security to raiding parties attacking the wall with picks and siege tools since the sappers were out of the reach of arrows and other missiles.

It should perhaps be noted that the thick walls of Dover and other medieval donjons and curtains consisted of a relatively thin shell of ashlar filled with rubble. Once a stone had been pried loose, the adjoining ashlars could easily be barred out, whereupon the loosely packed core could be excavated in full security and relative comfort, provided of course that the sappers could proceed without interference. A donjon with no machicolations for vertical defense could be defended against sapping only through sorties by the besieged garrison.

The obvious solution to the problem was of course the round tower surmounted by projecting fighting galleries from which the defenders could cover the whole circuit of the base; and such towers became the fashion from the thirteenth century onward. Château Gaillard and the French castles built by Philip Augustus were the first in Europe to be given round towers of impressive size, but in England no such keeps were erected until during the reign of Henry III. In England, the royal master masons evolved some transitional forms polygonal in shape whereby some of the advantages of the round keep were gained but not its strength. At Windsor, the upper bailey and parts of the middle and lower ones were enclosed by walls, and at Dover an enceinte was erected around the keep. Henry II also ordered the construction of

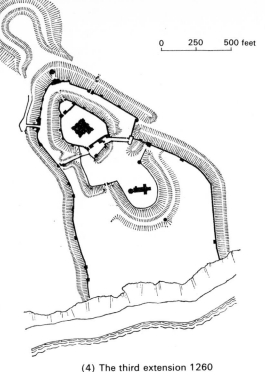

The military history of Dover begins in prehistoric times when earthworks were thrown up on the summit of the chalk cliff (1). The Romans built a lighthouse and the Anglo-Saxons a church within the ancient enclosure, both of which have survived. A square keep was erected nearby (1180–1190) (2). A second stone curtain around the keep was built in 1216 on the prehistoric works, and an outer bailey was obtained by throwing up an earth wall around the lighthouse and church (3). With the third and last extensions, completed in 1260, stone curtains with towers were brought to the shore escarpment following the ancient trace, the outer bailey was also enclosed by a stone curtain, and an earth barbican was thrown up northwest of the keep (4). Except for the demolition of the curtain around the outer bailey, the layout of Dover has remained unchanged since 1260.

0   250   500 feet

(4) The third extension 1260

an outer curtain wall, as the idea of concentric fortifications began to be accepted. Two gateways at Dover with flanking towers on each side of the entrance were also built; subsequently, the gateways came to be protected by a barbican placed so as to expose the attacking force to maximum flanking fire.

In the late twelfth century, the flanking tower was introduced whereby the defenders were enabled to protect the base of the curtain wall against breaching tools and engines. Since adjoining towers also covered each other, one defended the other from assault and escalade. The systematic development of mural towers during the late part of the thirteenth century permitted a castle built on a fresh site to dispense with the donjon. This type of castle became an integrated whole, in contrast to the keep and bailey castle, and was better able to stand up against the improvements in siege-craft coming into use—such as the trebuchet, crossbow, and Greek fire. The ditches were also widened to make the enemy keep his distance.

Within the walls stood the buildings, the *domus Regio in castello,* serving military, domestic, and administrative functions. Most important was the *hall,* the center of social life and administrative activity. Others were the chapel, the king's chambers, kitchens, barns and stables, the gaol, and the *camera clericorum.*

During the reign of Henry III (1227–1272), the number of Crown castles gradually declined to 47, mostly due to deterioration of the structures. The last wooden castle collapsed in 1270. It is surprising to learn from contemporary reports how quickly these seemingly impregnable structures fell to pieces from natural causes. Towers and walls kept falling down owing to faulty workmanship; dry rot attacked the rafters and caused the roofs to collapse. Although the castellans were allowed £5 a year for the upkeep of their castles, they usually embezzled the funds and materials intended for maintenance.

The Castles of Wales

Castle construction in England reached its culmination during the reign of Edward I (1272–1307). He spent altogether £80,000 on eight new castles in Wales and £20,000 on the reconstrucion of the Tower of London whereby it was turned into a concentric castle and given its present appearance.

The strategic thought behind the construction of the eight castles in Wales—Builth, Aberystwyth (in mid-Wales); Flint, Rhuddlan, Conway, Caernarvon, Harlech, and Beaumaris (in north Wales)—six of which could be victualed from the sea, was to impose English rule over a country whose native princes were in the habit of revolting as soon as an opportunity arose. The eight castles were the strongest ever built up until then, not excluding the London Tower and Dover. Five of them—Conway, Caernarvon, Aberystwith, Rhuddlan, Flint, and Beaumaris—were turned into *bastides* in Roman fashion, although more directly modeled on contemporary practice in Languedoc, by the addition of a

141

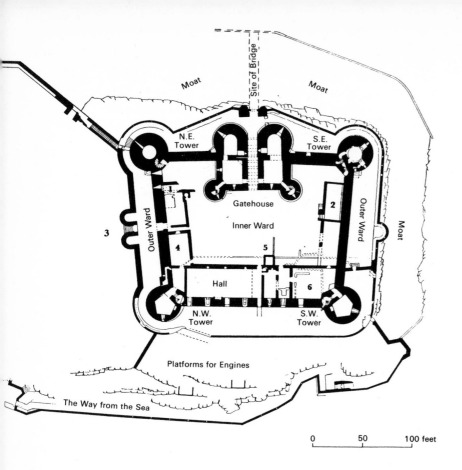

0 50 100 feet

Harlech Castle in Wales is a convenient sample for study for those interested in the details of medieval building operations, because all the details are documented. The work began on June 21, 1283, with the arrival of 10 masons and miners with their tools; in the middle of July came half a dozen timbermen and 15 more masons. Eventually, money arrived at the site— £260 before the end of October. The castle was designed by Master James St. George and constructed on contract. Master William Drydge held the contract for two towers, the north-western one for £117, with an additional £25 for the crenellations. By 1289, there were altogether 13 contractors working on the castle which had cost £9500 when it was turned over to the King in December, 1290. (1) well, (2) grain stores, (3) postern, (4) chapel, (5) staircase, and (6) kitchen.

fortified town adjunct to the castle. By the offer of economic and other privileges Englishmen were induced to settle as burgesses in the new towns and aid the castellans in their defense.

From contemporary documents it is surprising to learn at what small expense the English kings were able to pacify the previously so restless and troublesome Wales, once the castles were built. The garrison of Conway consisted in 1284 of a castellan with 15 crossbowmen, 15 men-at-arms, a priest, a blacksmith, a timberman, a mechanic, and 10 laborers. According to the payrolls for 1401 to 1404, the castellan of Caernarvon had 20 men-at-arms and 80 bowmen, while Harlech was held with only 10 men-at-arms and 30 bowmen.

The castles of Wales owed their inspiration to the symmetrical fortifications previously built in the Kingdom of Jerusalem, and in Sicily where Edward had seen them himself. The polygonal towers of Caernarvon with their decorative bands of sandstone were obviously copies of the flanking towers in the Theodosian wall around Constantinople. Similar design had been used in Savoy where the master mason responsible for six of the Wales castles, James of St. George, had been active. Characteristic of his six "patrons"—or designs—for the new castles was the replacement of the keep by rectangular or polygonal curtain walls protected by symmetrically placed polygonal or round towers, one inside the other to provide two integrated lines of defense; and a powerful, ingeniously designed gatehouse defended by a barbican. A postern was also added to facilitate sorties.

Although the principle of the design remained the same, the layout was accommodated to the site, and all eight castles differ in planning as well as in size. Of the eight, Caernarvon, sited at the mouth of the river Seiont and opposite the island of Anglesey, was conceived as the viceregal center and therefore built on a much larger scale than the others. It has a polygonal plan extending 570 feet east-west and 200 feet north-south. The curtain round the outer ward is defended by four towers; the two gates are provided with the characteristic flanking towers but lack barbicans. Hence, the defenses are not concentric as in Harlech and Rhuddlan. Work on the castle started in June, 1283, and was completed in 1330 after numerous interruptions, at a total cost of about £20,000.

The methods used by Edward I in building his Welsh castles were typical of the age, and serve to demonstrate how superior skills were diffused over Europe. His major problem was to recruit a core of skilled craftsmen competent to carry out his ambitious plans. How this was accomplished can be followed in considerable detail: In 1274, Edward's cousin, Count Philip of Savoý, had just about completed building his palace at St. Georges-d'Esperanche in Isère, designed and erected by Master James who when he came to England added St. George to his name. Master James was induced to enter into Edward's service on the promise of a stipend of 3s. a day, plus a pension of 1s. 1d. a day for his wife in the event of his death. When departing for his new service, he brought along practically all the key men who had been working with him on the palace, some 20 of whom have been recognized by name. Among them were the master carpenters Philip "Sente," William de Spinetto, and Theobald de Waus; the master masons John Francis, Giles of St. George, Adam Boyard, Albert de Menz, Gillot de Chalons, and William Seysel; Master Stephan the painter; Master Bertram de Saltu, the "ingeniator"; and the Master Fossatore Manassar de Vaucouleurs—practically all of them natives of Savoy.

Then the King issued orders to his sheriffs in all English counties to impress as many laborers as they could lay their hands on and deliver them under military escort to the sites. In 1282, for example, 150 masons, 440 carpenters, and 1,400 diggers and axmen from 23 counties were delivered to Chester for dispatch to the construction sites in Wales. On this particular campaign the sheriff in Yorkshire found it necessary to dress up the masons impressed by him in red caps and livery, "lest they should escape from the custody of the conductor." Once they arrived at the site, they were paid for their work: masons got 2s. a week, faulkners (hod carriers) 9d. a week, fossatores (diggers) 8d., women diggers 6d., smiths 2s., carpenters 1s. 8d., cart drivers 10d. The master mason was paid $7\frac{1}{2}$d. a day, and his deputy 3s. a week. Considering that a mailed knight mounted on a barded horse was paid 8d. a day and that the going rates for arbalesters and archers 4d. and 2d., respectively, these wages were not so bad.

Master George, commonly considered the most outstanding ingeniator and fortress builder in medieval times, died in 1309, but the work started by him was continued and brought to a conclusion in 1330, by which time something like £100,000 had been spent on the castles in Wales. But by then cannon had begun to be used, at Metz in 1324, Florence in 1326, and in Britain a year later. The cannon were as yet small and rather harmless; at the end of the fourteenth century a 150-pound gun was regarded as a "great cannon." It took nearly a century for the guns to be sufficiently developed in size and firing power to make a significant impression on the heavy stone curtains of the medieval castles. The first documented success in reducing a castle with a heavy siege gun dates from 1414 when the Elector Frederick I by means of "Faule Grete" managed in two days to pound a breach in the wall of the Friesack stronghold held by his feudal noble Dietrich von Quits. In England in 1464, the Earl of Warwick reduced the impregnable Bamborough castle within a week. The death blow to the old order of fortifications came that year with Charles XIII's campaign in Italy when his heavy siege guns knocked down all strongpoints encountered. Having lost their military importance, the medieval castles were turned into uncomfortable seignioral residences.

New Italian and French ideas of static defense

The development of the efficient siege gun forced reassessment of the entire system of static defense, at first in Italy where all the great men of the Renaissance—such as Leonardo, Machiavelli, Francesco di Giorgio Martini, and others —devoted considerable thought to overcoming the breaching power of the heavy guns. However, the ideas that came to have a lasting effect on subsequent developments were contained in the book *Trattato dell' architettura* written by Francesco Martini, where he deals with the entire subject of military engineering, and suggests the use of a star-shaped bastion consisting of a low masonry wall possessing an external batter in order to deflect the cannon balls. After his death in 1502, his idea was taken up by Francesco de Marchi, a professional soldier in the service of Pope Paul III. In his famous work *Della architettura militaire* published after his death in 1597, de Marchi enters into all the details of the design and construction of fortifications, including the star-shaped bastion. He concludes the work with the interesting observation that cities should not be fortified because too many strong fortresses cause too great destruction since they tend to prolong the sieges. In earlier times, he notes, wars passed more quickly, were less costly to the prince, and caused less destruction to the people.

Among the very first fortifications constructed in accordance with the new principles may be mentioned a triangular bastion in Verona's curtain built in 1520. Italian engineers built similar bastions on Cyprus in 1560, and when the Order of St. John was forced to evacuate the island and moved to Malta, it fortified Valetta with a crémaillière, or indented curtain provided with triangular bastions. The forts along the English Channel built by order of Henry VII were also constructed to resist heavy guns, although planned on modified medieval principles.

After de Marchi, the initiative in military engineering passed to France, where in 1633 was born Sebastian le Prestre de Vauban, the greatest military engineer in modern times. Vauban began where de Marchi left off; he took the simple polygonal or star-shaped bastion originally invented by Francesco Martini and turned it into a complex design of walls and moats with indented traces called crémaillère, the tenaille, star trace, or bastion trace. Inspired by Vauban, fortress engineering rapidly turned into an exercise in mathematics—in France, Holland, and Germany.

As eventually evolved, toward the end of the seventeenth century, a Vauban fortification scheme, regardless of its "trace," consisted of a masonry-faced wall called the "escarp," built with arches in the rear for increased resistance against cannon fire. The rear of the escarp was heavily banked with earth to strengthen it and provide a platform for cannon. This "rampart" was elevated to a "parapet" to protect the guns and their crews.

In front of the escarp was a wide and deep ditch which was provided with a masonry wall opposite the escarp. This was known as the "counterscarp," the top of which served as a fighting platform, the so-called "covered way," which was protected by a low earthwork—the "glacis"—sloping down toward the field. Since the glacis was lower in elevation than the rampart, it could be swept with fire from the guns placed behind the parapet and over the heads of the defenders lining the covered way, the first line of defense.

Such, briefly described, were the elements of a fortification

143

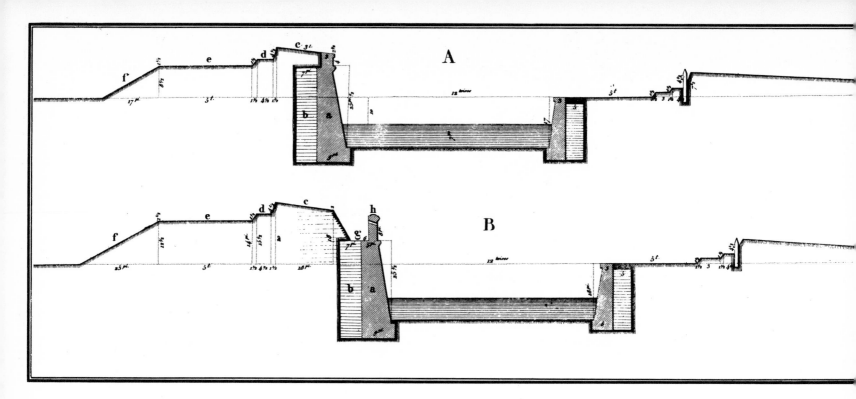

After the medieval fortification system had proved incapable of withstanding cannon fire, the sharpest brains of the Renaissance busied themselves with the development of static defense systems accommodated to the new missile weapons. Finally, 200 years later, in the middle of the seventeenth century, two military engineers—Vauban in France and Coehoorn in the Netherlands—developed, independently of each other, systems of defense that were capable of standing up to cannon siege. Above are two profiles of the works in a Vauban fortification. As encountered by the enemy, they were, from the right: the glacis, a sloping earth embankment, the covered way for the musketry, the masonry counter escarp, a deep and broad trench, the escarp with its parapet and rampart on which the guns were placed. In Profile B was a covered way also along the escarp. (From Vauban's book Défence des Places.) The trace on the next page, also from Vauban, illustrates how a siege of such a fortification should be carried out. The attack is directed against bastion No. 3, and the siege artillery has been brought up to the "première ligne," about 300 toises (600 yards) from the outer works. It keeps up enfilading fire against the ramparts of the bastion while sappers dig a set of zigzag approach trenches that cannot be enfiladed. Halfway to the works they dig a second ditch which is manned with musketry to protect the continuous advance of the sap up to the glacis, where a third and final concentric trench is established. The cannon is brought up to this "troisième ligne" and maintans breaching fire while the sap continues into the trench and the breach.

intended to stand up against heavy, flat-bored siege guns firing round iron balls, as developed by Vauban during a long life devoted to war. They incorporated his experience gained from 48 sieges and the construction of 160 fortifications. Throughout his career he kept insisting that "the art of fortification cannot be reduced to rules and systems but is solely a matter of common sense and experience," but long before his death in 1707, contemporary military writers had frozen military engineering into a Vauban mold in which it got stuck until the early nineteenth century. Throughout France and Continental Europe, every major city was fortified à la Vauban, and their existing parks and green belts are, more often than not, planted on the debris of Vauban works razed about a century ago.

As systematized by contemporary writers, Vauban was supposed to have employed three "systems" of fortification. His "First System" applied the principles developed by his Italian predecessors and was therefore no innovation. The curtain wall was short and protected by a so-called "ravelin," a detached work with double embankments placed in front of the wall and flanked by a bastion. The dimensions of the layout were accommodated to the range of musketry so that the works could be exposed to flanking fire. The covered way was laid out as a transverse to guard against enfilading fire. Bastions and curtains had an elevation 25 feet above the surrounding country, 17 feet over the glacis, and 8 feet over the ravelin. The ditches were 18 feet deep, and so was the thickness of the parapet.

In the "Second System," as used at Belfort and Landau, Vauban employed defense in depth by means of a second ditch and wall flanked by two-storied gun chambers, or lower bastions placed behind the outer works. These bastions commanded both the inner ditch and the outer bastions. Their thick walls and strongly vaulted roofs protected the cannon in a manner not possible on open ramparts. The idea of the layout was to bar the further progress of an enemy who

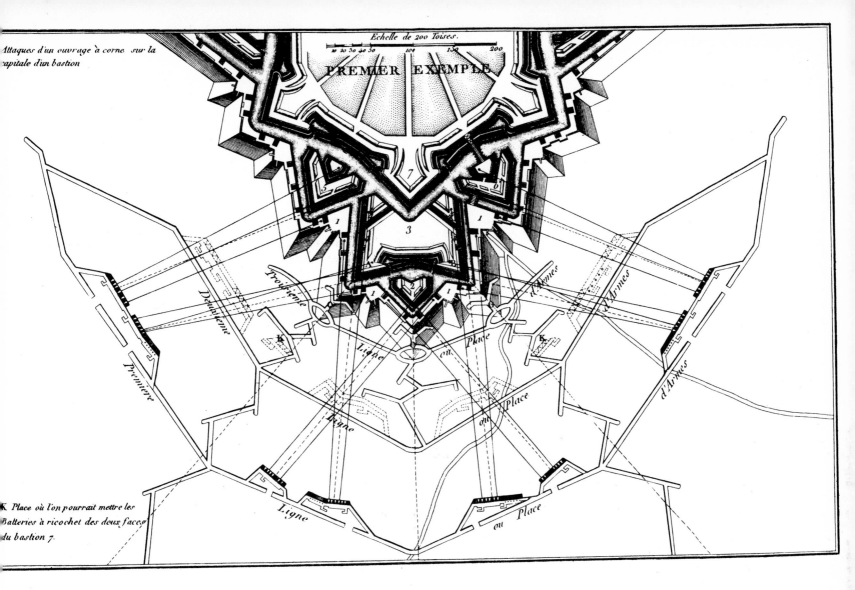

Echelle de 200 Toises.

PREMIER EXEMPLE

Troisième

Deuxième

Première

Ligne

ou

Place

d'Armes

d'Armes

d'Armes

Ligne

ou

Place

Ligne

ou

Place

K Place où l'on pourrait mettre les
Batteries à ricochet des deux faces
du bastion 7.

had succeeded in forcing his way into the outer bastions. The "Third System," finally, was a further elaboration of the Second and used by Vauban himself only once, at Neu-Breisach.

Sapping à la Vauban

It is not surprising that the great master of building fortifications was just as masterly in wrecking them. Of the 40-odd sieges directed by Vauban, all were successful. The principal siege method improved by him involved digging a system of trenches with the ultimate aim of bringing siege guns up to the glacis, from where they could breach the escarp. The trenching began beyond the range of the effective fire of the defensive guns, or 600 to 700 yards. There, a 15-foot-wide and 3-foot-deep trench was dug along a line parallel to the works. While the sapping operation was in progress, the siege guns

were placed so as to enfilade the faces of the works in front of the attack.

While the batteries of the first artillery position kept firing with the intent of silencing as many of the rampart guns as possible, the sap was carried forward with zigzag trenches, usually three sets for one bastion, which could not be enfiladed by musketry and cannon fire from the bastion. Halfway between the first artillery position and the bastion a parallel cross ditch joining the three approach trenches was dug and manned by troops to hold back sorties while the approach sap went on, still in zigzag fashion, toward the glacis. When close to the glacis the sap was joined by a third parallel trench which was manned to protect the sappers working up to and alongside the glacis. Guns were now brought up to the close position and kept on breaching fire against the escarp, while a mine gallery was driven under the covered way into the ditch. While the defenders were being held back by artillery and musketry fire from the advanced siege position, the mine

145

gallery was brought across the ditch and up to the breach.

This systematic trenching developed by Vauban gave rise to the idea that it was impossible to arrest the methodical and protected progress of the sap, and that no worthwhile resistance could be made until the besiegers had reached the covered way, which explains the complicated arrangements of the outworks. Siege warfare as carried on during the eighteenth century thus became a chess game regulated by the prevailing rules of war. A commandant was not expected to give up his fortress until a bastion or curtain had been breached and he had beaten back one assault through the breach. If he then surrendered, the garrison was treated with military honors. If he persisted in the defense and forced the besiegers to mount a major attack, he had forfeited his right to live and he and his men were slaughtered.

The entrenched camp

After the Napoleonic wars the Vauban fortifications fell into disuse, and new ideas operating with detached fortresses, using caponnières, casemates, and so on, came into favor, and became obsolete as soon as they were completed. So was the costly and fantastic idea of the so-called "entrenched camp," meant as a last stronghold of a national establishment in time of war, a refuge for the members of the royal house, the government, the central bank, and the remnants of armies defeated in the field. This idea was made obsolete by a series of experiments carried out by the French army in 1886 using the old fort Malmaison as a target for 8-inch shells containing a high-explosive charge. One shell was all that was required to destroy a casemate; a revetment wall was overturned and breached by two shells placed behind it. The military engineers thought they had the solution to that problem, and fortifications began to be constructed with steel-reinforced concrete of tremendous dimensions.

Of the "entrenched camps" built from 1850 onward, the largest and costliest were those constructed around Paris, Bucharest, and Antwerp. The Antwerp enceinte enclosing the strategic center of the national defense of Belgium, had a length of 9 miles with forts situated $1\frac{1}{4}$ miles apart. In 1914, it was captured in three days, after four of the main forts had been pounded to bits. The concrete forts around Liège, stated to be the strongest ever built, lasted 11 days. After these demonstrations the French government abandoned Paris; and Bucharest capitulated when its turn came. That was the end of the silly notion of the "entrenched camp," on which so many governments had drained their nations' resources since the middle of the last century. Those that remained after the 1914 to 1918 holocaust were turned into storehouses and military schools. But the lesson was not learned. In 1930, French military engineers embarked on the construction of a 314-kilometer-long subterranean Chinese wall extending from the Swiss to the Belgian border. Each one of the forts in the Maginot Line was sunk 325 feet into the ground and required the excavation of 750,000 cubic meters of earth and the pouring of 120,000 cubic meters of concrete. Its guns were enclosed in turrets of heavy armor plate. Only one of the forts was actually captured in 1940; the rest were outflanked and abandoned. In 1968, they were offered for sale cheap.

Of more lasting value was the development of scientific engineering to which the extensive fortification building gave rise, particularly in France, from the time of Vauban, as discussed in an ensuing chapter.

Town building on the northern marches

The fortified cities hitherto discussed have one feature in common: they began either as Roman legion camps or as Celtic tribal oppida that were turned into civitates by the Romans. East of the Rhine and north of the Danube reigned a rustic wilderness populated with heathens who had their local cult places which also served as markets for barter trade a few times a year.

The gradual conversion of these rustic centers into wall-enclosed towns follows a common pattern that is altogether different from that recognized above. They began as loci for the missionary work carried on by the Church. A missionary priest sent out by his bishop, who resided in some old Roman cité to the west of the Rhine, established himself with the heathens, and when he had succeeded in converting some of them, he built a wooden church. As the mission work became stabilized, the timber church was replaced by a larger one, usually built of brick. In time, a castle was added to house a missionary bishop who directed the work farther east and north.

The missions which eventually succeeded in converting the heathens residing around the Baltic were originally established by Charlemagne, who ordered a keep to be built on the Elbe estuary in 809, followed by a church a few years later. The missionary bishop established at what became Hamburg was given full administrative powers over the area and the town that soon began to cluster around the keep. It was from this keep that the conversion of Scandinavia and the Baltic was directed and, as a consequence, the area was opened up to German traders. In this manner one mission center after the other got its church and keep, nearly all of them built by German masons, and the towns growing up were settled by German burghers. Within the course of a century a whole string of German towns developed and were given a charter: Lübeck (1157), Rostock (1218), Wismar (1229), Stettin (1243), Königsberg (1255), Stralsund (1234), Riga (1158), Stockholm (1255). Those west of the Oder formed themselves into a protective association—the Hanseatic League—which held a monopoly on the Baltic trade for several hundred years and provided the capital and simple skills required to develop the untouched natural resources of this backward corner of Europe. But the building skills brought along by the recently urbanized carpetbaggers were not very impressive; distinguished building had to await the arrival of the Cistercian Brothers. Houses were built of wood with turf roofs, most of them lacking chimneys, and placed helter-skelter around the church fronting the market square.

In the Baltic basin, the island of Gotland with its fortified town of Visby forms an exception to this pauvre pattern. The islanders grew tremendously rich on the entrepôt trade in Russian furs and oriental produce with western Europe, since the southern route was closed by the Arabs, and were in a position to call on western master masons to aid them in erecting the keeps, churches, houses, and sculptured monuments enhancing their Christian status. But since no records have survived, these Gotland masters will forever remain anonymous ghosts.

With this digressive note on the deferred urban developments on the northern marches, of about the same insignificance from a technological point of view as the early American prairie towns, the remainder of this chapter will be devoted to the subsequent growth of the major cities in western and southern Europe as they began to expand beyond their medieval walls after 1500.

Tale of three cities

To begin with there was Rome, which during the fourteenth century had slid into a state of utter misery. The population had dropped to 17,000, and large parts of the city within the Aurelian wall had turned rural in character. Sheep grazed in the uninhabited valley between the Palatine and Aventine hills, and whatever houses remained among the ancient monuments consisted of hovels above which towered the battlemented keeps of the nobility.

After the return of the popes from Avignon, a program of reconstruction lasting nearly 200 years got under way. Pope Paul III began by straightening Via Flaminia around 1450 and turned this ancient road into Il Corso, running from the Arch of Marcus Aurelius (later Porta del Popolo) to the Church of San Marco. His successors continued what he had started. Sixtus IV ordered a master plan prepared, and after the turn of the century Via Giulia and Via Ripetta were laid out among the hovels. The work of reconstruction received a serious setback from the sack of Rome by the French in 1527, resulting in a rapid decline in population from 90,000 to 30,000. Thus, when Paul IV resumed the work in 1555, four of the seven hills were unoccupied, namely Caelian, Viminal, Esquiline, and Quirinal; and existing streets and houses followed the ruins of the ancient aqueducts, which

facilitated the scheme of his great successor Sixtus V who in a brief five years brought about the transformation of Rome.

He gave Dominico Fontana a free hand to go ahead and open up the city. Fontana laid out the main streets now converging on the great basilicas Santa Maria Maggiore and St. John Lateran, namely Via Sistina, Via Felice Quattro Fontane, and Via di Porta San Lorenzo (subsequently razed to provide space for the central railroad station), in addition to the Lateran and Esquiline piazzas. He also laid out Via del Babuino beginning at Piazza del Popolo, whereby a visitor entering the city from the north was offered three main streets, each one leading to a different part of the city. Finally, a succession of popes from Paul V to Alexander XII employed Lorenzo Bernini during the first half of the seventeenth century to lay out the colonnaded Piazza of St. Peter and Piazzas Navona, Colonna, and Berberini. This final spurt of papal effort gave Rome the baroque flavor that it still possesses.

While Berberini's work was reaching conclusion, London burned and, as a consequence, came close to being converted into the most magnificent baroque city north of the Alps. For six days, beginning on September 2, 1666, the fire raged and reduced the city of London to ashes. By historical accident the catastrophe was witnessed by two of England's greatest architects, John Evelyn and Sir Christopher Wren. While the fires were still smoldering, actually in less than a week, both architects presented excellent schemes to King Charles II for rebuilding the city on a baroque plan. Either plan would have vastly benefited London during the following centuries. But nothing came of them because there was at the time no ruthless Nero or Sixtus in power in England. The authorities bent to the frantic anxiety of the little men and shopkeepers in the city, the King and his commissioners were aghast at the expense and legal difficulties involved in settling thousands of property claims, and the unique chance was lost. The city was rebuilt on its old medieval plan. The only lasting benefit derived from the conflagration was the canalizing of the Fleet River, later turned into an underground sewer.

The difference in the response to a similar challenge, between the popes and the Royal Commissioners of Charles II is worthy of notice, but not much more because of the futility of taking sides. The lovely perspectives of Rome, the long broad vias lined with grand baroque palaces were bought with the proceeds of sordid graft, the plunder of pilgrims, the sale of remunerative papal offices to the highest bidders, and a callous disregard for the people made homeless by the razing of their shanties. To prevent the slums from once more creeping up to the new vias, the popes forced the corrupt officeholders to part with some of their loot and spend it on palaces lining the streets. One should not be too quick to condemn the Crown commissioners for their timidity. Social morals are always to be preferred to architectural splendors. Indeed, when the papacy reformed, the popes ceased to be

Antonius Tempesta's map, "Urbis Romae Prospectus," published in 1593, indicates how far the reconstruction of Rome had progressed by the time the Baroque popes had given the city its present layout by the middle of the next century. Via del Corso (1) had been completed by Pope Paul II from Porta del Popolo to the bottom of Capitol Hill, Leo X had laid out Via Ripetta (2) to the Tiber, and Fontana had pushed through Via del Babuino (3) to the large open place marked "Platea Trinitatis" (now Piazza di Spagna). The street named Via Trinitatis is now called Via dei Condotti east of the Corso and Via Borghesi west thereof. This northern sheet of the map ends to the north of Piazza Colonna.

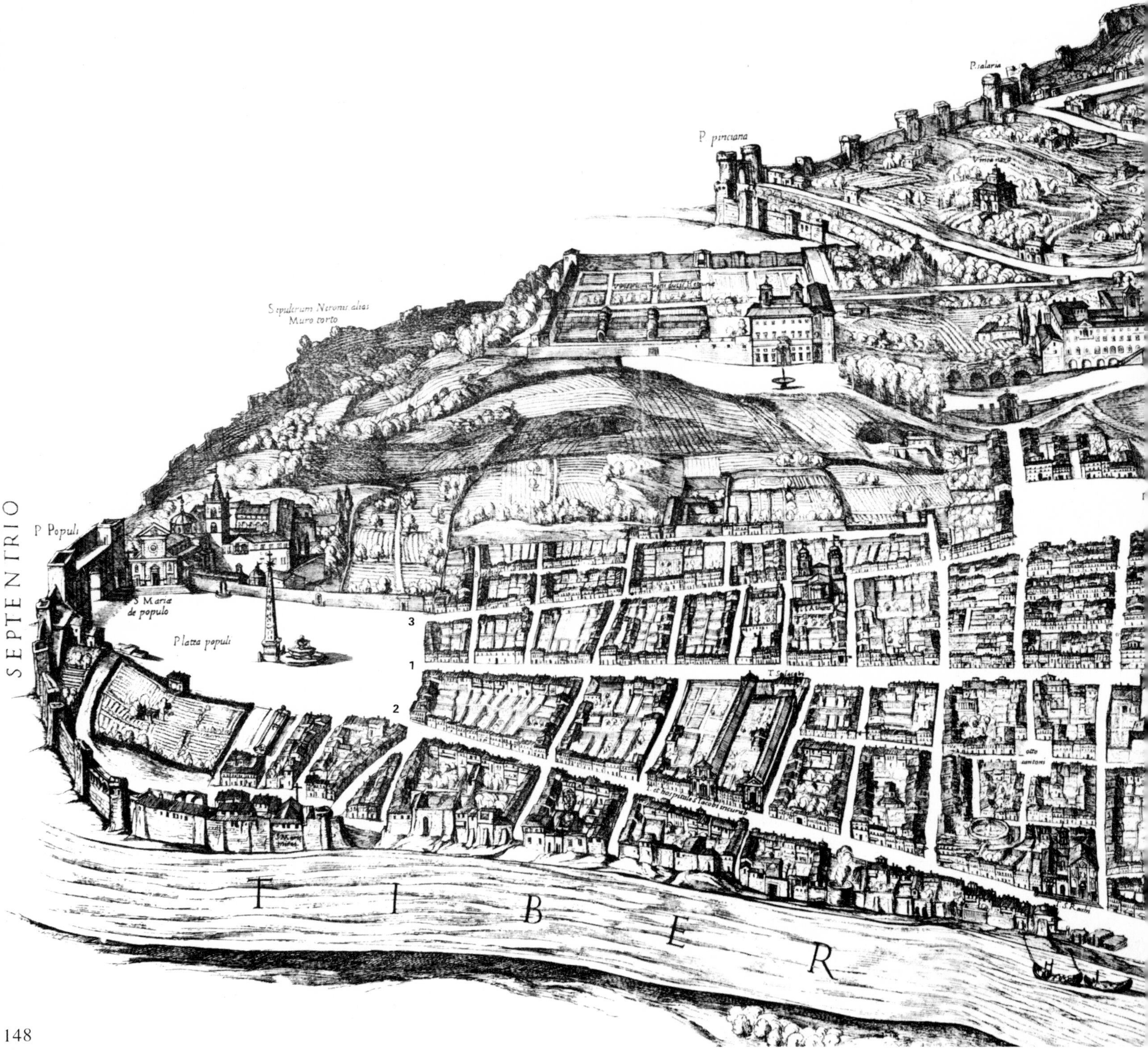
P.salaria
P pinciana
Sepulcrum Neronis alias
Muro torto
SEPTENTRIO
P Populi
S Mariæ
de populo
Platea populi
3
1
2
otto
cantoni
T I B E R

concerned with physical reconstruction. The slums returned; and for nearly 300 years nothing was done to stop the decay until a political scoundrel completed what the papal rogues had left undone.

Of the major European cities breaking out of their medieval enceintes, Amsterdam surely serves as a model for a moral, technically admirable, and well-nigh perfectly planned expansion, despite fearful odds. With the exception of Venice, there is not anywhere in Europe a site so ill situated for town building, and only the Dutch with their mastery of hydraulic engineering would have been able to deal with the formidable difficulties presented by a 50-foot layer of mud on top of 10 feet of sand. The entire city therefore rests on pilings.

Amsterdam began its civic life in 1204 with the building of a keep protected by a dam on the east side of the river Amstel where it flows into the Ij, an inlet of the Zuider Zee. In the usual fashion houses began to cluster around the keep, and when the city received its first charter in 1300, the opposite shore of the river had also been occupied. The built-up area was enclosed by a defensive ditch and wall—the Voorburgvallen—in 1342. Forty years later a second ditch was dug about 60 yards from the first one and connected to the Amstel and the Ij with sluice gates. In 1481 the wall was for the first time provided with towers, only one of which—the Schreijerstoren—now remains. Inside the enceinte, houses built on 40-foot pilings were erected along the canals.

During the ensuing century the city grew and prospered, partly owing to the hospitality offered to numerous refugees from Antwerp and Brabant, driven to flight by religious persecution. To obtain more space, the city in 1610 entered on the most ambitious town planning program undertaken in Europe up until then, whereby the building area was quadrupled. The bastions holding back the expansion were demolished and replaced by a new canal—the Heeren Gracht—forming a polygonal trace around the medieval city. Further beyond and parallel to it were dug two more canals—the Keizers Gracht and Princen Gracht. These three tree-lined canals with their fine old patrician houses still form the principal thoroughfares of Amsterdam.

Then more refugees arrived, this time Huguenots driven out of France after the revocation of the Edict of Nantes. They were settled to the west of the Princen Gracht, and around 1670 a 6½-mile-long ditch, known as the Singel Gracht, was dug around the greatly expanded building area. Beyond the ditch was constructed a continuous chain of bastions à la Coehoorn. These fortifications were razed a hundred years ago, and the area was laid out with gardens.

With this kind of civic enterprise it is not surprising that Amsterdam prospered and its influence spread beyond the seas. That the city became a world center also for the arts, music, and philosophy followed, as the inevitable blessings of inviting the victims of human beastliness to find refuge on its many man-made islands. But that is another story.

149

Roads and Bridges and Harbors

"They were drawn by his direction through the fields, exactly in a straight line, partly paved with hewn stone, and partly laid with solid masses of gravel."

PLUTARCH

When the Italian engineers surveying Strada del Sol brought their traverse from the hills down into the valley they unwittingly followed a route that their Roman colleagues had picked out 2,000 years earlier. When construction started, the Roman road bank and pave of a long-forgotten road were laid bare. This Roman construction could not be missed by a man riding a bulldozer; a more subtle investigation might have revealed a cattle trail prior to the Roman road, perhaps also a wild animal trail before that.

This is not to say that all existing roads have had their beginnings as animal paths, but a great many have for the simple reason that animals, particularly bovines, have an uncanny way of picking the most comfortable grade between two points, avoiding obstacles presented by water or precipitous ground. In Britain, these prehistoric tracks followed the downs and open country in a sinuous course, avoiding on the one hand the forest-covered sides of the valleys and on the other the swampy lowlands. In parts of Hampshire and Wiltshire, these prehistoric roads are locally known as "hollow ways" because after thousands of years of use they have turned into deep grooves. The Icknield Way, from Norfolk to Grimes Graves, must have possessed some commercial importance for the flint trade in Neolithic times.

However, some ancient roads were man-made from the outset. A number of them are on the island of Malta, which was an important center in the Mediterranean in Neolithic times, judging from its numerous megalithic tombs. On this island there are rut roads cut into limestone rock. The ruts are 10 to 20 inches wide at the top and 4 inches at the bottom and the track, i.e., the distance between the ruts, is 4 feet 6 inches. These Maltese rut roads most likely began as *travois* trails subsequently deepened by cart wheels, and have been interpreted as having been used as haulage ways for earth and water to plots on the sterile hillsides. But it is equally likely that the ruts were cut when the hills were covered with soil and served to bring down produce to the settlements along the shore. They possess considerable in-

terest from a technological point of view because they are the forerunners of similar roads in Greece as well as the "stone railways" in England prior to the Industrial Revolution which inspired the tramways.

The ancient empires of Egypt and Babylon were not road builders. They relied on water for transport. Whatever roads they possessed were short and were built for the purpose of hauling construction materials to major building sites. According to Herodotus, the Egyptian hierarchy spent 10 years building a five-furlong causeway from the western shore of the Nile to Gizeh. It was 10 fathoms wide and the highest elevation of the road bank was 8 fathoms. When the roadway after serving its original purpose was turned into a state funeral road connecting the shore temple at the river with the mortuary temple to the east of the Great Pyramid, it was enclosed by a wall of "polished stones decorated with carvings."

Similarly, there were parade streets leading up to the temples and palaces in the Euphrates valley; the market squares in the nome centers of Egypt were also paved with flagstones. The Hittite capital Hattusas had a processional road ca. 1200 B.C., and on Crete there was a track system, part of it graded and even paved, connecting Knossos with Phaestos and a port on the Gulf of Mesara. The origin of this strategic roadnet has been dated to about 2500 B.C. As for the rest, the roads of the early empires consisted mainly of tracks maintained and policed to enable runners to speed along them. The courier system of the Egyptian administrations was able to convey news of the Nile in flood at Elephantine to Memphis at an average speed of 11 kilometers per hour at the end of the Old Kingdom. Two and a half centuries later, Hammurabi ordered his man in Larsa, 200 kilometers distant, to appear before him within 48 hours. The bureaucrat was thus expected to travel behind a team of onagers pulling a chariot provided with solid wooden wheels at about half the speed of the professional Egyptian couriers. It could not have been a very pleasant journey.

Work on Via Appia began in 312 B.C., and it is therefore Italy's and Europe's first-built conventional road. Originally, it ended at Capua, 162 miles from Rome, but in 244 B.C. it was extended to Brindisi. The first mile from Porta Capena was paved in 191 B.C., and both sides of the road soon became burial grounds for Roman aristocrats who lined Via Appia with impressive funeral monuments.

Castorius' road map

On the following eight pages is reproduced a Roman road map showing the roadnet between Rome and the English Channel. The scale used in reproducing the original is 1:1.6. The map was compiled ca. A.D. 365 by a Roman cosmographer named Castorius of whom nothing further is known. He used as the basis of his map a "mappa mundi" drawn at the initiative of Agrippa and for many centuries used by the central government, provincial administrations, legion commanders, and other officials. Castorius added the roadnet and other features not included in Agrippa's map. The chief value of Castorius' compilation lies in the accurate distances given, the spelling of place names, the relative commercial and administrative importance of the towns, and the like, whereas topographical details have been largely neglected. The map contains altogether 68,651 distances given in mille passum (1,480 meters), of which about 70 percent are correct within one or two kilometers. It gives Alpine passes and 534 town symbols, of which 311 are in Europe. There are three imperial centers (Rome, Constantinople, and Antioch) and six empire fortresses, their importance marked by the number of turrets. Provincial capitals are denoted by two turrets, a button on a tower indicating that the town was also a commercial center. The garrison towns and the baths are also noted, and important harbor towns are represented by lighthouses. The map was originally in seven colors, but they have become faded and altered with time.

From the lettering, it appears as though the map shown was copied from the original Roman edition some time during the early eleventh century in a monastery in southern Germany. It has obtained its name "Tabula Peutingariana" from Konrad Peutinger, courthouse recorder and historian residing in Augsburg. In the summer of 1507, he was visited by Emperor Maximilian's librarian, Conrad Celtes, who entrusted a map roll to him for safekeeping and asked him to publish it after his death. How and when Celtes acquired the map still remains a mystery, but the likelihood is that he had removed it from some monastery and kept it instead of turning it over to the imperial library. Celtes died in 1508 and Peutinger neglected to publish the map. It vanished from view and was not found until 1714 when an inventory was made of Peutinger's old library. A bookdealer, Paul Küz, bought it at a modest price, and when he died, his heirs sold the map to Prince Eugen of Savoy for 100 ducats. Seventeen years later, Emperor Charles IV purchased Eugen's library from his widow, and the tabula was at long last incorporated into the imperial library. The complete roll, which measures 23 feet in length, was cut up into 11 sheets in 1861, and the four westerly sheets here shown have been photographed in the Vienna State Library.

151

Canterbury / Dover · Saintes · Leiden · Kassel · Rennes · Rouen

Bagnères-de-Bigorre · Bordeaux · Toulouse · Périgueux · Cahors · Limoges

Besancon Bonn Metz Trier Avenches Mainz

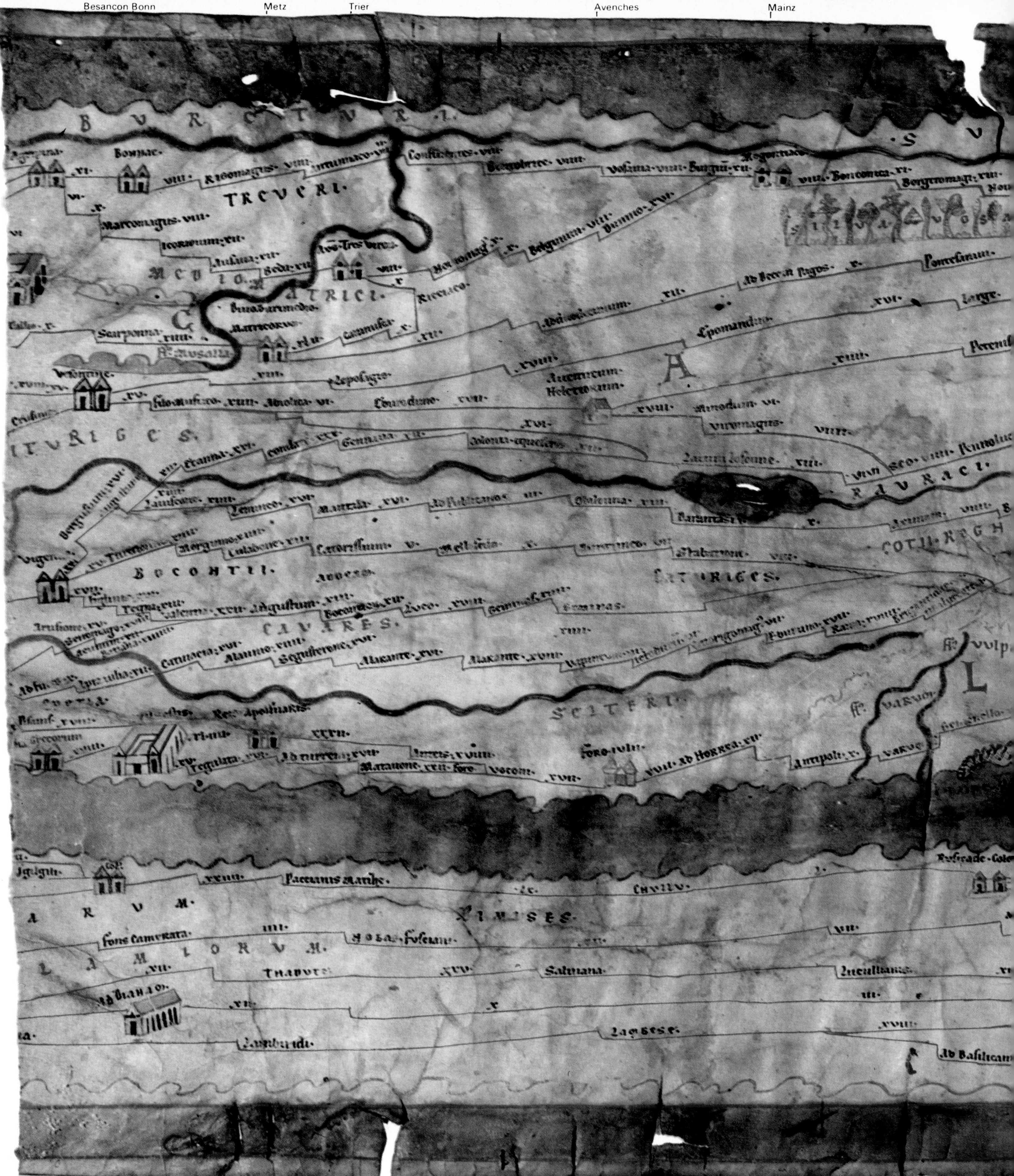

Vienne Marseille Aix Fréjus

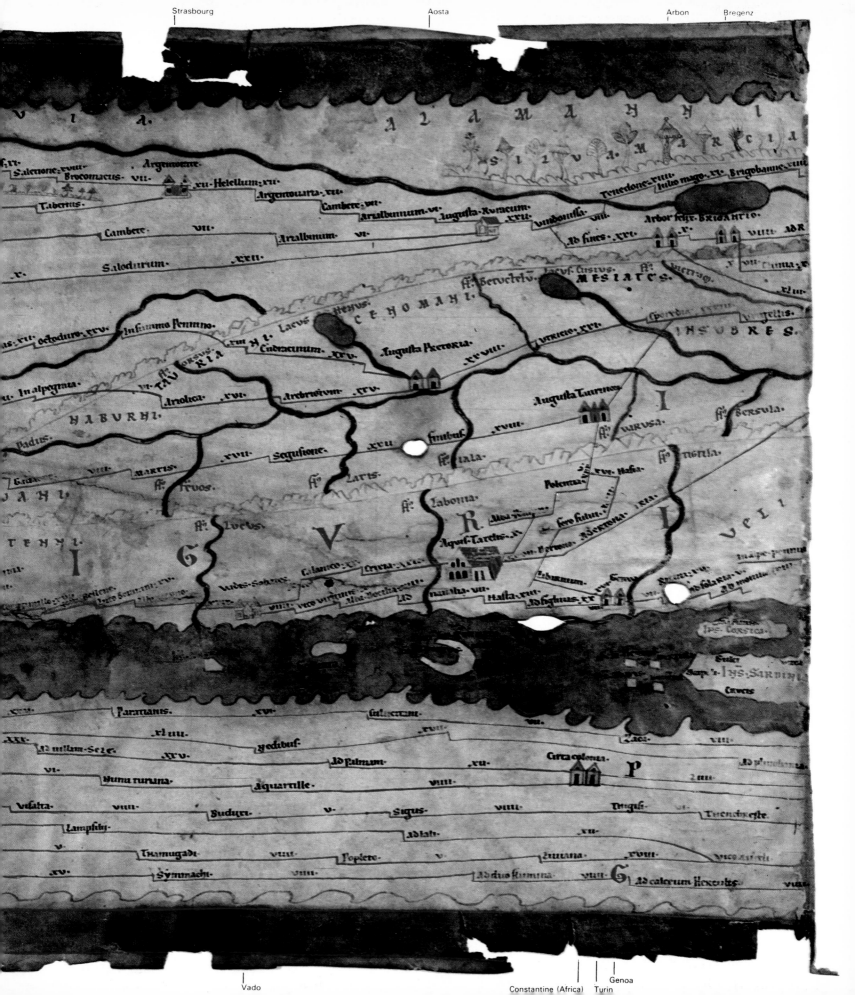

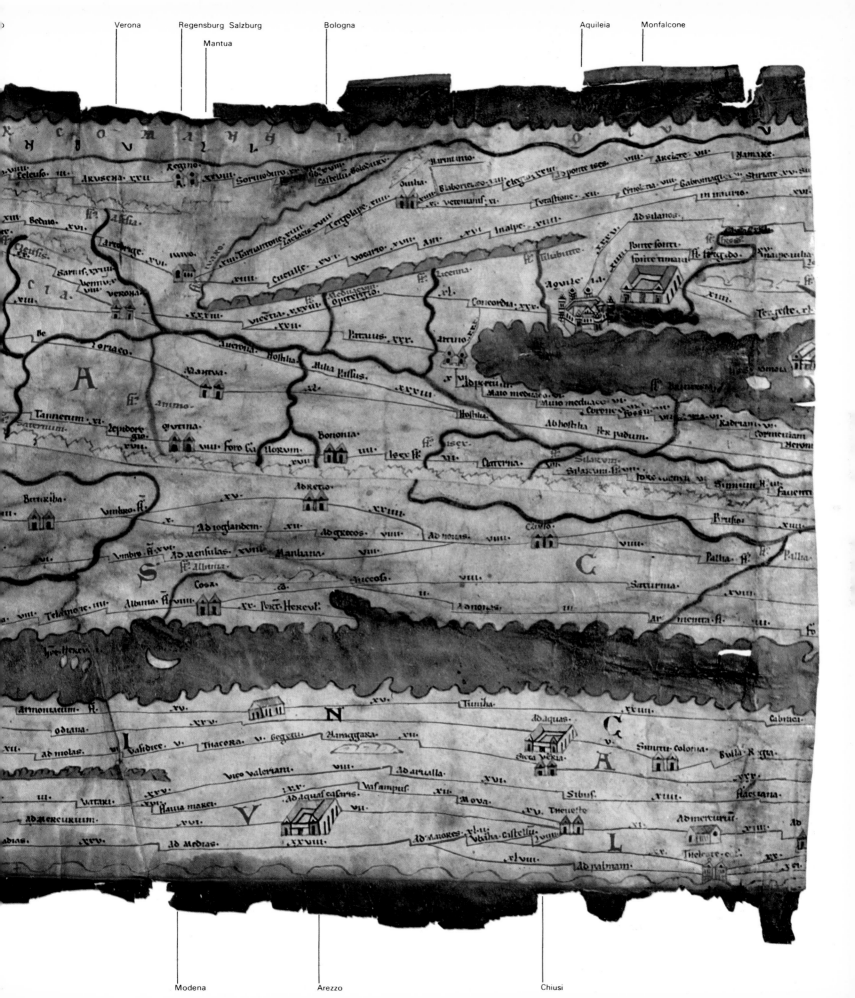

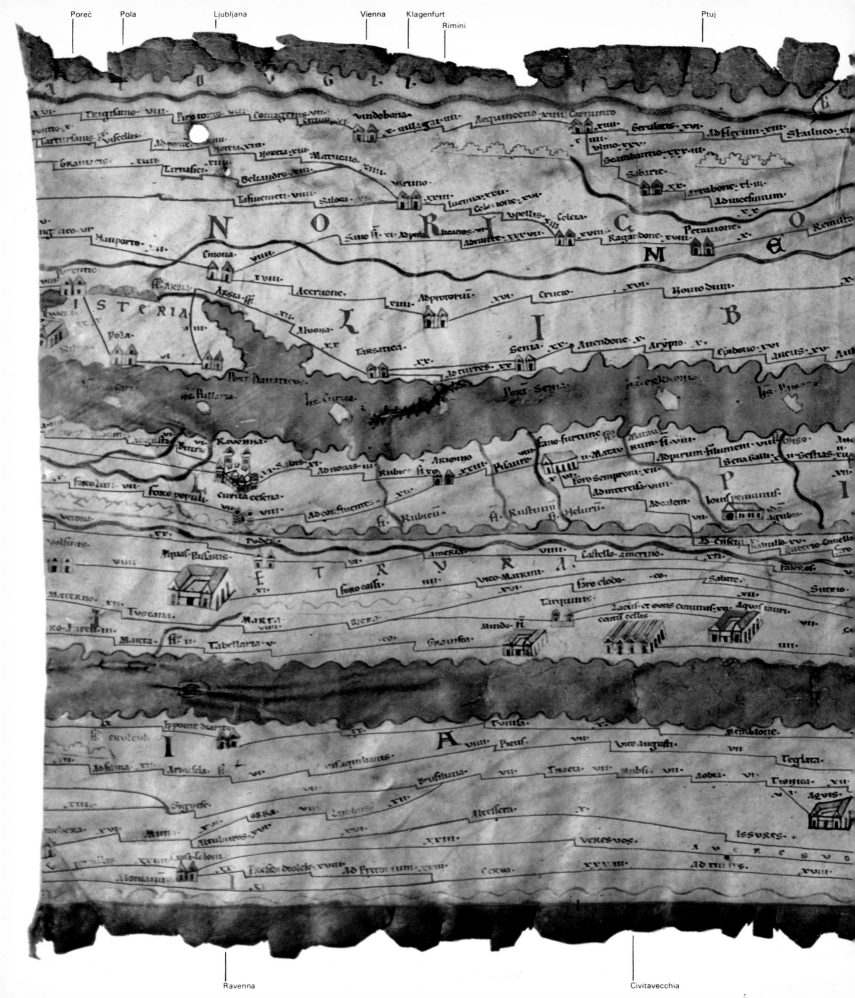

Poreč Pola Ljubljana Vienna Klagenfurt Rimini Ptuj

Ravenna Civitavecchia

Zadar
Spoleto

Potenza

Dubrovnik

Tivoli

Bagni di Stigliano

Acquaviva

Vaticano

Rome
Portus
Carthage

Ostia

Carthage was founded in 814 B.C. on a promontory on the coast of North Africa where the Phoenicians had maintained an emporium and factory since 1600 B.C. To the north and south of the point were two roadsteads for anchorage of large ships. These have now turned inte shallow lagoons. The town of Karthadshat ("the new town") was built on the point, while the citadel, called Byrsa after the 4,225-foot-long strips cut from an oxhide, was sited on a 180-foot-high hill about 660 yards from the shore, which was lined with quays protected by a mole. At present Le Kram, to the south of the citadel, were two man-made basins, one round and one square. The round one, known as Cothon, was reserved for naval vessels and had slipways for 220 ships. Of this only a shallow lagoon remains, as shown above. To the right is a harbor view of Paris from the Nürnberg Chronicles. Both pictures give—each in their own way—a false impression of the historical reality they purport to represent.

160

In respect to road communication, the western empires remained undeveloped long after road transport had been thoroughly organized in China some time prior to 1100 B.C. During the Chow Dynasty it was further improved by standardizing the roads into five classes: (1) pathways for men and pack animals, (2) roads for narrow-gage vehicles, (3) wagon roads, (4) two-lane wagon roads, and (5) three-lane highways. The road system was centrally administered with local highway authorities responsible for enforcing the traffic laws. The vehicles had to conform to standard specifications, speeding was forbidden, and special rules applied to road crossings. Passengers and goods traffic proceeded by poststages.

In the western empires the further development of the roads, as different from paths and tracks, was closely associated with the introduction of the horse into the Mediterranean basin. The ass had been used as a pack animal in Neolithic Egypt, and onagers were employed as draft animals in Sumeria. The domestication of the horse and the development of bridling and harnessing appears to have occurred on the steppelands of Asia; and the Egyptians, to their regret, did not know about the horse until they met him in the shape of fear-inspiring war stallions harnessed to the light chariots of the barbarian Hyksos suddenly appearing out of the dust clouds in the eastern desert. The Hyksos rule over Egypt lasted somewhat more than a hundred years, but it left an indelible imprint on Egyptian history.

From their temporary masters, the Egyptians learned the use of the horsed chariot which in their hands revolutionized military tactics and set the pharaohs of the Middle Kingdom on the path to foreign conquests and empire building. Inside Egypt the horsed chariot brought about an equestrian state, a corps d'élite to which only the wealthy aristocracy could belong. The chariot officers took on an increasingly greater social importance, and new guilds—cartwrights, saddlers, metalworkers—developed. Even the artists were influenced; movement and liveliness of composition became characteristic of the imperial art.

It was the same all over the Middle East as one empire after the other rose on the débris of the preceding ones. In Hatti, Babylonia, Assyria, Persia, chariotry became the principal instrument in the imperialistic wars that ravaged the Fertile Crescent during the Bronze Age and long afterward. The chariot also enabled the conquerors to organize and administer their vast realms owing to the immensely improved communications provided by it.

Although speedy couriers in the Emperor's service covered vast distances at speeds that were not exceeded until Napoleonic times, they drove their horses over narrow chariot tracks rather than roads. Even the "Great Road of Empire" from the Delta via Gaza, and Meggido to Mesopotamia, remained a track until the upstart Assyrians came upon the scene. The Assyrian kings building the first empire on iron had a need for roads to carry their clanking siege trains before which the ancient cities of the Middle East crumbled into dust. Tiglath-pileser I (1115–1102 B.C.) boasts of his campaign in the mountains of Elam: "I took my chariots and my warriors over the steep mountains and through their wearisome paths I hewed a way with pick axes of bronze to make a passage road for my chariots and troops." Elsewhere his engineers built pontoon bridges and widened and graded the ancient tracks for his wagons and siege engines.

What the Assyrian administrations started, their successors the Persians brought to early perfection. Cyrus (550–530 B.C.) ordered a survey made to determine just how far a horse and rider could travel in one day (horses had first been ridden in Assyria about three centuries earlier). On the basis of this survey he organized a courier service throughout his far-flung empire. Posting stations with relays of horses and grooms were built one day's ride from each other, or about 25 kilometers. When fully developed, the Persian "Royal Road" joined Susa and the other imperial palaces at Persepolis and Ecbatana and continued via Babylon, Nineveh, Nisibis, Tarsus, the Cilician Gates, Laodicea, to Ephesus and Sardis, the administrative outpost in the west. The distance from Nineveh to Sardis was 2,600 kilometers and was covered by imperial couriers in nine days, or on the average 12 kilometers per hour. Again, it was not a built road like the Roman ones, which makes Cyrus' courier service even more impressive.

To find roads in a true technical sense, one has to follow the Minoan line of descent, via Mycenae and Greece. The Mycenaeans, after having destroyed the Minoan power, became active road builders, and their capital was joined with the other commercial centers by an extensive roadnet extending from Nauplia via Corinth to distant Athens. These were *built* roads, part of which were paved with polygonal slabs in the Minoan tradition, with culverted causeways crossing narrow watercourses.

The Greek roads were, like all previous ones, very narrow with a few outstanding exceptions. Transport from the mines and quarries was facilitated by wider roads suitable for wheeled vehicles. But the most interesting of the ancient Greek roads were the so-called rut roads, resembling those on Malta but technically far superior. They were the sacred roads leading from Athens to Eleusis, from Sparta to Amyclae, from Elis to Olympia, and those converging on Delphi.

These roads have been called "railways in reverse," since the ruts were carefully cut to a width of 8 to 9 inches and a depth of 3 to 4 inches, sometimes deeper to obtain better grade. The gage between the inverted rails was 4 feet 11 inches, corresponding to the standard track of the carriages. At certain points the roads were provided with sidings and crossings. Two vehicles reaching a crossing at the same time frequently gave rise to violent quarrels such as the one between Oedipus and King Laius which inspired Sophocles to write his famous tragedy.

This Piranesi engraving shows the pave (summum dorsum) and the curbstones (umbones) of Via Appia. The polygonal basalt blocks were laid in mortar. The width of the driveway was only 12 feet on this road. Later ones, such as Via Salaria, were 20 feet wide along stretches with heavy traffic, and this width appears to have been standard in Syria. Via Appia had a total length of 360 miles, 120 of which were paved.

Roman road building

The Romans were, of course, the incomparable road builders of ancient times; indeed, only recent generations have been their equals and in some instances surpassed them. Under a succession of emperors, more than 50,000 miles of roads were built joining all parts of the empire, from Hadrian's wall to the edge of the Sahara, from Morocco to the Euphrates. Since western Europe in Roman times was sparsely populated (about 8 persons per square kilometer, as against 23 around the Mediterranean) the network of roads was much thinner in the west than in the east. The main traffic arteries consisted of 29 military and 343 other roads terminating in Rome. The system was catalogued in the *Itinerarium*

Antonini, compiled about A.D. 200, which gives the military stations on the main roads and the distances to Rome.

The roads were divided into five classes:

Via—a road permitting passage of two carriages, 14 feet wide

Actus—a single carriage road, 7 feet wide

Iter—a road for pedestrians and horsemen, 5 feet wide

Semita—a 2½-foot path

Callais—a mountain herder path

The first via, on which construction began in 312 B.C., was built by the censor Appius Claudius, from whom it obtained the name Via Appia. It ran from Rome to the armories of Capua, a distance of 162 miles. It was at first a gravel road, but in 295 B.C, it was paved with polygonal basalt blocks. During the reign of Julius Caesar it was extended from Capua to Brindisi, whereby the total length came to be 360 miles, 120 miles of which were paved.

The censor Gaius Flaminius connected Rome with the Po Valley by Via Flaminia, begun in 220 B.C. From this time on, the roadnet kept increasing until all Italy was crisscrossed by

The map below shows the Roman roadnet on the mainland and on the islands of Sicily, Corsica, and Sardinia. The well-developed net on Sardinia is due largely to its important mining industry. Although expanded and accommodated to motorized traffic, the layout of Italy's present road system agrees in the main with the Roman. Important vias *are indicated by heavy lines;* actus *roads with thin lines.*

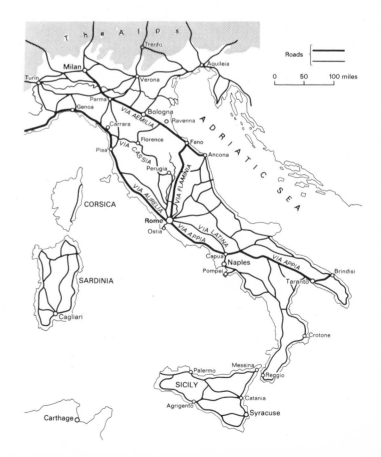

a net of roads feeding the vias, which were extended into Dalmatia in 145 B.C., into Asia Minor in 130 B.C., and into Gaul in 120 B.C. The roadnet kept expanding until A.D. 200, after which it began to shrink. By A.D. 400 the central road administration was no longer capable of maintaining even the existing roads, and landowners along the vias were charged with the duty of repairing roads and bridges.

The construction of a Roman road followed definite technical specifications. A *via* began with the digging of two parallel trenches, between which the topsoil was stripped down to firm ground. The excavated space was backfilled with fine dry earth which was well tamped to form a base layer called *pavimentum*. On this was placed a layer of fine-mesh rock aggregate, either laid dry or in mortar; this was the *statumen*. Then a concrete was poured, consisting of a gravel aggregate mixed with two parts of lime; this was the *ruderatio*. Upon this was laid another aggregate consisting of lime, chalk, broken tile, gravel; or a mixture of sand, gravel, and clay; in fact any suitable local material—even slag from nearby ironworks. This was the *nucleus*. Finally, the wearing surface, or *summum dorsum,* was put in, usually consisting of rectangular or polygonal stone slabs on arterial roads, or gravel and lime concrete on less important ones. The summum dorsum was cambered to shed water and provided with curbstones.

The five layers raised the road considerably from the ground on both sides of it, from which it came to be called a *highway*. On both sides were shallow trenches, sometimes 80 feet from each other, to drain the roadbed.

This was the best practice which appears to have been followed on the principal Italian roads. Elsewhere, as in Britain and Gaul, the roadbed was built up without stripping the topsoil, and the road was topped with gravel. Paving was only laid in the vicinity of the cities. This class of road was known as *via glarea strata*. In Africa and northern Syria, a third type of construction was used, utilizing the desert surface. These roads were called *via terrena*.

In mountainous country, Roman engineers followed the ancient practice of cutting wheel ruts in the rocky surface, no doubt to prevent laden wagons from slipping. Such Roman rut roads are found in France and Yugoslavia and vary in gage from $3\frac{1}{2}$ to $5\frac{1}{2}$ feet, thus reflecting local customs in wagon building.

Cuttings were of course resorted to, and some of these are impressive, even to a modern bulldozer driver. On Via Cassia, at Lake Barcciano near Rome, an unknown engineer made a 5,000-foot-long cut to a depth of 66 feet and a width of 20 feet; on Via Appia near Terracina there is a 120-foot cut where the engineers marked the rock face with the depth of the cut every 10 Roman feet. On a few occasions, the road engineers even drove tunnels such as the Petra Pertusa in the Furbo Pass on Via Flaminia. The tunnel is 131 feet long, 16 feet wide and high, and was driven in A.D. 78 by order of Emperor Vespasian. The longest road tunnel built in the ancient world was the 1,000-pace Pausilippo tunnel between Naples and Pozzuoli driven by Cocceius in 36 B.C. It was originally intended for pedestrians only, but was repeatedly enlarged until it obtained the dimensions of 25×30 feet.

A characteristic feature of Roman roads, in Britain and elsewhere, is their straightness. The London-Chichester road, for example, runs for 13 miles on a tangent leading to Chichester, after which it angles slightly to the left to avoid topographical obstacles. Then it runs on another long tangent before angling to the right on the last leg leading to Chichester. The staking of such a road through wooded country must have been exceedingly laborious. The 13-mile tangent suggests that after a number of trials, a straight line had been established between the two ends, separated by a distance of nearly 60 miles, after which the necessary adjustments imposed by topography were made.

A Roman via *bed consisted of four or even five layers. After the topsoil had been stripped to firm ground, the excavation was backfilled with well-compacted dry earth, the* pavimentum *(1). Then came a layer of finely crushed rock laid in mortar, the* statumen *(black line). On this was poured a concrete of gravel aggregate, the* ruderatio *(2); and then a layer of aggregate consisting of lime, broken tile, gravel, even slag from nearby ironworks, the* nucleus *(3). The cambered wearing surface, or* summum dorsum, *consisted of polygonal stone slabs (4) on roads in Italy, whereas elsewhere in Europe it was of gravel. The road was lined with curbstones,* umbones *(5).*

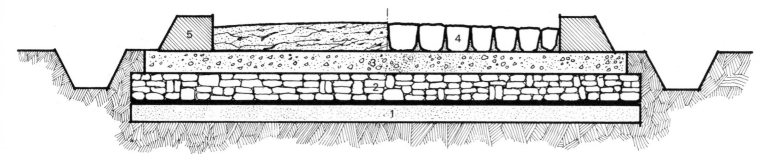

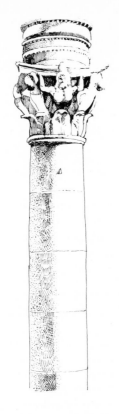

The end of Via Appia in Brundisium (Brindisi) on the Adriatic was marked by two sculptured columns, one of which remains in good condition.

The Roman predilection for straight roads developed early. Gaius Gracchus (153–121 B.C.), who introduced the practice of placing milestones spaced 1,000 paces apart, had his roads staked in straight lines through the fields. As put by Plutarch, "When he met with a valley or deep watercourse, he either caused them to be filled with rubble or had bridges built over them, and so well leveled that all being of equal height on both sides, the work presented one uniform and beautiful appearance."

Via Aemelia, running in a straight line from Parma via Bologna to Fano—a distance of 150 miles—served as an axis for the homestead scheme designed for the settlement of veterans on the Lombardy plains, but elsewhere in the Italian heartland, the roads, although run on long tangents, were always laid out in conformity with the topography, in order to avoid bridges and excessive cuttings. In the eastern Mediterranean, the Roman engineers followed the ancient routes and merely improved the existing roads, although former fords were spanned by bridges.

In peacetime, roads were built by the army which needed good surfaced roads capable of carrying its heavy siege trains of ballistae and catapultas. This was work detested by the legionnaires, and many mutinies had their origin in this enforced labor. Slaves and captives of war were also put to road building. Provincial roads were built by local contractors supervised by local administrative authorities. Such contracting work inevitably became shot through with graft and fraudulent practices.

Road financing took many forms. The early roads—like the aqueducts—were financed by booty taken from defeated enemies. A law of 111 B.C. made all property holders along the roads liable for the cost of their construction and maintenance. In some instances emperors and prominent citizens footed the bill. Augustus paid for Via Flaminia but left one stretch to be built by the general Calvisius Sabinus. Some roads were financed by tolls, but by and large the vias were paid for by the imperial treasury.

All roads lead to Rome

From the point of view of communications, the center of the far-flung Roman roadnet was the miliarium aureum, a stone pillar placed in the Forum of Rome. It carried the distances to the main cities in the Roman world. Major provincial centers were also provided with a similar miliarium. However, in respect of road building and maintenance, distances were reckoned from the gates in the Republican wall to the city limits of the provincial centers.

Traffic on the extensive Roman roadnet was centrally administered by the cursus publicus, a post and courier service which originated in the third century B.C., a service restricted for the use of government couriers and officials. The dispatch riders changed horses at post stations placed at intervals of 6 to 16 miles. Officials traveling in two or four-wheeled carriages stopped over at mansiones, or rest houses, spaced 20 to 30 miles from each other. To be admitted to a mansione the traveler had to show a passport which stated the quality of the transport and the accommodations to which the bearer was entitled by reason of his office.

The regulations recognized five classes of transport:

1. Clabularia were heavy transport wagons carrying 1,500 libras or Roman pounds (1,085 lb avoirdupois) and drawn by a team of 8 oxen in summer and 10 in winter. Lower officials, legionnaires, and sick persons were allowed to climb aboard and ride on the cargo.

2. The raeda was a four-wheeled cart loading 1,000 libras (728 lb avoirdupois) and pulled by 8 mules in summer and 10 in winter. It was used for express goods and the transport of precious metals.

3. The carrus was a four-wheeled wagon loading 600 libras (437 lb avoirdupois).

4. The vereda loading 300 libras (218 lb avoirdupois) was drawn by four mules and took two or three passengers.

5. The birota was a two-wheeled cart drawn by three mules and loading 200 libras (146 lb avoirdupois) in addition to one or two passengers.

The courier service used only riding horses and pack horses carrying 100 libras (73 lb avoirdupois).

The smallness of the weights pulled by the draft animals was due to the poor method of harnessing which choked the animals when they tugged at the breastband. With such a harness, a horse pulled only 137 pounds, or one-quarter of its capacity when provided with a proper harness. The animals were not arranged in tandem but only abreast of each other, whereby their pulling power was further reduced.

The draft animals served only between specified mansiones and were fed by enforced contributions from the local farmers who had also to provide one-fourth of the horses needed.

The average speed of the imperial post was 50 Roman miles, or 47 statute miles per day. The *tabellarii*, or official couriers, were expected to travel twice as fast. The record speed attained was 149 statute miles in 24 hours. Private travelers using the roads had to provide their own transport and spend their nights in inns.

The imperial post system worked satisfactorily for many centuries, although always exposed to abuse. During the reign of Emperor Constantine (324–337), it finally broke down, mainly because of the heavy tax load that ruined the entire Roman economy, more directly because of the system of compulsory exactions from the local population who by now had been turned into paupers owing to excessive taxation. Fraudulent demands by government officials for free transport also contributed to the collapse of the system.

Roman bridge building

The Assyrian military engineers were proficient in using pontoon bridges, still a favorite military means of crossing a river, and the technique was further developed by the Persians. Xerxes even succeeded, after two initial failures, in bridging the Hellespont by means of 614 ships lashed together and anchored in the strong current coming down the Straits. He had to chop off the heads of a succession of engineers until in the end he found one with better luck than his unfortunate colleagues, and the Persian king managed to get his army across to the Greek mainland in 480 B.C. This was the greatest bridge building feat in history until the Romans became interested in the technology of bridging rivers.

Roman bridge building began with timber constructions quickly built by military engineers to get troops and engines across a river fronting the advance. In distant provinces these timber bridges built by the military were left standing and not replaced by stone bridges. This was true of Newgate Bridge in London and—the most famous of them all—Trajan's bridge over the Danube near Orsova, a trussed timber construction carried on stone piers. The bridge was designed by Appolodorus, a Greek engineer from Damascus who also designed Trajan's column which is still standing in his forum in Rome. The Danube bridge built in A.D. 99 consisted of 20 stone piers carrying a trussed timber construction for a total length of 1,170 yards. Caesar also had his engineers span the Rhine with a trestle bridge in only 10 days.

Such proficiency in bridge construction takes a long time to develop. A couple of centuries before Xerxes managed to get a pontoon bridge across the Hellespont, the Etruscans had built a timber bridge—the *Pons Sublicius*—across the Tiber; and around 500 B.C. they built a stone bridge at Bieda, north of Rome, using wedge-shaped blocks in the arches, one of which, spanning 24 feet, is still standing.

Numerous Roman bridges remain today, some of them from the second century B.C. when they began to be built of stone, subsequently also with an inner core of concrete faced

Piranesi's engraving of Pons Aelius Hadrianus, now Ponte Sant'Angelo, in Rome, gives a good impression of Roman bridge building of the first order. The bridge was completed in A.D. 136 to serve as a stately approach to the mausoleum of Hadrian sited on the right bank of the Tiber. The view is from the upstream side on the right shore.

Piranesi's scale drawing of the central pier of Hadrian's bridge gives an idea of the enormous stone masses used in the bridge piers. The pier rests on piling and is built of ashlar held together with iron clamps set in lead. His dimensions are given in Roman palmi, which converted into present-day measures give a length of 233 feet and a height of 82 feet, extending to the high-water level of the river. Enclosed by the stone ballast and built with larger dimensional stone is the pier proper which measures 102 × 39 feet. Above high water it was constructed of sperone. The pier ends 33 feet above water in a broad corniced platform carrying the 34-foot-wide vaults. The plan shows the different shapes of the pier upstream and downstream.

with stone. *Ponte Cestio* in Rome, however, is constructed throughout of stone blocks joined by iron clamps to increase the strength. The present Ponte Molle *(Pons Mulvius)* spanning the Tiber about 1½ miles north of Porta del Popolo has two of its four central arches built with the original travertine. Yet the most magnificent bridge ever built by Roman engineers was the Ponte d'Augusto at Narni during the reign of Augustus; it was destroyed by a flash flood in 1304. It had four spans carried on circular arches varying in width from 52 to 104 feet. A 64-foot shore span still exists. The Roman bridge at Alcantara crossing the river Tagus in western Spain has a central span of 89 feet. These two spans were apparently the widest ever accomplished by Roman bridge builders.

With so many ancient bridges still standing, it may be assumed that Roman engineers had mastered the most difficult problem in bridge building, namely, the sinking of piers so as to obtain a solid unyielding foundation. When bedrock could not be reached, the piers were built on piling consisting of charred alder, olive, or oak timbers driven into the soft ground by means of a pile driver. The piles were driven together as closely as possible and the interstices filled with ashes.

Cofferdams were used in pier construction. According to Vitruvius, "A dam is formed in the water by oak piles tied together with chains and driven firmly into the bottom. In the enclosure formed by the piles and below the level of the water, the bed is dug up and leveled, and the work carried up with stones and mortar until the walls fill the vacant spaces of the dam."

Harbors and docks

The ridiculously poor efficiency of Roman road transport made it wholly inadequate for the flourishing Roman economy. Nor could the ancient empires before the Roman conquests be maintained with the limited volume of goods carried by pack animals. Goods moved by sea and river, the cheapest way of moving bulk cargoes hitherto invented.

Moving a cargo in a ship does not entail more capital than that invested in building the ship. But when the ship reaches its destination, it is essential for the orderly development of trade to invest heavily in harbor and dock installations. When the Phoenicians began to roam the Mediterranean in search of trade and piracy around 1200 B.C., they found an inland ocean lacking natural harbors, where the waves broke on unsheltered shores. Their home ports, at Sidon, Tyre, and Byblos, were to some extent sheltered by islands or rocks, to the leeward of which they beached their ships. This pattern was followed when they began to establish their factories on the opposite shores of the ocean, at Carthage, Cadiz, Motya, Sulcia, and elsewhere around the Mediterranean. They looked for a small island near the coast where they could beach their ships without risk of having them pounded to bits. It was at these sheltered spots that the Phoenician merchant-pirates subsequently built their harbors.

The Phoenician harbors were generally obliterated by the larger works of the Romans, and it is now impossible to get an idea of their construction. Carthage, however, has been described by Appian, albeit a few hundred years after the destruction of the city. There were two harbors, an inner and an outer. The entrance from the sea was 70 feet wide and could be closed by chains. Merchant vessels used the outer harbor, while the circular inner port—still remaining as a

circular basin 328 yards in diameter—was lined with ship-yards and dock installations capable of accommodating 220 vessels.

The Greek ports were sited by natural indications, in sheltered bays and behind promontories, but as the shipping grew, these natural harbors required considerable works in order to handle the increasing business. Peiraeus, the out-port of Athens, was created out of three natural bays around the rocky isthmus of Akte. About 330 B.C. the engineer Philon developed the three basins, of which Cantharus was the largest and was provided with projecting jetties leaving an entrance 55 yards wide. Near the entrance was the naval yard, while farther in was the *emporium* where merchantmen were berthed. The two small harbors Zea and Munychia to the east of the promontory were provided with slipways accommodating 372 vessels under roofs carried by colonnades. The slipways could accommodate a ship of 16 feet beam and 100 feet length, and were apparently planned for war galleys because the merchantmen at this time were much larger—up to 200 feet in length and 40 feet of beam. The out-port was joined with Athens by a nearly straight road 25 miles long and enclosed all the way by a defensive stone wall.

As previously noted, all the ancient harbor works were enlarged by the Romans who also built numerous new ports where needed, whether or not natural conditions invited their construction. The first one was Ostia (from *ostium* or mouth of the Tiber) which according to legend was originally founded by Ancus Martius, fourth king of Rome, to guard the estuary. However, the oldest existing remains in "Ostia Antica" derive from the fourth century B.C.; and the city—the first colony established by the Romans—dates from 335 B.C. It retained its status of colony until well into Imperial times, to judge from the 12-inch lead water mains under Decumanus Maximus, the main street, which bear the legend COLON . . . AQUA OSTENSIE cast in the pipe. Ostia's first industry was the extraction of salt, but at the time of the First Punic War (264 B.C.) it was also a naval base, at which time it had grown into an *urbs,* or commercial city. All the grain and other vital supplies consumed by the metropolis about 15 miles inland had to pass through the harbor, and Via Ostiense, the road linking the city with its outport, was for centuries the busiest one in the ancient world.

However, despite the continuous expansion of Ostia as a city and harbor during the Republic, natural conditions—such as the silting up of the estuary and the gradual accretions to the shoreline—eventually put a stop to further growth. For this reason Augustus planned a new harbor in the vicinity, but it was left to Claudius to start the work in A.D. 42. The harbor was finished in A.D. 54 during the reign of Nero, who issued a series of coins bearing the inscription "Portus Augusti" to commemorate the occasion. For this reason the new harbor became referred to as Portus.

The harbor of Claudius was wholly artificial, having been created by building two moles into the open sea so as to

167

Aside from the inevitable artistic interpretations and the faulty accentuation of detail resulting therefrom, this Roman engraving, dated 1575, gives a good impression of Portus, Rome's man-made port to the west of Ostia. The harbor basin to the right was completed in A.D. 54 and had an area of 247 acres protected by two moles. But this— Portus Claudius— soon proved to be too small, and Trajan ordered a hexagonal basin to be built measuring 766 yards along its largest diagonal. This second basin had an area of 115 acres and a maximum depth of 16 feet. It was completed in A.D. 100. What is sometimes interpreted as a drydock between the two basins, the artist depicts as a supply canal from Fiumicino with swiftly flowing water.

CA·DEL·TEVERE

MARE·TYRRHENO

obtain a sheltered basin of about 247 acres. A large barge used to bring an Egyptian obelisk to Rome was included in the construction of the western mole on which a lighthouse was built. The moles were constructed by dumping quarry stone and concrete blocks weighing up to 3 tons as ballast, as described by Pliny: "Huge stones are transported hither in broad-bottomed boats, and being sunk one upon the other, are fixed by their own weight, gradually accumulating in the manner of a rampart."

The original Portus was enlarged by Emperor Trajan who in A.D. 100 ordered a hexagonal basin excavated out of the marshland. The basin, which measured 766 yards across with an area of 115 acres, was completed six years later. It had a depth of 13 to 16 feet. The bottom was paved to facilitate removal of silt washed in from the outer harbor. Near the entrance to Trajan's harbor was a 130×820-foot basin which may have served as a drydock. The quays lining the basin were built of concrete blocks made with pozzolana cement which sets under water. This hydraulic concrete has resisted the physical and chemical aggression of seawater for nearly 2,000 years.

Owing to the silting up of the coast, Ostia is now situated 2 miles from the existing river estuary; but other Roman harbors, such as Civitavecchia, Pozzuoli, Terracina, and Anzio, are working ports retaining substantial remnants of the old Roman works. At Anzio they were badly damaged by high explosives during the invasion in 1943.

There is no point in being sentimental about the wreckage caused to ancient Roman harbor works during the Second World War. The lethal damage was done long ago by natural causes—by pounding waves and sand. For a thousand years or longer, no one was capable of maintaining the harbor works, and during the Middle Ages one port after another had to be abandoned, since medieval shipping used only harbors that could be kept open with minimum or no work. None of the Roman ports except Alexandria have retained their ancient importance.

The Roman harbors were provided with lighthouses to mark the entrances between the moles. In other words, shipping economy had developed to such an extent that wasteful coasting could not be tolerated, and ships sailing on direct tracks made their landfalls from the open sea. These Roman lighthouses ran into large numbers; but most of them collapsed and only a few survived the Middle Ages. The one at Boulogne, known as "The Old Man," was 197 feet high and was erected on the order of Caligula during his abortive attempt to invade Britain in A.D. 39. It collapsed in the middle of the seventeenth century. On the British coast, at Dover, the Romans erected two lighthouses, one of which, near the church of St. Mary-in-Castro, still survives. It is an octagonal stone structure with each side measuring 15 feet and a present height of 43 feet; originally, the tower must have been twice as high. In the center is a 14-foot-square shaft that at one time held a staircase.

169

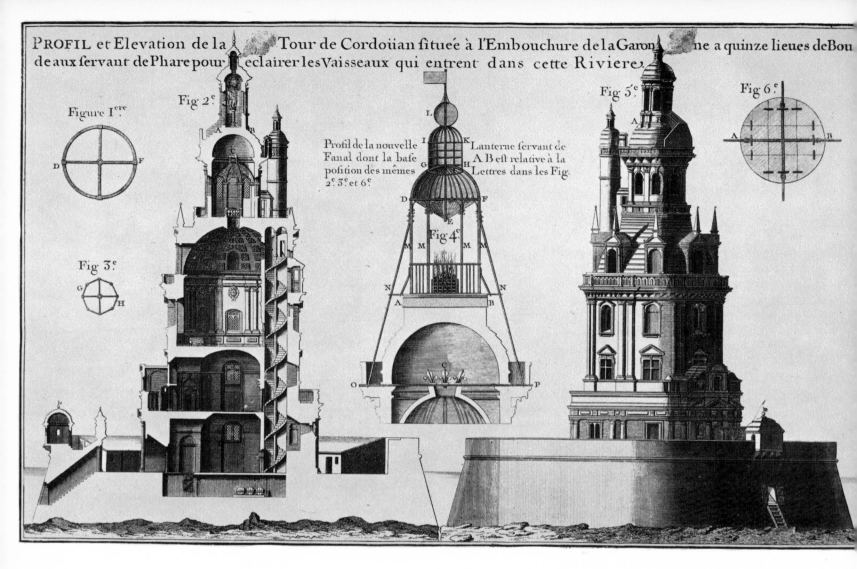

PROFIL et Elevation de la Tour de Cordoüan située à l'Embouchure de la Garonne a quinze lieues de Bordeaux servant de Phare pour eclairer les Vaisseaux qui entrent dans cette Riviere.

The Cordouan lighthouse at the mouth of the Gironde was completed in 1584 and was the first lighthouse to be built on a rock in the open sea. The engraving showing details of its construction derives from Bélidor's Architecture Hydraulique *published in 1737–1753, the first technical handbook produced north of the Alps. It remained for nearly a century the only standard work on civil engineering.*

These tall structures were built for the purpose of projecting a wood fire high enough to be seen far out at sea. But these navigation aids were by no means Roman innovations. Such guidance fires were lit to assist Phoenician ships to make landfalls, and when Ptolemy Philadelphus, in 280 B.C. or thereabouts, ordered his engineer Sosastrus of Chidos to erect a tall tower on the island of Pharos to protect the two harbor basins of Alexandria, lighthouses had become commonplace around the Mediterranean. The only difference was that Sosastrus' tower was built on a much more ambitious scale. It was at least 280 feet high and rested on a base nearly half as high. The base carried an octagonal tower, on top of which was a cylindrical one terminating in a lantern. Here a wood fire was kept burning, and by means of polished metal mirrors the firelight was concentrated so that it could be seen

at a distance of 300 stades (35 miles). From this seventh wonder of the world all subsequent lighthouses were called *pharos*. The original one was shaken down by an earthquake in the fourteenth century. The Alexandria pharos appears to have been the only one provided with mirrors to concentrate the light of the resinous wood fire kept burning after nightfall. Actually, the pharos was not originally intended as a lighthouse. It served as a landfall beacon for several centuries and did not carry a fire until well into the first century A.D.

Construction in the Middle Ages

The vast inheritance left by the Romans was managed altogether differently in the east than in the west. Byzantines and Arabs maintained the existing road system and its

170

bridges; the last Roman bridge spanning the Tigris did not collapse until some time in the tenth century. At about this time the posting service maintained by the Caliphs included 930 stations.

It was indeed different in the West. The new rulers were too busy slitting one another's throats to find time for anything else. The fine Roman roadnet disintegrated; the pave was mined for building stone; the right-of-way was taken over by greedy landowners. However, since everything else was in a similar mess, the condition of the roads did not really matter, because what traffic existed was local in character and modest in scale. The transport needs could be satisfied by an occasional train of pack animals picking their way along narrow tracks, furtively pulled along by poor wretches expecting every moment to be robbed and slain.

Nonetheless, long-distance communications did not cease altogether. The church and the religious houses maintained a postal service using messengers traveling on foot. During the reign of Charlemagne the economy improved temporarily, at least in some parts of his vast realm, so as to require wagon transport for moving local surplus. But many of the old Roman bridges had collapsed, it would seem, judging from a contemporary regulation that wagons be covered with tarpaulins of leather to prevent the goods from being damaged by water when fording rivers. Three centuries later, goods moved across the Alpine passes, from the Po valley to the Rhine, in wheelbarrows and on porters' backs. It was not until the thirteenth century that the St. Gotthard pass was opened up by a rough pack trail.

Opening the St. Gotthard pass necessitated the bridging of the forbidding Schöllenen defile by a stone bridge, later known as the "Devil's Bridge." But to reach the bridge in the gorge, a suspension bridge had to be built, hanging in chains along the precipitous rock wall. Times were indeed changing elsewhere, too. In England, four roads—Watling Street, Ermine Street, the Fosse Way, and the Icknield Way—all of them inherited from the Romans, were protected by the "King's Peace" so that anyone could travel on them un-molested. The maintenance of the roads became, under the influence of the Church and the monastic orders, a pious duty, and indulgences and benefactions were granted for contributions toward the restoration and maintenance of roads connecting centers possessing religious importance.

Kings also took a hand, and in 1353 Edward III paid for the paving of the Westminster road, while in France the *strata publica* were built and maintained by the Crown. These improved carriage roads were paved either with cobbles laid in sand or, in France, with stone blocks laid with mortar. Over these cobbled roads goods moved at a speed of 14 to 22 miles a day, owing to the more efficient horse harness. Passengers traveling by cart made 30 miles a day in flat country and 25 miles in mountainous areas. Commodities such as wool increased in price by about 1.5 percent, and grain by 15 percent per 50 miles traveled. A King's Highway

This engraving taken from "Le diverse et artifose machines" (Paris, 1588) shows a cofferdam constructed of sawed piles provided with tongue and groove. The piling is so technically advanced that it is tempting to regard it as science fiction. How-ever, it appears to be historically established that such a piling was used by John Perry in a cofferdam built along the Thames in 1715, and Perry explained that he had seen it used elsewhere. Otherwise, cofferdams continued to be constructed in the Roman fashion, i.e., rows of piles with the space between them filled with rammed clay.

in the better parts of Europe carried on the average about 1,000 tons of freight a year, although three salt roads in the Salzburg area are reported to have carried as much as 7,000 tons in 1370.

Sea transport also began to improve, as evidenced by the report that wine from Gascony could be brought to England at only 10 percent of its laden cost. The beacons guiding the ships to harbor were lit again. Genoa obtained its harbor light in 1139, and the 161-foot-high stone tower still standing at the entrance to the Leghorn Harbor was put in service in 1304. About 200 years later the very first lighthouse to be built on an exposed rock in the sea—the Cordouan at the mouth of the Gironde—was completed. At least one thing not previously attempted by the engineers of the ancient world had at long last been achieved in Europe north of the Alps.

European Reclamation and Canal Building

"In this land twice each day and night the ocean sweeps in a flood over a measureless expanse covering up nature's age-long controversy and the region in dispute between land and sea."

PLINY

There lived, still according to Pliny, "this miserable race occupying elevated patches of ground above the level of the highest tide, living in huts like sailors in ships when the waters cover the surrounding land."

What the Roman author referred to were the *terpen,* or low mounds in the Low Countries, which at high tide rose something like 3 feet above the water. At one time these patches of dry ground were only about 100 feet in diameter, but when Pliny visited this area along the coast of northern Holland, there were about 1,200 such patches varying in size from a few acres to 40 acres. The four sea provinces of the Netherlands (Zeeland, Holland, Friesland, and Groningen) were like the fens in England without doubt miserable areas for human habitation, but so were other vast tracts on the Continent. Everywhere, arable land constituted high patches of ground surrounded by marshes and bogs. Europe is to a large extent a man-made continent whose fertile "bottom-lands" have been recovered with the aid of the spade.

The Roman inheritance

The marshy valleys of Italy were reclaimed and made habitable as the Roman conquests brought them under a civilizing administration; and as the colonization proceeded on the other side of the Alps, reclamation work continued unabated throughout the centuries. In England, Agricola even tried to reclaim the fens—the lowlands to the south of the Wash. It was an ambitious exercise in civil engineering which included the digging of a 60-foot-wide canal extending from Peterborough on the Nene to Witham south of Lincoln, for a total distance of 25 miles. From Witham, the 7-mile-long Foss Dike led to the river Trent. In Holland, Claudius Drusus built a mole in the Rhine west of Cleves to prevent the water of the Waal from flooding the area between the two rivers. He also had a canal dug between the Rhine and the Yssel to deviate the excess waters of the Rhine. This large-scale reclamation and flood control program was started in A.D. 12; later, in A.D. 45, Dominitius Corbulo dug a ships' canal joining the Rhine with the Meuse. To the east and north of the Rhine the savages remained content with their bogs.

The Romans were aided by nature in their reclamation work because the climate remained fairly dry during the many centuries that they were occupied with expanding the arable land area of their conquests. By A.D. 500 the climate worsened, and on the Continent the rainfall increased. In the Low Countries, the terpen sank about 4 inches per century. To save them from being inundated at high tide, they had to be provided with dikes. Also in France the land became waterlogged, in some areas such as Normandy and Marne aided by deforestation.

As in so many other fields—in building, agriculture, horticulture, metal smelting and road building—the early medieval reclamation schemes throughout Europe were started by the religious orders—Benedictines, Cistercians, Premonstrantians—who, as stated in a previous chapter, were instrumental in carrying on the Roman engineering tradition. Under their guidance, land in Flanders and Zeeland was gradually recovered from the sea, at first by the natural compaction of seaweed, later by the planting of marsh samphire and marram grass until the flatlands could be used for the grazing of sheep. As the land slowly rose, it was drained and eventually turned into ploughland. The counts of Flanders kept granting the religious orders mud flats and submerged land to recover in this slow natural fashion.

Gradually, a more rapid method of land recovery was developed. A tract of waterlogged land was enclosed by dikes to prevent it from being submerged at high tide. Two large ditches were dug at right angles to each other across the submerged land to carry off the water drained by a number of small meandering ditches. A sluice gate discharged the accumulated water at low tide.

172

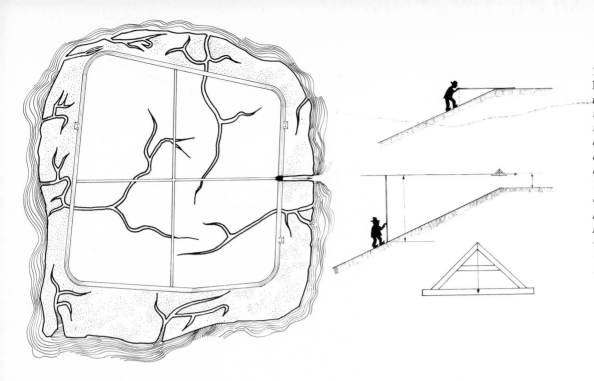

These dikes against the sea were 100 feet wide at the base and constructed of clay made waterproof by being trodden by oxen or horses. The inland side of the dike had a slope of 1:1.5 and was lined with clay or planted with grass. The sea side of the dike sloped 1:0.75 and was banked with several rows of seaweed bundles, piled in steps up to and somewhat over the top. The seaweed compacted under its own weight, and the heat generated by its putrefaction turned it into a solid mass. Piling was not used until the fifteenth century, and fascines—the technical term for bundles of faggots—until even later.

After centuries of struggling with the sea, a corpus of hydraulic engineering know-how was accumulated by the Orders, and by the tenth century the reputation of the Netherlanders as experts in land reclamation had been firmly established. For many centuries, all major reclamation works in Europe were bossed by either Netherlanders or Italians.

La bonifica in Lombardy

In Italy, land reclamation was conducted along altogether different lines from what it was north of the Alps, in the Netherlands, England, and France. The smoldering fires in the ruins of the western Roman Empire had hardly been extinguished before the Benedictine and Cistercian Orders began in a small and uncoordinated way to drain the marshes of Lombardy. From the seventh century onward, the Benedictine abbeys of Palazzolo, Monteverdi, and Salvatore became increasingly occupied with recovering waterlogged wastelands. Here, then, in Lombardy is found another technical junction linking up Roman hydraulic engineering with the present.

Out of this early work developed a comprehensive scheme of river control and land reclamation that eventually became known as *la bonifica*. It was concerned not only with health and agriculture but with the entire technology of river control and hydraulics in the widest sense. Some of the men employed on la bonifica belong to the roster of early European scientists. Galileo Galilei was engaged in hydraulic work and was at one time superintendent of the waters of Tuscany. His pupil Benedetto Castelli, when professor at Padua, became involved in flood control in the valley of Pisa and later superintendent of the areas around Bologna and Ferrara, between the rivers Po and Reno. This is where he wrote his famous work *Della misura dell' acque correnti* ("On the Measurement of Running Waters") published in 1628. This was the first important work on hydraulic technology.

Galileo's most distinguished pupil Vincenzo Viviani was also a practical hydraulic engineer and the founder in 1657 of the first great scientific society—Académica del Cimento. Finally, at the end of the seventeenth century, Domenico Guglielmini published his *Della natura de fiumi* ("The Nature of Rivers"), and Giovanni Battista Barattieri his *Architettura d'acque* ("Hydraulic Engineering"). From these works emanates everything worth knowing about the behavior and control of rivers.

But all that came later. In the Low Countries as well as in northern Italy, the work of the ancient religious orders was taken over by brotherhoods and chartered companies. In the

173

Low Countries they were succeeded around 1200 by a central organization known as the *Hoogheemradschappen,* or Main Polder Boards, while in Italy the councils of Milan, Florence, Verona, Venice, and others formed special commissions charged with the duty of administering the comprehensive la bonifica schemes.

The droogmakerij

In the early part of the sixteenth century, a new technique of land reclamation was developed for draining the large lakes in northern Holland. The drainage of these largely man-made lakes—formed by excessive peat cutting—was accomplished by the so-called *droogmakerij* and involved the digging of a *ringvaart,* a wide ditch around the entire area to be recovered. The soil was thrown up on the inside of the ditch and compacted to form a *ringdikj,* or bank. The enclosed area was then crisscrossed by wide channels at right angles to one another, and between them were excavated a number of small parallel ditches. The drainage water flowed by gravity toward the ringdikj where wind-driven drainage mills lifted the water into the ringvaart from which it flowed, through sluices if necessary, to a river or drainage canal.

This new development in land reclamation was made possible by the Dutch invention around 1450 of the *wipmolen,* or hollow-post mills in which the top (or cap) carrying the sails could be turned to face the wind. This invention widened the application of wind power and set off a chain reaction of mill and pump designs.

The droogmakerij required large-scale capital investments which were raised by companies of "adventurers" who speculated in the profits to be made from the recovered land. After a few modest initial attempts, Count Van Egmond began the drainage of the Egmondermeer in 1556, which was quickly followed by even more ambitious undertakings. In total, however, the droogmakerij during the sixteenth century was of small importance compared to the ancient method of gaining land from the sea by diking. The major expansion of the droogmakerij took place in the following century when during the 25 years from 1615 to 1640, 63,046 acres of new land was obtained in this manner; but this record dwindles in comparison with the land recovered from the sea by the Polder Boards which during the 150 years from 1540 to 1690 gained 413,318 acres from the sea by traditional methods.

One of the most outstanding hydraulic engineers associated with the droogmakerij was Jan Adrianszonn, or Leegh-water as he later called himself. He was born in 1575 and his early fame rested on the construction of the technically superior mills used to drain the Beemster in 1608. It was here that he developed a system of multistage waterlifts using two to four scoop wheels working in tandem. In 1629, he presented his famous plan for draining the Haarlemmermeer, the

T. *Brueck by Houte wael wyt 18 Roeden*   Y. *Buiten Dicks Lande*
Y. *Stucken wt den Dyck gebroecken.*   Y. *De Nieuwe vaert*
W. *Den niew ghemacten Dam.*   Z. *Wech nae de Diemer meer.*

largest ever, which required 160 drainage mills of his improved type and an estimated capital of 3.6 million guilders. That was too much for his time, and the successful completion of the scheme had to wait until 1852.

Draining the fens

In the sixteenth and seventeenth centuries the fame of the Dutch hydraulic engineers brought them far afield. In 1628, Leeghwater was invited by the Duc d'Epernon to drain the marshes of Gironde, and in 1584 a Brabanter from Bergen op Zoom with the English-sounding name of Humphrey Bradley was brought to England, originally as a consultant for the construction of the Dover harbor. In a brief issued by the Privy Council in March, 1588, to the commissioners of the Fen country, Bradley was recommended as being competent "to make viewe and platt for the several fenns, the true dyssentes of waters, and the qualities of soile through which waters should be carried."

The "several fenns" mentioned in the brief referred to the 700,000 acres of marshy lowlands, some of it subject to

periodic flooding, extending inland some 35 miles from the shores of the Wash and traversed by four sluggish rivers—the Ouse, the Nene, the Welland, and the Witham. Of this immense area the part lying between the Nene in the north and the uplands of Norfolk in the south and known as the "Great Level" had been drained piecemeal by the local abbeys up until 1540 when the work ceased.

In December, 1589, Bradley presented his "Treatise" to Lord Burghley wherein he made the claim that "practically the entire surface of the land is above high-water level and the only way to redeeme the land from the waters is to draw off the waters by directing them along the shortest track, in canals dug with such a width and depth as can serve to make the waters run into the seas." The entire project could be accomplished by gravity without recourse to embankments, machinery, mills, and "inestimable expense."

Bradley's Treatise raises the question of how he went about surveying and above all—leveling—the flat fenland with sufficient accuracy to obtain the slight gradients needed to get gravity flow over the long distances involved. Mapping the area to be drained should not have been particularly difficult because the plane table had been in use for several

One of the numerous but now forgotten dam catastrophes in the Netherlands took place on March 5, 1651, when the St. Anthoniedyk in Amsterdam was swept away. This time however, the artist Pieter Nolpe was on the spot when the rescue and reconstruction work got under way. Small boats loaded with fill and, at a distance, a piledriver are being brought to the site.

The sketch below is a patent drawing of a drainage mill filed in 1589 by Simon Stevin, Dutch businessman and mathematician. This particular mill design, one of 102 similar inventions for which patents were applied for, became of great importance because it was capable of lifting four times as much water as other mills. The Stevin patent claim included alternative drive with a horse whim.

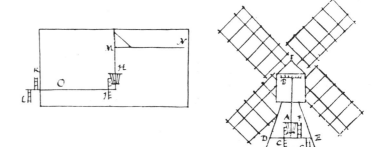

175

decades, the method of triangulation had been devised in 1533 by Gemma Frisius, professor of mathematics at Louvain; and in 1513 a regional map of the upper Rhineland had been drawn on the basis of a field survey made with the aid of an angle instrument known as the "polimetrum." Finally, in 1571, the English mathematician Thomas Digges had published the description of an instrument called "theodolitus," orginally suggested by his father. There was at the time also a linked steel chain for measuring distances.

Therefore, the actual mapping of the fens was no problem, but leveling must have been a slow and demanding job because the only means available for determining small differences in elevation was the old Roman chorobates. Although the survey instruments mentioned above could also be used as levels, they were unsuited to the fine work required in the fens where the slope can be less than 5 inches per mile.

There is no record of the Bradley survey, and the accuracy of the leveling remains unknown. But nothing came of the Bradley project, and the reclamation of the Great Level had to wait until 1653 when Cornelius Vermuyden from Zeeland succeeded in reclaiming 307,000 acres under a charter granted by Charles I. This giant drainage project was financed with capital raised in the Netherlands, and its completion in about 25 years still remains something of a record. The gravity scheme worked for some years, but the shrinkage of the peat lands and the lowering of the elevation due to efficient drainage eventually put a stop to the gravitational flow, and the Great Level had to be provided with drainage mills to keep it viable.

After four years of frustrating attempts to get official permission to go ahead with the Great Level plan, Bradley went into the service of Henry IV and obtained a virtual monopoly on land reclamation in all of France. He formed an association organized on the pattern of the Hoogheemradschappen and raised capital in the Netherlands for successful drainage schemes in Poitou, Normandy, Picardy, Languedoc, Provence, and elsewhere in France. Thus, one and the same man became the pioneer in modern land reclamation in both England and France.

The Pontine Marshes

However, the greatest drainage scheme in all of Europe, the Pontine Marshes, an area of 300 square miles to the southwest of Rome, proved too much for the efforts of the hydraulic engineers both north and south of the Alps. The area had been reclaimed by the Romans but had returned to its original state of malaria-infested swamp during the Middle Ages. In 1514, Julius de Medici obtained a concession from the Pope to drain the area, and employed Leonardo da Vinci to prepare a plan. Leonardo proposed to reopen the Roman canal running along Via Appia for the purpose of catching the waters flowing from the Lepini mountains. At right angles to this canal he planned to cut another to discharge the waters into the sea. But the Pope died and so did Leonardo's project.

Seventy years later—in 1586—the Italian engineer Ascanio Fenici presented Pope Sixtus V with a drainage scheme that was to be financed by the merchants of Urbino. The Pope granted a concession stipulating that $5\frac{1}{2}$ percent of the reclaimed land be deeded to the papal treasury. Fenici's project came very close to succeeding. He put 2,000 men to work digging the main canal—the Fiume Sisto—plus a number of smaller ones, and sluices were constructed to control the water flow from the mountains. The first phase was completed in three years, but upon the death of Pope Sixtus in 1590 the works were abandoned.

In 1623, the Netherlander Nicolaas Corneliszoon de Witt came up with a new Pontine project and obtained a concession 14 years later from Pope Urban VIII. The de Witt scheme was financed by wealthy merchants in the Netherlands and Italy, but when the promoter died the scheme aborted.

Finally, in 1676 another Netherlander, Cornelis Janszoon Meijer, presented Pope Innocent XI with a drainage plan which the Pope turned over to an Italian engineer for technical review. On the basis of his *relazione,* Meijer reworked his original project and published the new one in 1683 as a memoir entitled "On the way to drain the Pontine Marshes," whereupon he died. Work on his scheme actually got under way but had to be abandoned owing to opposition by the local population in the marsh country. That was the end of a century-long splurge of enthusiastic activity. The marshes were left in their pristine and malaria-infested state until Mussolini's government carried through their drainage in the 1930s.

Canals and river clearance

This book has up to now been largely concerned with tracing the development of building and engineering as it evolved in the ancient civilizations around the Mediterranean, particularly in the Fertile Crescent. The high-water mark in hydraulic engineering and all that it implies was reached in the twin-river valley centuries before the Romans began their more modest attempts at taming rivers, draining marshes, and building canals. But it should perhaps be recalled in this context that in distant China, wholly isolated from the Mediterranean, a parallel growth of civilization was taking place, which owing to the relative political stability could keep on developing without violent interruptions for much longer periods than in the disruptive West. The economy kept expanding, the inventions necessary to resolve the problems generated by a complex society accumulated, production grew in numerous ever-increasing fields.

The Chinese canals

The numerous and mighty Chinese rivers were cleared for navigation at an early time, apparently simultaneously with similar work in the Euphrates. Just when they were joined by man-made canals is not known, but in 215 B.C. the Ling Ch'u canal in the Kuangsi province was built, and in 133 B.C. the Han capital Ch'angan was linked with the Yellow River by means of a 90-mile-long canal. The Han emperors and the T'ang Dynasty that followed them after a 400-year interregnum of political unrest were active canal builders. Their work culminated in the 600-mile Grand Canal, the first section of which was completed in A.D. 610. Two centuries later, when hardly a river in Europe was capable of navigation, the Pien Canal carried about two million tons of goods a year.

The Chinese canals were dug across a flat country, and their gradients were therefore modest; however, owing to their great lengths, the grades kept accumulating and creating undesirable currents. For this reason the major canals were provided with weirs of stone and timber and placed about 3 miles apart. In the weir was a gate through which the barges could pass. In places where the topography produced a considerable difference in level, there was a break in the continuity of the canal, and the upper and lower sections were joined by a slipway over which the barges were hauled.

Navigational weirs offer the simplest and cheapest means of overcoming differences in level over a stretch of river or canal. Unless the difference between adjacent sections is slight, opening a gate to equalize the levels of the two sections results in a waste of water and time. If instead of placing the weirs a mile or more apart, the distance is reduced to a few hundred feet, the difference in water levels can be equalized much more rapidly since the volume of water is far less.

This is the principle of the "pound lock." But such a lock is costly and difficult to construct owing to the water pressure on it. It has to be built of stone and made watertight, while some arrangements must be made to open and close the gates. This tremendously difficult construction problem in hydraulic engineering was resolved in A.D. 984 when the Chinese engineer Ch'iao Wei-Yo constructed two pound locks with gates 250 feet from each other on the Pien Canal. The gates were raised and lowered by means of a windlass. These were the first pound locks ever built. Later, during the Sung Dynasty, Chinese engineers succeeded with the difficult problem of building a summit canal across the watershed dividing the two rivers.

Ch'iao Wei-Yo was assistant commissioner of transport in the Huainan province, and the reason for his becoming engaged with the problem of the pound lock was the increasing theft of tax rice upon damage to the barges when they were pulled over the slipway at Huai-Shin on the Pien Canal. According to an official report of 984, the slipway laborers at the five stations between An-pei and Huai-Shib, each one with five upstream and downstream slipways, were in cahoots with local bandits and carried on a systematic sabotage of the barges to get at the rice loads. It should perhaps be added in this context that the Chinese canal system was developed largely for the purpose of carrying tax rice to the capital.

River clearance in Europe

While these Chinese developments were going on, post-Roman Europe had not yet begun clearing the rivers, for the simple reason that there was nothing to ship by inland waterways. It took something like 700 years before there was a local surplus large enough to require water transport, during which time the rivers had become littered up by mill weirs to accumulate water for driving corn mills. When eventually river navigation began to be contemplated some time in the twelfth century, the numerous mill weirs presented obstacles that had to be overcome in some fashion or other. The early solution to this problem was the "staunch" or gate placed in the weir and consisting of a number of boards set in a framework of vertical timbers. When a barge was to pass through a staunch, the boards were removed one by one, and when the rush of water had abated, the hinged timber frame was swung aside to let the boat through. Going downstream the boat rode on the flood of water released by the staunch; going upstream was more difficult, and the boat had to be winched up against the strong current.

The first staunch appears to have been built in 1116 on the river Scarpe in Flanders, but the first documented one was constructed in 1188 to 1198 by Alberto Pitentino on the river Mincio in Italy. Here the gate was of the so-called portcullis type; i.e., the entire gate was lifted by means of a windlass. The next development was, of course, to place the gates close to each other, thus obtaining a pound lock. The earliest documented pound lock was the one constructed in 1373 at Vreeswijk, at the junction of the Utrecht Canal and the river Lek. According to regulations issued in 1378, the lock was opened at 2 o'clock P.M. three days a week.

The first summit canal

Before we turn our attention to the main stem of European canal developments, some mention should be made of one of the more interesting of the early canal projects in northern

177

Elevation vuë de côté de la Machine à creuser les Ports pareil à celle dont on se sert à Toulon.

Echelle de la Machine.

Many of the machines, perhaps most of them, used on old-time working sites were driven by thread wheels. The engraving, from Bélidor, shows a dredge used for deepening the harbor of Toulon, but there is reason to believe that similar contrivances were employed for river clearance at the end of the Middle Ages. The scale is in old French toises, equal to 1.95 meters.

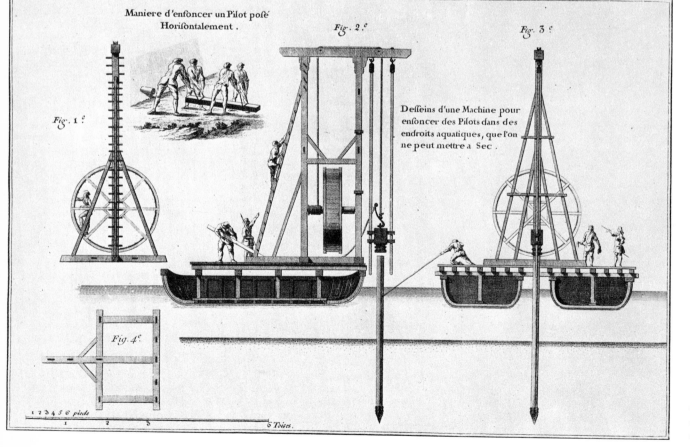

Maniere d'enfoncer un Pilot posé Horisontalement.

Fig. 1.^e

Fig. 2.^e

Fig. 3.^e

Desseins d'une Machine pour enfoncer des Pilots dans des endroits aquatiques, que l'on ne peut mettre a Sec.

Fig. 4.^e

Piling in water has always been important in hydraulic engineering and essential for cofferdam construction. The Bélidor engraving gives an idea of how a seventeenth century pile driver was used. The hammer was, of course, lifted by a tread wheel.

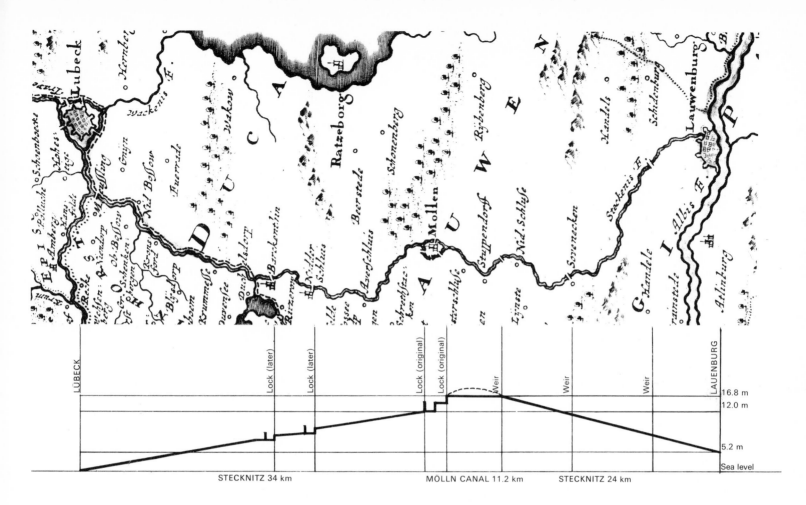

The only technologically interesting medieval hydraulic development in northern Europe was the clearance of the Stecknitz and Delvenau Rivers south of Lübeck and the summit canal at the Mölln Lake. The work was done 1391 to 1398, whereupon the Lübeck merchants were enabled to run barges from the Baltic to the North Sea via the Elbe. Stecknitz was indeed the very first deux mers canal in Europe, because all earlier canals were used only for inland transport. On the seventeenth century map the two rivers are given as one and several more locks and weirs have been added.

Europe, whereby the town of Lauenburg on the Elbe was connected with Lübeck, center of the Hanseatic League. Indeed, the Stecknitz Canal established an inland waterway between the Baltic and the North Sea.

In the early fourteenth century, the river Stecknitz had been cleared of obstacles to navigation from Lake Mölln to Lübeck and the Baltic, a total distance of 43 miles. The 40-foot drop of the river from the lake to Lübeck was taken up by four weirs with staunches. Then in 1391 work began on a canal extending southward from the lake to join the river Delvenau on the other side of the watershed. The canal rose 16 feet in less than half a mile, after which it ran level across

the summit through a 12-foot cut until it joined the river Delvenau at a point 15 miles north of the town of Lauenburg. The fall from the canal junction to the river Elbe at Lauenburg was 42 feet and was taken up by eight weirs.

The 7-mile stretch between the lake and the Delvenau appears to be the first summit canal built in Europe, and like all subsequent summit canals, it posed the problem of how it should be supplied with water to keep it navigable. The designer relied on water seepage through the sandy soil in which the 12-foot cutting was made, and to retain the water he constructed two pound locks at the lake end, known as the Hahnenburger Kistenschleusen, each one taking up 8

179

feet of the total 16-foot difference in levels. Together the two locks accommodated twenty 35-foot barges.

The name of the engineer who planned and executed this scheme is forgotten, but most likely he was somebody from the Netherlands. No one in Europe except a Netherlander, or an Italian, was at that time capable of carrying out such advanced hydraulic works.

The canals of Lombardy

The other major developments in canal construction at this time took place in Lombardy, as part of or closely related to la bonifica. Between 1179 and 1209 a 31-mile irrigation canal was dug tapping the Ticino River at Cassa della Camera, from which it ran to Abbiate and then turned east to Milan. In 1260, the canal was widened for navigation, and the fall of 110 feet was taken up by numerous weirs. This was the beginning of the famous Naviglio Grande.

When in 1387 work began on the Milan cathedral, the marble was taken from the quarries near Lago Maggiore and brought to the city on the Naviglio Grande, which ended in a basin outside the west wall. A small canal was therefore dug joining the basin with the old moat which passed near the building site. Where the new canal joined the moat, a staunch was built because the level of the water in the moat was several feet higher, being fed by the Seveso River. To get a marble barge to the building site, the staunch was opened, releasing a rush of water into the canal; and when the two levels had become equalized, the barge was brought into the moat. But having lost water, the moat had become too shallow for the barge to continue, and so the staunch was closed to permit the entire moat to fill up again so that the barge could proceed to the masons' yard near the cathedral.

This troublesome and time-consuming navigation was immeasurably improved in 1438 with the building of a second lock in the canal. This was the first pound lock in Italy, and the engineers were Filippo da Modena and Fioravante da Bologna. In 1451, Bertola da Navate was appointed ducal engineer of Milan and was ordered to investigate the possibility of enlarging an old irrigation canal between Milan and Binasco, with the view of extending the Naviglio Grande to Pavia. Instead, he recommended building a canal southward from Abbiate to Bereguardo, to be connected by land portage to the Picino, which was to be cleared for navigation to Pavia.

Work began in 1452, and six years later the 12-mile-long Bereguardo canal was completed. The difference in elevation over this distance was 80 feet which was taken up by 18 locks. This was the first time in history that such a difference in elevation had been taken up entirely by means of pound locks. Before the job was concluded, Bertola had plans ready for a canal joining Milan with the Adda River to the east of the city.

Work on the Martesana canal began by constructing a weir across the Adda River at Trezzo, whereupon the canal was dug parallel to and separated from the river by a masonry wall running for 5 miles to Gropello where it turned westward across the plains to Milan. The 24-mile-long Martesana required only two locks, but it had to cross the river Molgora on a masonry arch, while the Lambro stream was brought under the canal in a culvert. In Lombardy, hydraulic engineering was beginning to take on sophisticated overtones.

Of the technologically important work going on in Lombardy during the fifteenth and sixteenth centuries only one detailed accomplishment will be recorded, partly because of its future implications, partly because of the famous name associated with it. Upon Bertola's death, Leonardo da Vinci was appointed ducal engineer in 1482, and after spending 10 years on other activities, such as designing masks and sets for the ducal balls, he turned his attention to hydraulic engineering. The *Naviglio Interna,* i.e., the enlarged moat of Milan, had a number of old locks provided with portcullis gates which needed replacement. The new lock gates designed by Leonardo were finished in 1497 and became the prototype for all subsequent gates, including those used today.

The sketch of one of them, that of San Marco, below the terminal basin of the Martesana canal, has survived among Leonardo's papers, subsequently assembled in the *Codex Atlanticus.* It shows a 95 × 18-foot masonry lock provided with so-called miter gates folding back into recesses in the walls when open. The water inside the lock was discharged through a hinged-wicket sluice door placed at the bottom of the large gate and occupying about one-sixth of the gate area. It was tripped by a rope connected to a pulley on the canal bank.

Thus, by 1500 the technology of canal building was complete; the principle of lockage to overcome difference in levels was mastered; the design of the gates had been perfected; the problem of supplying a summit canal with water had been solved; a canal could be brought over a river crossing its way; indeed Leonardo and others had contemplated bringing a canal through an intervening hill by means of a tunnel. The invention of the miter gate capped several centuries of development in canal building, and it was rapidly adopted all over Europe in the canal building craze during the latter half of the eighteenth and the beginning of the nineteenth centuries. Throughout this long period the technology remained pretty much the same as that developed by 1500.

Numerous canals, such as the famous Brussels Canal, were built in Flanders in the sixteenth century by Dutch engineers who became the acknowledged experts in canal construction, as well as in drainage. Northern Renaissance princes eager to catch up with developments abroad kept these engineers busy preparing canal projects. But many, indeed the majority

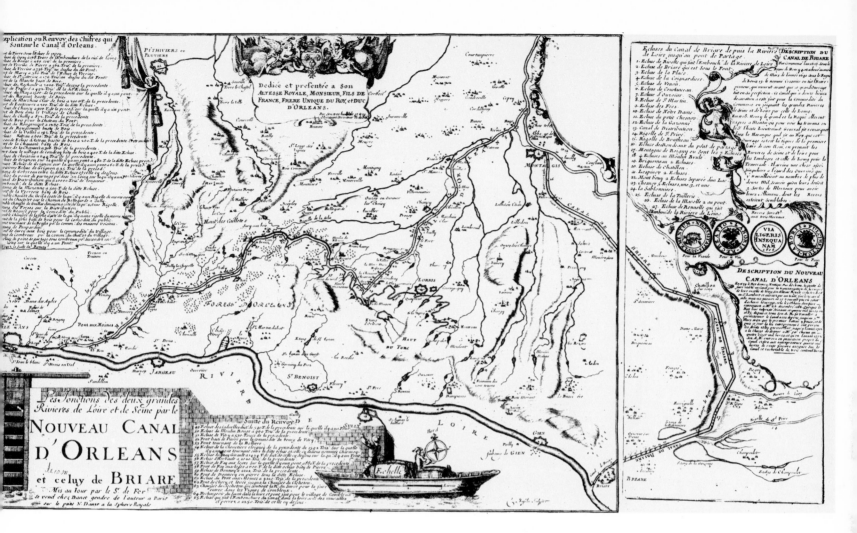

The Briare Canal was built 1605–1642 and has a length of 34 miles. To overcome a difference in elevation of 266 feet, the canal was provided with 40 pound locks. The so-called Orléans Canal between the Loire and Montagis is 46 miles long, including a 4.5-mile summit stretch. The canal rises 98 feet from the Loire by means of 11 locks, originally constructed of timber. From the summit it descends 132 feet by way of 17 locks. The map is contemporary with the canal.

of these projects, were too advanced for the meager economic resources and nothing came of them—until a century or more later. In those instances where work was actually started, it ceased for various reasons, mostly because of war and inability to pay the bills.

The canals of France

One of the most ambitious canal schemes dreamed up by a Renaissance prince was the *Canal des deux mers,* an inland waterway joining the Atlantic and the Mediterranean. The idea of such a canal fascinated Francis I, and he invited Leonardo to Paris in 1516 to draw up plans for the gigantic undertaking. Leonardo died three years later without having accomplished anything, but in 1539 a survey was actually carried out by Nicolas Bachelier who proposed that the river Aude be joined at Carcassonne with a canal crossing the divide and emptying into the Garonne just above Toulouse.

King Francis died and nothing further happened to the grand scheme until 1559 when Adam de Craponne made another survey of the route; this time the outbreak of the religious wars put a stop to the work. Then, in 1598, King Henry IV and his minister Duc de Sully engaged Humphrey Bradley to prepare a scheme. But Duc de Sully wisely ab-

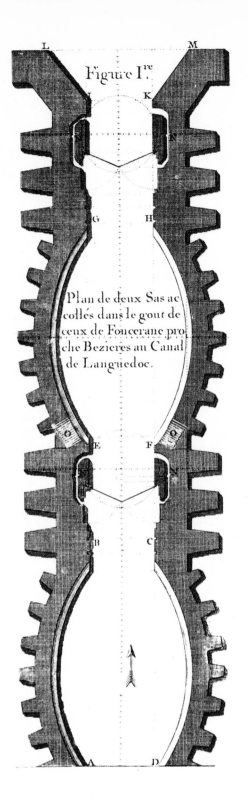

Plan of two locks in the stairway near Béziers on the Languedoc Canal. (From Bélidor.)

stained from the deux mers canal in favor of more practical and commercially more promising projects such as the 34-mile Briare Canal linking up the Loire and the Seine. The work on this canal was started in early 1605 by the contractor Hughes Cosnier, who obtained 6,000 drafted troops from Sully for the digging. When upon the assassination of the King in 1610 Sully was forced to resign his office, a commission appointed by the new regime found that 26 miles had been dug and 35 locks completed. Although the commission, which included old Bradley, was impressed by what had been done and recommended that the work be finished, the project was abandoned for political and financial reasons, and nothing more was done for 17 years. When at long last authorization was given in 1629 to start work again, Cosnier was dead, and since there was no one else capable of doing the job it was abandoned.

Finally, in 1638 Cardinal de Richelieu granted letters patent to a company which undertook to finish the work and pay the land compensations still outstanding in return for complete ownership of the canal. The company restored the neglected canal works and completed the remaining 8 miles and the last five locks so that the Briare canal could be put in service in 1642.

This first long French canal, when operated by a private company, proved a good investment which produced 13 percent in dividends in the following century. The annual tonnage carried rose to 200,000 tons in the eighteenth century, doubled during the nineteenth, and trebled in the present century.

Later, between 1682 and 1692, the 46-mile Orléans Canal was dug to accommodate the trade coming upstream the Loire to Orléans. The work is memorable for the exceptionally fine leveling performance by Sébastien Truchet of the 20-mile Rigole de Courpalet, the feeder canal bringing water to the summit of the main canal. The difference in elevation between the terminals of the feeder is only 4 feet, or 2.4 inches per mile, i.e., nearly 1:26,400.

The Languedoc Canal

While this canal construction was going on, the deux mers canal was held in abeyance. It required a minister of the caliber of Colbert, a king like Louis XIV, and a great engineer to get the grand project launched. However, calling Pierre-Paul Riquet, a 58-year-old retired tax collector, a "great engineer" is somewhat of a misnomer. Riquet does not fit into the distinguished lineage of the Low Countries' and the Lombardian hydraulic masters, and before 1662 he had done nothing to indicate his innate engineering gifts beyond showing an academic interest in mathematics. It is to the credit of Colbert and the King that they reacted positively to a letter addressed to the minister and dated November 26, 1662. It read in part as follows:

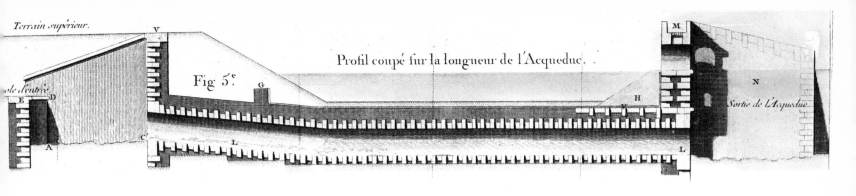

The drawing shows the construction of a type of culvert used to bring small streams under the Languedoc Canal. It was used at Quarante and completed in 1680. (From Bélidor.)

"Monseigneur,

I am writing from this village about a canal which could be constructed here in the Province of Languedoc, joining the two seas. You will be surprised to hear me speak of something of which I apparently know nothing—one does not expect a tax gatherer to go about with leveling instruments. But you will excuse my daring when I explain that I am writing under the orders of Monseigneur the Archbishop of Toulouse. Some time ago the Archbishop did me the great honor of visiting me, and he asked how the canal could be constructed, for he had heard that I had made a particular study of the problem. I told him what I knew and promised to meet him again on my return from Perpignan, and then I could show him the ground so that he could see exactly what the possibilities were; and I did this, and the Archbishop in company with the Bishop of St. Papoul and various other dignitaries have all enquired about the matter . . . ."

After the introduction Riquet goes on to explain his project for a 144-mile canal 50 feet wide and 9 feet deep, running from Toulouse to Étang de Thau. The route of the canal was essentially the same as the one surveyed by Bachelier in 1539, but the original feature of the Riquet scheme was his proposal to feed the summit canal with water according to a plan developed in collaboration with François Andreossy the previous year and tried out in models in his garden. The feeder canal was 26 miles long and tapped three mountain streams, the Alzau, Vernassone, and Lampy.

Colbert being familiar with the problems of the canal was suitably impressed and succeeded in imparting his interest to the King. A royal commission was appointed to investigate and report on the scheme, and Riquet was encouraged to work up the details. The commission reported favorably in 1664, but recommended that instead of making the rivers Lers, Fresqel, and Aude navigable, the canal should traverse the entire distance between Toulouse and the Mediterranean, ending at Étang de Thau.

The report was approved by Colbert and the King, but some doubt remained about the validity of the plan of feeding water to the summit at Naurouze. Riquet now offered to dig this feeder canal at his own expense, to prove to the satisfaction of his royal patron that it could be done. The 20×9-foot feeder canal with a length of 26 miles was dug in a surprisingly short time, from May to October, 1665, with complete success, whereupon Colbert ordered the work to continue. Money was assigned from the royal treasury, but most of the cost was to be borne by the taxpayers of Languedoc. The Estates refused to approve a special canal tax, and so Riquet proposed that existing taxes be assigned to him for six years. The amount raised by this tax-farming venture amounted to about one million dollars in gold.

Riquet was appointed general contractor in October, 1666, and in January of the following year he had 2,000 men digging. In March, the force was doubled, and by 1669 he had more than 8,000 men and several hundred women at work. The canal was divided into 12 sections, each one with a complete organization under an Inspecteur-Général. Under the contract, Riquet was to complete the work in eight years.

This was obviously an overoptimistic estimate, and when the eight years were up the work was far from finished and had been slowed down for lack of money, opposition from landowners, and the continuous quarrels with the two canal commissioners appointed by the government. There was also a scarcity of brick suitable for lock building. With inadequate funds forthcoming from the treasury and mounting costs, Riquet had to dig into his own private fortune, and when that was gone he had to sell his shares in the canal. His always precarious health, which forced him at times to be carried in

183

a litter to inspect the work, finally collapsed and he died in October, 1680, seven months before the official opening of the canal by Louis XIV in May, 1681.

Although the Languedoc Canal was far from complete when opened—it was not finished until 1692—Voltaire did not exaggerate when he declared that it was the most useful and glorious monument of the reign of the Roi Soleil. From the Garonne (at Toulouse) the canal rises 206 feet over a distance of 32 miles to a summit level having a length of 3 miles, after which it descends 620 feet to the level of the Mediterranean over a distance of 115 miles. The rise from Toulouse is taken with 26 locks, and the descent with 74 locks. The length of the canal is 150 miles, its width at water level 64 feet, and the depth 6½ feet. The sides slope 2.5:1, to provide a bottom width of 32 feet. The oval-shaped locks had a length of 115 feet and were provided with 21-foot gates.

When a difficult obstacle in the form of a hill was encountered near Bézier, Riquet after a violent quarrel with the commissioners drove at his own risk a 180-foot tunnel through the hill. Three aqueducts were needed to take the canal over the rivers Repudre, Orbiel, and Cesse; and numerous small streams were taken in culverts under the canal. In order to store water for keeping the summit section navigable throughout the year, Riquet built a 105-foot-tall earth dam with a masonry core at St. Ferréol, in the valley of Laudot, capable of storing 250 million cubic feet of water for use in the summer.

The final cost of the scheme came to 17 million livres, to which the Treasury contributed 7.5 million, the Estates 6 million. The rest, or 3.5 million, represented Riquet's private fortune. He was completely ruined when he died.

The Languedoc Canal, or Canal du Midi as it is now called, was the largest and most important civil engineering development concluded north of the Alps. There were to be many more canals built in Europe during the following century, but none can compare with this masterpiece of hydraulic engineering, designed and built by a retired tax collector turned amateur engineer in middle age.

Viaduct over the river Trebe on the Languedoc Canal.

Milan, or Mediolanum as the city was called during the best part of
its long history, was founded by the Celtic Insubrians after they swept
down the Adige Valley and defeated the Etruscans at Ticino. Having
settled permanently in the Po Valley, they left indelible genetic imprints
on European civilization. Many of Rome's major authors, such as
Virgil, Livy, and Pliny, are of Celtic ancestry. It is therefore no accident
that the area shown on the map is the heartland of the technological and
economic rebirth of Europe at the end of the first Christian millenium.
Here is also the source of European hydraulic technology, and here the
first major canals were constructed—such as Naviglio Grande from
Ticino to Milan (1179–1209), Martesana (1470), and Naviglio Interno
(1495). It was in the third canal that Leonardo da Vinci built the San
Marco lock and provided it with miter gates, which became the prototype for all subsequent pound locks, including those used today.

185

At Béziers, about 12 miles from the Mediterranean, are found most of the technical innovations on the Languedoc Canal. Above is Pont Aqueduc that takes the canal over the river Orb, and below is one of the eight locks that takes the canal down to the coastal plains (Béziers in the background). To the right is the junction of the 26-mile feed canal, Rigole de Montagne (now Rigole de canal du Midi), at the very summit of the divide. The junction is at Col du Nauroze at an elevation of 268 feet above Toulouse and 620 feet over the Mediterranean. Without the feed water, the two canal stretches could never have been joined, and it would have been necessary either to port the cargoes over the divide or, as on China's early canals, pull the barges over a slipway. The canal is 150 miles long and has a width at water level of 64 feet.

Deferred Birth of Engineering Sciences

"For although the Barbarian peoples derive no power from eloquence and no illustrial rank from office, yet they are by no means considered strangers to mechanical inventiveness, where nature comes to their assistance."

DE REBUS BELLICUS (A.D. 370)

All the works of antiquity and the Middle Ages—the pyramids, temples, drainage schemes, bridges, cathedrals, harbors, and all the rest—had one thing in common: they were built by master craftsmen devoid of theoretical knowledge, except possibly simple mathematics and geometry. Nevertheless, these masters possessed a tremendous amount of empirical know-how which was kept secret by the building brotherhoods. The monastic orders, such as the Cistercians and Benedictines, were the depositories of the construction knowledge of the Romans, and throughout the Middle Ages the body of craft mysteries kept expanding by additional influences from the far more sophisticated and technically advanced builders of Byzantium and Islam. The sudden development of the pointed arch was, as previously stated, due on the one hand to the Crusaders' encounter with it in Syria, and on the other to the translations from the Arabic of the ancient writings on geometry. A Gothic cathedral is an exercise in geometry, but it is definitely not one in theoretical statics.

Such masters as Geoffrey, Richard the ingeniator; Urri, Elgas, Nicholas, et al., who served the Angevin kings; or cathedral builders like Magister Odo; the monks Caozo and Aezilo who were responsible for the Cluny Abbey church; and Magister Gerardus Lapicida, rector fabricae of the Cologne Cathedral, were steeped in the mysteries of their craft, but beyond that they were men of genius with native instincts and feeling for material, of what could and could not be done. They had no need for science and book learning; they knew all that was necessary about the jobs they undertook.

These men would be masters in any age, competent to deal with any task confronting them, be it a railway, superhighway, or underground nuclear power plant. But such men have always been rare. There were apparently enough of them at times when there was a need for them, when the economy flourished so as to permit the undertaking of ambitious building works to the glory of God and temporal princes. But when the tempo of the western economy began to accelerate about 150 years ago, it was no longer possible to rely entirely on native engineering talent; some way of imparting their genius to the lesser breeds of men had to be invented to cope with the growing bulk of work generated by the industrial economy. The means of dividing the native engineering genius and imparting it in ever smaller bits, spreading it thin over a wide field, was at hand—the scientific method. But it was not an easy process, and the inheritance of ugly and misconceived structures left by the nineteenth century bears witness to the early failures resulting from substituting scientific engineering for native building genius.

Contributions of ancient writers

Science, theoretical knowledge and analysis, still retains traces of its ancient character of being sacrosanct and in danger of being adulterated when applied to practical ends. When Eudoxus and Archytas confirmed certain hypotheses by experiments, Plato abused them for corrupting and debasing the excellence of geometry. Even Archimedes did not think the invention of engines worthy of serious study. Plutarch wrote: "Mechanics were separated from geometry and were for a long time despised by philosophers."

In Greece, then, as in Rome as well, the great works may have been bossed by a member of the equestrian estate, but the *technites* actually responsible for the design and execution of the work were men of a lesser breed, many of them slaves, the best ones freedmen. This ancient stigma attached to being involved in construction still adheres to the engineering profession. No engineer carries official rank; he is not seated at the King's table at an official banquet. A former miner turned politician is so seated, as indeed is a captain of His Majesty's bodyguard, his chaplain, his personal physician, professors of philosophy—but not the engineer or architect responsible for the structure wherein the regal feast is being held.

187

Although Archimedes disdained the practical applications of geometry, he nevertheless produced the first book on statics, *De planorum equilibriis,* dealing with the theory of the lever and the determination of the centers of gravity of the parallelogram, triangle, trapezoid, etc. His works on hydrostatics are also well known: the upthrust on a body immersed in a fluid equals the weight of the fluid displaced by the body. In pure mathematics he invented the "exhaustion" method which enabled him to calculate the area of a circle, and which may be regarded as the germ of infinitismal calculus.

With Archimedes the Golden Age of Greek science came to an end, although two great mathematicians were to follow him: Appolonius of Pergamos (240–170 B.C.) who found a theory of conic sections, and Diophantus, the inventor of algebra, who lived in the second century A.D. Also worthy of mention in this context are Ctesibius, Philo, and Hero, whom Vitruvius recommends for study by anyone aspiring to become a master builder. Hero's "Mechanics" is an engineering manual describing a number of elementary machines, such as the capstan, lever, pulley, wedge, worm gear, plus a few more complicated ones.

However, the ancient work that came to have a profound influence on European building practice when published in 1484 were the 10 books *De architectura libri decem,* written by Vitruvius, a Roman master builder in the service of Augustus. The books contain a wealth of technical minutiae on building materials, rules of design, dimensioning of columns, temples, palaces, theaters, harbors, building in water. The tenth book deals with machines: hoisting gear, scoop wheels, pumps, catapults and ballistae, etc. When these books were made available by the process of printing, they had a tremendous influence on such sixteenth century leaders of the classicist movement as Palladio and Vignola.

But even such an outstanding authority as Vitruvius has only the following to say about static loads: "When lintels or beams are loaded, they are apt to sag in the middle and cause fracture of the work above; but when posts are introduced properly wedged up, this is prevented." Or "the weight of the wall may be carried by arches formed of wedges, correctly anchored . . . . In building where piers and arches are used, the outer piers are made wider than the rest, in order to resist the thrust of the arches."

The long struggle with statics

The scholars of the Middle Ages lived in a world of their own and contributed nothing that could be applied on a building site. Jordanus de Nemore, who appears to have lived in the thirteenth century, wrote a number of treatises on geometry and statics and attempted to formulate the principle of the rectilinear and angular lever, but did not succeed. A contemporary but unknown writer discussed the principle of the inclined plane with better success. Finally, toward the end of the Middle Ages, Blasius of Parma, a physician who taught at Bologna, Padua, and Pavia, concerned himself with the lever and inclined plane in his *Tractatus de ponderibus,* which came to exercise an influence on the mathematicians of the Renaissance, among them Leonardo da Vinci.

But these were problems that the master builders knew by instinct long before the learned professors and mathematicians began to search for rational explanations and proofs. Many of the assertions put forward were rejected out of hand as false by the craftsmen—which indeed they were.

Two of the major problems of fundamental importance to engineering which occupied the scholars of the sixteenth and seventeenth centuries were (1) the composition of forces, and (2) bending. Neither of these problems was of any particular concern to the builders who knew how to deal with them, but they occupied the minds of such giants as Leonardo, Galileo, and many others. It is interesting to note that these two simple problems, familiar to any schoolboy, required for their final solution such men as Newton, in respect of forces, and Coulomb, who came up with the final answer to bending.

Of the numerous Renaissance scholars occupied with these and related problems in statics Jean Battista Alberti (1404–1472) was the first to reconcile theory and practice. In his magnum opus *Ludi matematica* he offers guidance to the surveyor and engineer for the measuring of heights, calculations of the width of a river by observations from one bank, river regulation, etc. Leonardo's investigations included tests on the strength of materials, analysis of pulleys and levers, forces acting on an arch, strength of beams, structural forces, and so on. But his work was not published in his lifetime and had no influence on further developments, although it seems possible that Galileo may have known about it.

Outside Italy, the Netherlander Simon Stevin (1548–1620), merchant and mathematician, busied himself with the same static problems as his Italian contemporaries: center of gravity, the lever, the inclined plane. He made correct use of the parallelogram of forces and the notion of static moment. He was also the first to represent the magnitude of force graphically as the length of a straight line in the direction of the force. He thus laid an early foundation for graphic statics, which during the latter half of the nineteenth century became such an important designing tool.

In the design of domes and bridges, intuition and a feeling for statics were all that was required. Conduction of water in canals demanded a certain knowledge of hydraulics, as well as accurate field surveys that could be carried out only by mathematically trained surveyors. But beyond that, the early Renaissance builders were also "engineres dilletantes" in Italy as elsewhere.

But then Italy produced the greatest scientist of all time in the field of mechanics, Galileo Galilei (1564–1642). He relied on direct observation and experiment rather than on blind faith in ancient authorities, including Aristotle. Galileo was

Contemporary copper engraving of Galileo Galilei (1564–1642), professor of mathematics, and the first natural philosopher to apply practical tests in his researches in statics and mechanics.

magnitude of the bending strength in relation to the tensile strength of the beam. Nonetheless, his *Discorsi* initiated the study of the properties of materials and the mechanics of elastic bodies, although Galileo himself treated his materials as inelastic bodies.

During the seventeenth century there was a marked shift from Italy to France and England in the development of the natural sciences. René Descartes (1596–1650) published in 1637 his great work, "The Method," whereby he introduced his system of coordinate geometry by which the visually minded could see and grasp the implications of mathematical expressions. His contemporary Robert Hooke (1635–1703) observed that the force with which a spring attempts to regain its natural position is proportional to the distance by which it has been displaced. He emphasized that this proportion applies not only to coiled springs, but also to any "elastic body"—metals, wood, stone, silk, glass, etc., and to tensile as well as comprehensive stresses. Hooke's proposition "Ut tensio sic vis" (as the pull, so the stress) laid the basis for the further development of the classical theory of strength and elasticity of materials.

Mariotte (1620–1684) reverted to "Galileo's problem" and applied Hooke's new method of approach to the fibers of a beam subjected to bending. He correctly placed the bending moment at the middle of the rectangular cross section, rather than at the lower edge in the manner of Galileo ; and although

appointed professor of mathematics at the University of Pisa at the age of 25, and while there occupied himself with mechanics, such as finding the law of falling bodies and the behavior of the pendulum. After lecturing on astronomy in Padua and Florence, he was called before the Inquisition Tribunal in Rome in 1633 to answer for his cosmic heresies. After that unpleasant experience Galileo devoted the rest of his life to less dangerous fields of research, such as mechanics and statics. He published his great work *Discorsi e dimonstrazione matematiche intorno a due nuove science* in Leyden in 1638, and it was immediately put on the Index Expurgatorum and prohibited in all Catholic countries.

In this book he uses the term "statical moment" for the first time, as applied to the product of a weight and the speed with which it moves. To the lever he applied the statical moment proper, i.e., the product of force and distance from the fulcrum. At the same time he also made himself pioneer of an entirely new field of science, namely, the theory of the strength of materials.

Galileo begins his discourse with a cantilever beam subjected to a load at one end. To the beam he applies the principle of the angular lever. He arrives at the correct conclusion that the bending strength of a rectangular beam is directly proportional to its width and to the square of its height. But he does not consider elasticity and therefore errs in the

René Descartes (Renatus Cartesius), mathematician and philosopher, born 1596 at La Haye in Touraine. Tempted by Queen Christina of Sweden to leave his native habitat, he froze to death in Stockholm in 1650. Copper engraving of a painting by Frans Hals.

189

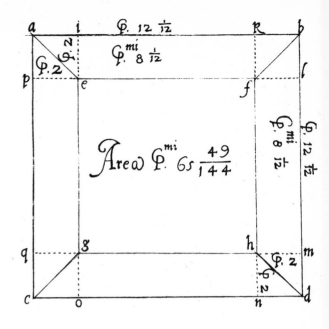

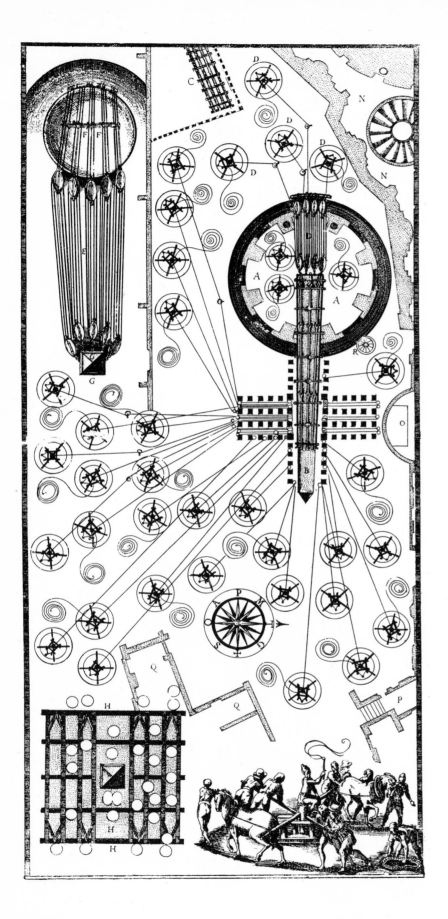

his calculations were in error, the basic principle of calculating bending stress was established. He also noted, as a result of his work of designing the pipelines supplying the Palace of Versailles with water, that the fibers in the lower half of a cantilever beam are in compression and those in the upper half in tension. This became the "Theory of Mariotte and Leibnitz."

The men who laid the foundations for the theory of statics and strength of materials included no architects or bridge builders. They were physicists, mathematicians, geometricians—many of whom were drawn into the natural sciences through their study of medicine. They were professors at universities or else found a living as "court mathematicians."

Domenico Fontana

It may therefore be appropriate in this context to devote some space to the career of a practical engineering genius of the Renaissance who also used mathematics to check on his intuitive solutions of mechanical problems. His name was Domenico Fontana (1543–1607), papal engineer and architect, who was responsible for the vaulting of the $137\frac{1}{2}$-ft. dome of St. Peter's in Rome, begun and left unfinished by Michelangelo.

The sketch near left (as well as the other engravings on this spread) is from Fontana's own book, and illustrates how he went about calculating the area of the base of the obelisk and the total volume and weight. To lift the 327-ton monolith from its plinth on Circus Maximus and lower it to its transport ramp, Fontana used 34 capstans, each driven by four horses, or two horses and four men (left). The obelisk was enclosed in a corset of iron before the lift (below). The most critical moment of the entire operation was the lift from the plinth in the Circus, which strained the technical resources of the time to the limit. This inevitably raises the question of how Caligula's engineers went about it when they lowered the obelisk in Egypt and erected it on Circus Maximus. No doubt they used the ancient method of an enclosed sand ramp.

Domenico Fontana (1543–1607).

given a number, and a trumpet blast started the operation of one or more capstans while a bell signal stopped them. Stringent discipline was maintained, and a public executioner was on hand to deal instantly with any obstructionists. The carpenters working on the scaffolding were provided with iron helmets to protect them from falling objects.

The two sites had been connected by a timber-reinforced level roadway over which the monolith was rolled. Owing to the poor subsoil, the new site had to be founded on piling, but since the site had been prepared in advance, the erection of the obelisk was relatively simple because it did not involve a vertical lift. The monolith was raised by means of rope tackle to a vertical position and secured to the new base. Subsequently, the entire operation was described by Fontana himself in a profusely illustrated volume.

Academicians and court mathematicians

In France, the advance of mechanics was influenced by the royal encouragement given sciences and higher education. The Académie des Sciences founded by Colbert in 1666 enabled many scientists to publish works which would otherwise have gone unnoticed. Before the existence of the Académie, the Franciscan Mersenne conducted an extensive

However, the job that made Fontana world-famous was not the dome of St. Peter's but the removal, transport, and re-erection of the great Egyptian obelisk from its old place on Circus Maximus to its present position in the piazza fronting St. Peter's. This operation was regarded at the time as the most amazing technical achievement of the century. Fontana was entrusted with the difficult and hazardous undertaking upon winning a competition against the outstanding masters of Italy. He was 42 years of age at the time.

The preparatory work took seven months. In order to determine the number of ropes and capstans, Fontana needed first to establish the weight of the obelisk. To determine the volume of the truncated pyramid, he dissected it into a prismatic core, four wedge-shaped lateral pieces, and four corner pyramids. Having laboriously calculated the separate volumes using vulgar fractions, he added them together to obtain the total. He then weighed a specimen of the stone and found the specific gravity to be 86 libras (Roman pounds) per span, equivalent to about 160 pounds avoirdupois per cubic foot. In this manner the total weight was calculated to $963,537\frac{35}{48}$ Roman pounds, or about 327 long tons. On the basis of this weight he decided to use 34 horse-driven capstans and five huge levers.

After a large and heavy scaffolding had been erected around the obelisk, and a similar one on the new site in the piazza, the work began April 30, 1586. The 327-ton obelisk was lifted from its base and placed in a horizontal position on a bed of rollers. This critical operation was planned to the smallest detail. Each capstan with its associated tackle was

Isaac Newton (1642–1727), natural philosopher and mathematician, set physics on a new track with his basic research, and by his invention of infinitesimal calculus facilitated the mathematical analysis of materials and structures. It took a long time, however, before his mathematics was put to practical use for engineering.

correspondence with all contemporary geometricians for the purpose of collating and publishing as much of their works as possible. However, the royal encouragement of the sciences was negated by the subsequent religious persecution of the Huguenots which drove untold numbers of gifted Frenchmen from their homeland, among them Nicolaus Bernoulli who settled in Basel where he became the ancestor of the most prolific and continuous line of mathematicians known to history. Of his nine male descendants not less than eight became eminent mathematicians. One of them, Jacob I (1654–1705), made important contributions to the theory of bending by his concept of a beam consisting of filaments capable of being stretched and compressed and the effect of resistance arising therefrom. His nephew Daniel (1700–1782) laid the foundation of mathematical physics and made important contributions to the science of hydrodynamics.

Descartes died in 1650, eight years after the birth of Isaac Newton and four years after Gottfried Wilhelm Leibnitz was born. Both of the latter invented independently of each other the most important mathematical tool available to the engineer and scientist, namely, *infinitesimal calculus*. Leibnitz arrived at the concept of calculus by pondering over the notion of continuity of mathematical functions, whereas Newton reached it through his work in dynamics. He had formulated his three laws of motion on which the entire science of dynamics came to rest, but his second law required the development of special mathematics to determine the rates of change.

Then arose the problem of calculating the total effect in a given time of a variable that is changing from instant to instant. Newton solved this problem by the invention of integral calculus, after which he discovered the reciprocal relationship between differential and integral calculus. The actual dates of Newton's discoveries of calculus—as well as universal gravitation and the nature of light—are not known, except that all this magnificent work was accomplished during the two years 1665 and 1666 which Newton spent at Woolsthorpe, since Cambridge was closed owing to the Great Plague.

Among the numerous other scientific contributors to the development of statics, Leonhard Euler (1707–1783), prominent member of the Basel school of mathematics, founded by Bernoulli, occupies a central position. He divided his career between the newly established academies of St. Petersburg and Berlin, being a favorite of both Catherine the Great and Frederick the Great. We are not here concerned with his scientific work in a wider sense, but only with his contributions to engineering theory. He worked out Jacob Bernoulli's ideas of the elastic line in the bending of beams of constant and variable sections under different conditions of loading. He succeeded in proving that short columns fail by compression only, while long ones fail by bending. His famous "Euler Formula" published in 1757 is still used to express the buckling strength of struts and columns.

Leonhard Euler (1707–1783).

The link between study chamber and building site

It should perhaps be noted at this point that the long parade of great scientists and mathematicians and their contributions to structural theory, as briefly sketched in the previous pages, are of purely theoretical interest. The principles and laws, the formulas presented after long mental anguish, the papers published, the academies established by their royal patrons, and all the rest of the scholarly apparatus had no bearing on contemporary building and construction. From the point of view of the building masters, all this scientific rigmarole was conducted in a vacuum, as far removed from the workaday realities on the building sites as the discourses of the Greek philosophers.

This is not to say that their efforts were in vain; they added up eventually to a codex of structural science that, stripped of scholarly esoterics, became eminently useful to the engineering practitioners. But it took 500 years of mental effort by the outstanding brains in Europe to arrive at the simple laws of statics that any intelligent teenage schoolboy can master in a few weeks. One important link, although by no means the last one, bridging the vacant space between the study chamber and the construction site was forged by Augustus Coulomb (1736–1806), who succeeded for the first time in summarizing the static behavior of building elements in a definite and final manner. By his *Essais sur une application des règles de maximis et minimis à quelques problèmes de statique relatifs à l'architecture* published in 1773 he became the founder of structural analysis, or building statics. His Académie paper was based on his work as a Génie officer on the island of Martinique where he had occasion to investigate the solidity of such building elements as walls and vaults. He dealt with the laws of equilibrium,

193

the resolution of forces, friction, cohesion, the behavior of beams under different conditions of load. He was the first one to recognize the vertical components acting on a beam, the so-called shearing force. From his investigations he summarized the following familiar laws of statics:

1. The sum of the tensions must balance the sum of the compressions.
2. The sum of the vertical components of the internal forces must equal the load applied.
3. The sum of the moments of internal forces must balance the bending moment produced by the load.

Coulomb, then, was the first to state unequivocally that the forces acting on a beam must be in equilibrium. Although he applied exact scientific methods in arriving at his conclusions, Coulomb was conscious of their application to building practice. "I have endeavored, to the best of my ability," he wrote in his essay, "to make the principles I have used sufficiently clear so that a technician with some training can understand and use them." Alas, in this he clearly overestimated the capacity of his fellow engineers and indeed also of his fellow mathematicians. The rich contents of the essay were presented in such concise form and concentrated to so few pages that they escaped the notice of the experts in statics for 40 years.

Jean Baptiste Colbert (1619–1683), powerful minister of Louis XIV, founded the French Academy of Sciences and established the Corps de Génie, the engineering corps of the French army, whereby he contributed to the introduction of a mathematical and scientific education for engineers.

In a memoir dated 1779, Coulomb makes concrete suggestions for the use of compressed air for work under water. The wooden caisson proposed by him consisted of three chambers, the central one being the working chamber provided with an air pump and lock. The two lateral chambers served to give the caisson buoyancy when floating it out to the site.

Coulomb also investigated the effect and efficiency of human labor. To make rational use of the human body, the most versatile of all machines, one must, according to Coulomb, "augment the effect without augmenting the fatigue." He analyzed observations of carrying heavy loads when climbing mountains, and statistics compiled by Vauban on the transport of materials in wheelbarrows. He ends by insisting on the importance of breaks and recommends that no more than seven to eight effective working hours be permitted for heavy work.

Beginnings of scientific engineering training

As previously stated, the mathematical and physical investigations into the behavior of building elements and strength of materials had no impact on contemporary building practice. The statement holds largely true when applied to Europe as a whole, but in respect to France it must be modified substantially. In that country mathematics and structural theory began to be applied in a definite manner within the military establishment during the last quarter of the seventeenth century.

The man responsible for the early French predominance in what subsequently became known as the engineering sciences was Jean Baptiste Colbert, the brilliant minister of Louis XIV. After persuading the King to establish the Académie des Sciences and the Académie de l'Architecture, he acted on a suggestion by Sebastien le Prestre de Vauban, France's foremost military engineer, and constituted in 1675 the Corps de Génie, a body of military engineers trained by Vauban. The main reason for setting up the Corps was to improve the status of the Génie officers and put a stop to the high-handedness and abuse to which they had been exposed by temperamental field generals. The Génie cadets were to be given a scientific education, with special emphasis on mathematics, at existing state schools. Later they were trained at the École des Ponts et Chaussées.

This school, which as the name implies was part of the department of roads and bridges, was founded in 1747 for the express purpose of training sorely needed engineers. It was reorganized in 1793 by Jean Perronet, foremost French bridge builder throughout the centuries, when he became Premier Ingénieur of the department of roads and bridges. The instruction of the school was very informal; notes were made and circulated among the young men in training who were also encouraged to take part in ad hoc discussions with experienced engineers and visiting scholars. The students had access to Bernard de Bélidor's famous textbook *Science des Ingénieurs,* first published in 1729 and repeatedly reissued until 1830. It contained in a concise and handy form a wealth of information in the field of civil engineering, rules for construction and dimensions, elementary physics, specific gravities of building materials, and all the rest usually found in an engineering manual.

Thus the École des Ponts et Chaussées became the first channel through which physics and, beyond all, mathematics was directed to educational and, ultimately, practical ends; and as a consequence French engineering science reached an early flowering during the eighteenth century that came to have a profound influence on the development of civil engineering in the next century, too. It could have been of even greater and more enduring importance had not the French Revolution put an end to the supremacy of the French sciences. Professors and students were suspected of counterrevolutionary activities, and schools and universities were closed. As a consequence, there soon developed a serious lack of military engineers in the revolutionary armies.

For this reason Napoleon authorized his personal friend Gaspard Monge, a prominent mathematician, to found a new type of school that would supply the Génie Corps with scientifically trained engineers. This school, established in 1795, was given the name *École Polytechnique.* Here future engineers were given a broad education in the engineering sciences, and the graduates were prepared for the advanced design studies at the École des Ponts et Chaussées.

In addition to being responsible for the establishment of the first specialized engineering school in the world, Gaspard Monge, while professor of mathematics at the military school at Méziers, invented a new branch of mathematics called by him "descriptive geometry" which immediately became of tremendous importance for all branches of French engineering. Indeed, it was regarded as having such military value that it was kept a top secret by the French government for a quarter of a century. To later generations of engineering students there is nothing particularly strange about Monge's invention; it was simply what subsequently became known as mechanical drawing, a method of placing plans, sections, and elevations of three-dimensional objects on a two-dimensional sheet of paper. Projection drawing was taught to cadets of the French services, whereby a young naval officer by the name of Marc Isambard Brunel became acquainted with it. Later, when a refugee from the revolution, Brunel settled in England and introduced this most important engineering tool to his adopted country.

In this context mention should be made of one of the most distinguished lecturers at the École des Ponts et Chaussées, Louis Navier (1785–1836), who joined the staff in 1807 at the age of 22 and became a full professor of applied mechanics in 1830. Besides designing a number of the major bridges crossing the Seine, Navier made it his principal task in life to apply discoveries and methods of theoretical mechanics to practical construction as well as to the teaching thereof. His lectures at the École were published in 1826 under the long title *Résumé des Leçons données à l'École des Ponts et Chaussées, sur l'Application de la Mécanique à l'Établissement des Construction et des Machines* and constituted a general survey of structural theory and strength of materials. In this book Navier for the first time collected the isolated discoveries of his predecessors in mechanics and allied subjects into a single unified system of instruction. He taught how to apply known laws and methods to practical engineering, using them for the calculation of structural dimensions. A large number of methods still in daily use go back to Navier, indeed are named after him, such as the theory of flexure, buckling, eccentric loads, the solution of statically indeterminate structures, continuous girder on three or more supports, the two-hinged arch, the theory of flat slabs and plates.

L'École Polytechnique inevitably became the model for the numerous engineering schools started on the Continent during the nineteenth century, such as the Eidgenössisches Polytechnicum in Zürich established in 1855, the Polytechnic schools at Delft in 1864, and similar ones in Chemnitz, Turin, and Karlsruhe, to mention but a few. In the United States, the Massachusetts Institute of Technology was founded in 1865.

The British way of learning

In England, however, scene of the greatest and most rapid engineering developments that the world has ever seen, neither architects nor engineers placed much value on a formal university training for their juniors. The young man aspiring to practice was articled to a practitioner; he paid a fee and drew no salary; his training consisted in picking up what he could from salaried assistants to whom he made himself useful. His progress depended on his keeping his eyes and ears open and studying in his spare time. In an office handling important and interesting work, a keen pupil could learn a good deal, and with good connections and adequate funds he could eventually buy himself a partnerhip in a firm or open an office of his own.

In this way a fundamental difference developed between Continental and British engineers. The young Continental engineer after finishing his formal education was far more advanced in both scientific theory and design than his British counterpart, but unlike the Briton, he had no experience whatever of commercial and professional practice.

Still, Britain was not altogether lacking in theoretical engineering education. From its foundation in 1828 the University College in London offered lectures on engineering, and a Regius Professor of Engineering was appointed at

Thomas Young (1733–1829) introduced the idea of the modulus of elasticity in 1807, but it took nearly half a century before engineers knew what he was talking about.

Glasgow University in 1840. This chair was held by William McQuorn Rankin from 1855 onward, and his *Manual of Civil Engineering* published six years later provided English-reading students for the first time with the equivalent scientific data that previously had been available only in French. From then on there was no need for English engineers to learn French in order to read Bélidor and Gautier, as had been the case with such great engineering pioneers as Thomas Telford and John Rankin.

However, the lack of engineering colleges on the Continental model no doubt contributed to the long delay in acknowledging and adopting Thomas Young's idea of "a modulus of elasticity" which laid the capstone on the centuries-long struggle in learning how to understand the behavior of materials. Thomas Young, professor of natural philosophy at the Royal Institution, introduced the idea in a lecture in 1807, but it was ignored for several decades and was not applied until about the middle of the century.

Graphic statics, on the other hand, which at one stroke did away with the elaborate and time-consuming calculations hitherto needed for stress analysis, was more or less simultaneously developed by professors at the newly established schools at Zürich, Dresden, and Milan. Karl Culmann, professor of engineering sciences at Zürich, first published his *Graphische Statik* in 1864, which was further simplified by L. Cremona in Milan in 1872, and added to by Otto Mohr of the Dresden Technical College. Culmann's magnum opus *Anwendungen der graphischen Static* in four volumes published posthumously (1888–1906) by his pupil and successor to the chair at Zürich, Wilhelm Ritter, still remains the classical work on graphic statics, the most powerful engineering tool hitherto invented. But the full use of the method had to wait until the twentieth century.

The great British engineers of the nineteenth century, who set the world on a new track, had no need nor knowledge of such scientific designing tools. They were native geniuses, apprenticed as instrument makers, millwrights, or masons, and more related to the ancient masters of the Middle Ages than to the contemporary philosophers of natural science.

But no matter how handicapped they may have been by their lack of scientific training, it was to them that the future, at least the immediate future, belonged. In 1771, a small group of them, frequently called upon to give evidence before parliamentary committees on canal and harbor projects, gathered around John Smeaton to form the Society of Engineers, the first professional engineering body. John Smeaton was the first to call himself *civil* engineer—to accentuate the nonmilitary nature of his work. On the threshold of the railway age, Thomas Telford and H. R. Palmer turned the modest society into the impressive *Institution of Civil Engineers* which was incorporated under a royal charter granted June 3, 1828.

The British Century of Engineering

"They look to their own hands for a living . . . without them there is no building up a common-wealth."

ECCLESIASTICUS

On the night of December 1, 1755, the pitch-saturated timber of the Eddystone lighthouse caught fire, and the light which since 1698 had warned mariners approaching the port of Plymouth on the Cornish coast of the nasty gneiss reef 14 miles offshore was extinguished. That night marks the beginning of the great English tradition in civil engineering.

The private proprietors of the lighthouse asked the president of the Royal Society for advice as to who might be entrusted with the design and erection of a new lighthouse. The president, the Earl of Macdesfield, recommended a 31-year-old former instrument maker and fellow of the Society by the name of John Smeaton as eminently suited for the job.

Smeaton accepted the assignment and decided to erect the new structure entirely in stone. He designed the tower in the shape of an oak trunk, in order to spread the base over the largest possible area, and realizing that the structure would be exposed to heavy buffeting by storms coming straight from the Atlantic, and that transport of megaliths was out of the question, Smeaton had each stone cut to dovetail with its neighbor. The dovetailed masonry of each course was pegged to that above and below, to give additional solidity to the fabric. In this manner, Smeaton hoped that his tower would be the equivalent of a solid megalith structure.

Portland stone was used for the interior, with granite for cladding. The stones were cut with templates and each course fitted on the shore. The first foundation stone, weighing $2\frac{1}{2}$ tons, was placed on June 12, 1757, and by the end of the month two complete courses were laid. On October 16, 1759, the Eddystone tower was finished under the supervision of Josias Jessop, a shipwright from Plymouth.

Neglecting Smeaton's other major contributions to engineering, such as his steam pump, diving bell, and so on, we shall direct our interest to two important and far-reaching initiatives taken by him while working on the Eddystone. One was his research into the properties of lime in order to find a hydraulic cement to bond the masonry. He collected limestone from all parts of England and tested innumerable mixtures of lime, clays, lias, and pozzolana. From his

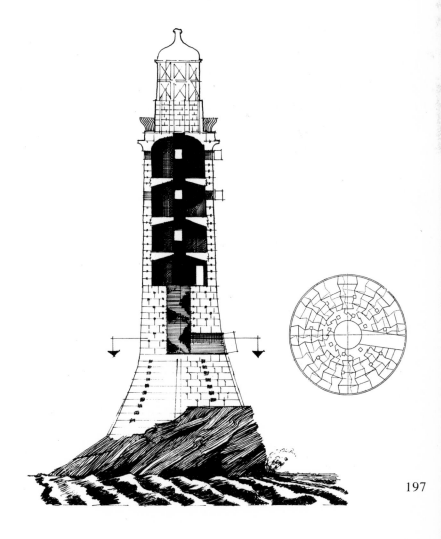

The Eddystone lighthouse, designed and constructed in 1757–1759 by John Smeaton on a reef outside Plymouth, can be regarded as the takeoff of pragmatic English civil engineering. The sketch below shows how Smeaton bonded the stone masonry in dovetail fashion to attain a structure resembling a megalith. The plan refers to the fifteenth course of the structure.

mechanical and chemical tests he concluded that a cement capable of setting under water must consist of limestone containing appreciable amounts of clay. The natural cements made from minerals in which calcareous and argillaceous constituents were present in suitable proportions would, after calcination and grinding, make an "eminently hydraulic lime."

In the end, Seaton chose Italian pozzolana from Civitavecchia mixed with Aberthaw lias for the mortar used to bond the Eddystone structure, but his pioneering work led eventually to the development of artificial cements made by mixing limestone and clay in what were for many years empirically determined proportions. The further development of the cement industry will be discussed in the following chapter.

His second initiative was to take care of young William Jessop upon the death of his father and train him in civil engineering. The latter, in his turn, by his generous and unselfish service helped shape the professional development of Thomas Telford and John Rennie, and through them the following generations of British civil engineers.

Before entering on the heroic age of British civil engineering, the Smeatonian line of descent is worthy of notice owing to its implications for the future. After his successful work on the Eddystone, Smeaton became the acknowledged leader of his profession, but with the training of William Jessop, the latter became as prominent as his master by the turn of the century. Among Jessop's major works are the two West India Docks built between 1799 and 1802 on the Isle of Dogs in Poplar. The Import Dock is 2,600 feet long and 500 feet wide, while the Export Dock has the same length but is only 400 feet wide. The depth in both is 24 feet. Both docks share a common basin at each end; the Blackwell Basin at the eastern end is oval-shaped and covers 6¾ acres. The basins were provided with lock gates at the river and dock end. Owing to the excellent workmanship, these vast hydraulic structures are as good today as when built.

In the burst of dock building in London, following upon Jessop's successful conclusion of the West India Docks, were included the London Docks at Wapping and Shadwell, on which work began in 1802. The London Docks were engineered by John Renni, a former millwright trained by Jessop, and they consisted—in the first building phase—of the Western Dock 1,200 by 690 feet and a 3-acre basin closed by dock gates. The depth was 20 feet. The lock was built within a cofferdam of timber piling driven with the aid of an 8-hp steam engine, the first time steam was used for pile driving. A 25-hp engine drove the pumps. The docks were completed in 1805; and 10 years later, work began on the Tobacco Dock, of similar size and layout.

Thomas Telford, former stone mason, was also active in dock construction on the north shore of the Thames, although a quarter of a century later. He cleared 27 acres of a slum region east of the Tower of London and built there in the short time of somewhat more than three years—from June, 1825 to October, 1828—two dock basins and entrance locks, together with warehouses and other dockside facilities.

However, dock work along the Thames was hardly more than a minor incident in the busy lives of the Smeatonians. These self-taught men engaged in an incredibly wide range of engineering, from the design of machines, integrated ironworks, bridges, canals, harbor works, roads and, eventually, railways. They were all of them native geniuses, who in a country emerging rich and powerful from the Revolutionary Wars happened to be available when there was a need for them. It is interesting to note that as the industrial revolution progressed in ever-widening circles, similar men emerged to make, at any rate, local engineering history. A generation earlier they would have remained hidden in the amorphous mass of wheelwrights, masons, carpenters, and similar artisans. A few generations later they would have been stopped in their personal development by the insistence on formal educational standards for entry into scientific engineering schools. They might have become rich and powerful as manufacturers and contractors, or they might have gained influence as gray eminences in the backrooms of engineering consultants, but being self-taught men, they would not have been accepted either by the professional societies or by government departments in need of their services.

The native British geniuses who snatched the initiative from the French engineering establishment and whose work left an indelible imprint on further development without their writing any scientific papers or textbooks were followed by subsequent generations of other self-taught men who acted as midwives to the railway age and steel age—the Stephensons, Brunels, Locke, and numerous others. But as the nineteenth century advanced and the newly established engineering schools began to pour out graduate civil engineers, and the projects multiplied by geometric progression, the individual efforts became lost in the mass; and it requires laborious and not particularly rewarding research to sieve out the engineering personalities involved, except in the most outstanding accomplishments. In the gigantic engineering schemes of the twentieth century, the individual contributions have become altogether lost, and the developments have become associated, in the ancient manner, with names that have had nothing or little to do with the work involved—dictators, presidents, politicans who in some way or other managed to become associated with them in the public mind. Thus we get the Hoover Dam, Nasser's High Dam, Hitler's Autobahn, Mussolini's drainage of the Pontine Marshes, Cape Kennedy's rocket launching sites, and so on.

Road building in England

In all European countries, the roads were in a deplorable state well into the eighteenth century and beyond. Sporadic attempts at road building and maintenance by various authorities had failed, and although France took pride in a roadnet of some 17,000 miles, the royal roads were paved only a few miles beyond the major towns, and road traffic consisted largely of pack and saddle horses, pigs and cattle, plus a few wagons and carts.

In England, the roads were in a thorough mess, and in the dusk it was impossible to distinguish the road from the heath or fens on the side of it, as told by Samuel Pepys who got lost on his way between Newbury and Reading. Between Reading and Beaumaris the travelers had to proceed afoot, because their carriages were taken apart and carried by peasants to the Menai Straits, as described by Macaulay.

To the north of York and the west of Exeter, goods were transported by trains of packhorses. Wealthy people rode in their own carriages drawn by six horses to prevent getting stuck in the mire. But often even six horses were unable to pull out a carriage sunk over the hubs in mud.

The poor state of the roads naturally had some evil economic effects, because the expense of road transport became excessive. In 1685, the cost of hauling a ton of goods from London to Birmingham was £12, or 15 pence per ton-mile. Coal could not be shipped inland, and was not used or even seen except in places where it was mined or where it could be shipped by sea. For this reason the fuel that provided the energy for the takeoff of the industrial revolution was generally known as *sea coal* in the south of England.

The reason for the poor state of the roads was a law which forced farmers and peasants residing along them to devote six days each year—without pay—to repairing them. The Great North Road linking up London with West Riding passed through poor and thinly populated areas that could not possibly supply the labor essential to keep this busy road in good shape.

True, in an attempt to encourage road improvement the Turnpike Act was passed in 1663, whereby the principle was established that the road users should pay a toll or road tax. "Companies of Proprietors" were given 25 years to collect tolls on the roads built and maintained by them. Although always detested and avoided whenever possible, the turnpike roads kept increasing in length to about 30,000 miles, administered by approximately 1,100 turnpike trusts by the time the first railway was put in operation in 1830. However, by then the road companies were sunk under a heavy burden of debts and were eventually replaced by the General Highways Act of 1835.

But 50 years after the passing of the Turnpike Act the roads were obviously in a far from passable condition, judging from an eye-witness report. Daniel Defoe made a journey through Kent some time in the second decade of the eighteenth century and described his experience in part as follows: "I left Tunbridge and came to Lawes through the deepest, dirtiest, but in many ways the richest and most profitable country in that part of England. The timber was

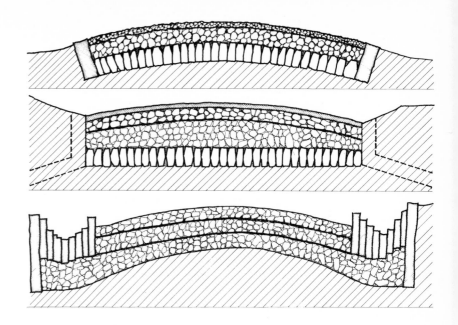

The sketches illustrate three early methods of reducing the cost of road building. The top section is that of a Trésaguet road bank (1764) consisting of a cambered foundation layer of large stones placed on edge, on top of which was a layer of smaller stones hammered into place. The wearing surface was comprised of walnut-size stones. The middle section shows a Telford road which also consists of a bottom layer of large stones placed edgewise, but on undisturbed ground, and carefully set and leveled by hammer. It was followed by two layers of 2½-inch stones placed with a 7-inch camber. On top of this was a 1½-inch layer of washed gravel. The macadam bed (bottom) was constructed entirely of stones hammered to a 2-inch mesh and placed in three 4-inch layers. The stones were compacted by the traffic, and the iron tires of the wheels ground off chips that filled the interstices between the stones.

prodigious, sometimes I saw one tree on a carriage drawn by two and twenty oxen. But it was carried only a small distance and then thrown down and left for others to take up and carry on, and sometimes it takes two or three years before the tree gets to Chatham. For when the rains come, it stirs no more that year, and sometimes the whole summer is not dry enough to make the roads passable."

The primary reason for the inability to build and maintain satisfactory roads was the same everywhere, namely, the high cost. Well into the eighteenth century, roads were built on Roman technical precepts, and although, or perhaps because, conscript labor was widely used, the cost of road building was prohibitively high. It may be said with some

199

measure of truth that all major efforts in subsequent road building have been aimed at reducing costs.

The first independent, i.e., non-Roman, method of road building in Europe was invented by Pierre Trésaguet (1716–1794) and was applied in France in 1764. Trésaguet's road consisted of a foundation layer of stones placed on edge to a height of 6 to 7 inches. On top of this foundation he placed smaller stones to about the same thickness, and hammered them into place. The top layer, or wearing surface, consisted of a 3-inch layer of walnut-sized stones and was given a moderate camber to shed water into the side ditches. The Trésaguet type of road could be built at less than half the cost of the previous ones, and the wearing surface lasted up to 10 years if properly maintained.

This, then, was the type of road that was subsequently adopted as standard all over Europe. In Russia, for example, 450 miles of Trésaguet roads were built connecting St. Petersburg with Moscow, and in 1781 work began on the great Siberian Highway leading from Moscow to Irkutsk.

In England, road building under the Turnpike Act followed a slightly different line of development. The first to apply sound principles to road engineering was a blind man, John Metcalf of Knaresborough, who built 180 miles of turnpikes in Yorkshire around the middle of the eighteenth century. His roadbed consisted of large stones covered with a layer of smaller stones to form a camber shedding water into ditches on both sides. The strange thing about "Blind Jack," as he was called, was that he selected and surveyed the route all alone, using only a long staff. According to contemporary witnesses, he possessed the uncanny knack of choosing the most economical route "by a method peculiar to himself which he cannot well convey to others."

However, the great technical advance in English road building was brought about at the turn of the century by Thomas Telford when he was working for the turnpike trusts in the county of Salop. He modified the Trésaguet road in numerous details which improved its lasting qualities but also increased its cost. A typical Telford road was 34 feet between the fences, while the so-called "metaled section" was 18 feet wide surrounded by an 8-foot graveled roadway. The slope of embankments and cuttings was 1.5:1. The longitudinal gradients never exceeded 1:30, and ascents and descents were joined by smooth curves. Filled sections were built up so that the side with the higher fill had a greater elevation than the opposite side, to allow for future consolidation of the road bank.

The 18-foot metaled section in the middle was formed by placing stones edgewise upon the undisturbed ground to a height of 7 inches. The stones were carefully placed by hand, and the height was adjusted by hammer. Two layers of broken $2\frac{1}{2}$-inch whinstone were laid on top of the foundation. No stone could exceed 6 ounces in weight, and each one had to pass through a ring with a diameter of $2\frac{1}{2}$ inches. The two layers were built up to a total height of 7 inches and given a

6-inch camber. At each hundred yards there was a lateral drain leading to the side ditches. The roadbed was finished off with a $1\frac{1}{2}$-inch layer of clean gravel.

This was the road that Telford introduced when he was put in charge of all road building in Scotland in 1803. During the next 18 years something like 920 miles of such high-quality roads were put down. The work was done by contract, and to prevent swindles and more innocent corner cuttings, Telford developed a legal instrument embodying rigid specifications from which modern construction contracts derive.

The Telford road was definitely not the last word in road building; indeed as mentioned in passing, Telford's insistence on treating the roadbed as a masonry job made his roads too expensive. Something else was needed if European and other countries the world over, sorely in need of a passable road-net, were to be able to undertake the costs involved. Again, as so frequently happens in engineering history, the man who came up with the idea that set the technology of road building on a new course was not an engineer, nor did he possess any experience whatsoever of road building. John Loudon McAdam was a businessman who after amassing a fortune during his 13 years in New York returned to his native country in 1783 and was appointed Deputy-Lieutenant and road trustee in Ayrshire. In this capacity he became fascinated by the economic aspects of road construction. For Telford, road building was but a passing phase in his rich and varied life as civil engineer; for McAdam, road building became the one passion of his life.

The McAdam principle of road building was to let the native soil support the weight of the traffic. McAdam was convinced that provided the roadbed could be preserved in a dry state, it would support any weight without sinking. Thus, he dispensed with the heavy stone foundation and built up his cambered roadbed on the dry natural soil with three 4-inch layers, using stones passing a 2-inch mesh and not exceeding 6 ounces in weight. No graveled wearing surface was needed, he reasoned, because the iron tires of coaches and wagons would grind off chips to fill the interstices. But it was important that the stones be free of clay, earth, and chalk, all of which hold water. The stones were compacted, layer after layer, by the traffic, whereby the roadbed was eventually turned into a solid impervious mass.

McAdam came to have a tremendous influence on road building everywhere in the world, and his many writings on his favorite subject were translated into many languages. His road building method also brought about a new source of income for the rural proletarians who could now put their women and children to breaking stone meshed to McAdam's specifications. By the end of the nineteenth century, about 90 percent of the highways in Europe were macadamized; in 1870, Britain alone had 160,000 miles of macadam roads. In the early 1830s, coaches drove over his new highways at an average speed of 10 miles an hour, and the mail was carried from London to Birmingham in 12 hours and to Exeter in 17 hours.

By that time stones mixed with hot tar and compacted after cooling had been tried out in Nottinghamshire, while in France bitumen had been experimented with as a surface for macadam roads. The Place de la Concorde was paved with bitumen in 1835. Some 20 years later, a few stretches of concrete pave had been laid out in Austria; a decade later similar experiments were made in England and the United States. But nothing came of these early trials with paved roads until the arrival of the motorcar whose pneumatic tires sucked out the fine binding material of the macadam roads and churned them into a mire. But that is another story.

Then came the canals

The reason for the canal building craze that set in about 1750 in England is therefore not hard to find, and is excellently illustrated by the heavy oak logs which Daniel Defoe noted lying beside the rutted Tunbridge-Lewes road waiting to be picked up after the rains by an ox team in order to bring them a few miles further toward their final destination, the Chatham naval yard on the Thames. It took a minimum of two years to move a log from the site to its destination 50 miles away. It required 22 oxen, in innumerable relays, to

This engraving from 1828 shows a peaceful yet economically fruitful scene at Paddington on the Regent's Canal. Here one horse pulls a heavily loaded barge without excessive effort. A bargeman is fishing while his mate is at the rudder and is keeping an eye on the lit stove in the cabin. Canal transport is still, under certain conditions, the most economical and pleasant way to move a surplus of commodities.

move it foot by squeaking foot. But put the log, or rather a number of them, on a barge pulled by one horse along a canal, and they would be at the naval yard within a few days.

In the ancient economies—and by "ancient" is meant in this context conditions up to a century and a half ago—road transport doubled the price of a commodity per 100 kilometers, not counting the tolls taken by highwaymen and robber barons. On the "metaled roads" built from 1750 onward, one horse could pull a load of 2 tons, while the same horse could pull a barge loaded with 50 tons on the sheltered water of a canal.

This tremendous difference in transport performance explains, for example, why people in one area could literally starve to death as a result of a disastrous harvest, while in a province not far off peasants and farmers were embarrassed by overabundant crops which wrecked the prices. It also explains why the ironworks in such countries as Russia and Scandinavia had to be supplied with ore and charcoal in winter when the frozen bogs and lakes provided the only roads capable of carrying heavy and bulky loads at less than prohibitively high costs.

Thus we find that as soon as industry and commerce in an area populated with enterprising people develop beyond the scale requisite for satisfying purely local needs, the transport bottleneck can only be broken by undertaking the heavy investments needed to develop inland water transport, for clearing rivers or digging canals. Romans and Phoenicians moved their bulky cargoes by water; in the Euphrates Valley bulk commodities were also moved either on the river or over the finely meshed canal net. Coastal and island people appear predestined to be become "shopkeepers of the world," and when they find it necessary to develop their uplands, they resort to the most efficient means they know: they build canals and let water carry the heavy loads.

By the middle of the eighteenth century the English economy had developed to the point where canals were being considered; but actual canal building did not set in until 1755 when some Liverpool merchants financed the digging of the short Sankey Brook Canal to join St. Helen's coalfield with the Mersey. Before the canal was finished they engaged a self-taught surveyor and former millwright, James Brindley, to investigate the possibility of uniting the rivers Trent and Mersey with a canal. Nothing came of this canal venture, but Brindley's survey, as well as his previous work of designing a flint mill for Josiah Wedgwood's potteries at Stoke-on-Trent, attracted the attention of Francis Egerton, third and last Duke of Bridgewater. On his "grand tour" of the Continent, the 23-year-old Duke had seen for himself the benefits brought by the canals, and when he returned to England, he set about developing the coal mines on his estate at Worsely and, in order to move his coal to Manchester, he proposed to cut a canal from the colliery to the city. Hearing about James Brindley, he appointed him to conduct a survey and report on the feasibility of a canal.

"Water is a giant; therefore lay him on his back," was the way James Brindley, the English pioneer of canal construction, expressed his engineering philosophy of keeping level grade without recourse to locks. He followed this precept when laying out the Worsley Canal along a traverse that followed a contour. As a consequence, he had to carry his canal on a 595-foot-long stone viaduct crossing the Irwell River. To prevent leaks the viaduct was lined with "heeled" clay.

Brindley returned from his "ochilor servey" and proposed digging a level canal without locks, but in order to do that he had to cross the Irwell River by means of a large stone aqueduct, which also meant that the approaches had to be lifted to grade by the construction of high embankments.

The Duke obtained the necessary enactment for his daring enterprise in 1760, despite opposition from leading technical opinion which regarded the scheme insane. Work on the 10½-mile canal began immediately, and the canal was opened for traffic a year later, on July 17, 1761. It was extended into the mine by means of a tunnel to facilitate loading the barges directly from the stopes. To make the canal bottom, the high embankments, and the viaduct waterproof, Brindley used puddled clay, but instead of employing cattle in the ancient manner, he put his men to "heel" the clay, by tramping it with the heels of their boots. Heeled clay is still an excellent way to waterproof hydraulic structures—provided of course that cheap labor with boots is available.

It has been claimed on good grounds that the Worsley Canal marked the beginning of the new age of industrialization. It broke the monopoly grip of the owners of the Irwell River who had charged exhorbitant tolls for its use for navigation. Unlike the river, the canal was navigable throughout the entire year. With the canal in operation the millowners in Manchester got coal in unlimited supply at 3½ pence per hundredweight, the lowest price in England. As for the Duke, he spent altogether £200,000 on the Worsley and some other canal ventures on which he obtained revenues amounting to £80,000 per year.

The tremendous commercial success of the Worsley Canal set off the canal boom in England. Five years later, on July 26, 1766, Josiah Wedgwood cut the ceremonial first sod of the 139.5-mile-long Grant Trunk Canal, up until then the greatest civil engineering enterprise in Britain. It was engineered by James Brindley in collaboration with John Smeaton, and it joined the Duke's canal at Preston-on-the-Hill with the Severn, where a canal port was built and named Stourport. The Grand Trunk like all other canals planned by Brindley was laid out on an easy gradient and followed the contours in a meandering way, without embankments and cuttings. When a watershed had to be crossed, Brindley preferred to tunnel through a hill rather than make a large cut. He used five tunnels on the Grand Trunk Canal: Harcastle (2,880 yards), Hermitage (130 yards), Barnton (560 yards), Saltenford (350 yards), and Preston-on-the-Hill (1,241 yards). The tunnels were low and narrow, being only 13 × 17 feet—indeed, the long Harcastle was only 12 × 9 feet—and they became from the outset serious bottlenecks since they permitted only one-way traffic. Therefore, a second larger Harcastle tunnel had to be excavated—by Telford in 1827—and provided with a towpath.

The canal boom culminated in 1792 with a flurry of speculation to raise money for not less than 30 canal schemes. Up until that year the canals had been promoted and constructed by local companies that had difficulties in raising finance, and they had to be built as cheaply as possible. This is reflected in the layout of the early English canals which without exception meander about the countryside avoiding

Mont Blanc, the 15,782-foot-high peak on the grandest mountain massif in Europe, was climbed for the first time in 1786 by Jacques Balmat of Chamonix. But it was the third climb made on August 1 to 2, 1787, by the Swiss scientist H. B. de Saussure with 18 guides and porters that is interesting from the point of view of engineering. The event inspired the exploration of the Alps and the scientific study of their complex geology—a necessary prerequisite to the driving of the long railroad tunnels that began 60 years later. Jacques Balmat served as the chief guide on this expedition, and is shown in the upper right-hand corner in this hand-colored engraving published in Basel in 1790. In his field notes, de Saussure made the observation that Mont Blanc eventually would have to be pierced by a tunnel to improve communications between the people north and south of the massif. His prophecy did not come true until 1965 when the vehicle tunnel between Chamonix and Entrèves was opened after six years of superhuman efforts. The Italian portal of the tunnel is shown at the right.

The West India Docks, London's first non-tide harbor basins, were built from scratch in the incredibly short time between January, 1800, and August, 1802, when on the 17th the first two ships passed through the dock gates. William Jessop designed the docks and managed the construction, while the architect Gwilt designed the warehouses. The West India Docks have two harbor basins, namely, the Import Dock measuring 2,000 × 500 feet and the Export Dock having the same length but a width of 400 feet. They share a common basin at each end, and the depth was originally 24 feet. The east one, called the Blackwell Basin (right), is oval in shape, has an area of 7.2 acres and was used for ocean shipping; whereas the Junction Dock (Limehouse Basin) is 1.3 acres and was intended for barges. The masonry walls of the basins were constructed on a bed of gravel and have a rounded profile buttressed at the back. The curved profile was chosen as the most natural shape for the lower part of the walls because it resembled a clay bank. Owing to the good foundation and the solid masonry work, the docks are still in the same shape as when they were built. The area was expanded in the 1860s by the addition of a third basin engineered by John Hawkshaw. The West India Docks initiated a construction boom along the Thames, during which most of London's existing docks were built. The contemporary engraving made by William Daniell was commissioned by the West India Dock Dompany for the opening in 1802.

costly locks and cuttings. Instead, they had about 45 miles of narrow canal tunnels, and since there were no towpaths, the barges had to be propelled through them by "leggers" who lying on their backs pushed the barges ahead in a zigzag course by placing their feet against the tunnel walls.

As more capital became available at the end of the eighteenth century, the canal companies eliminated many of the time-consuming long loops by shortcuts requiring deep cuttings or locks. The canals built by Rennie, such as the Kennet and Avon, the Rochdale and Lancaster, are built almost on Continental principles, with such costly engineering features as deep cuts, embankments, aqueducts, locks, and storage dams for water to ensure navigation throughout the year. At Devizes, on the Kennet and Avon Canal, Rennie put in a tier of 29 locks.

Toward the end of the century there was, as a result of the speculative wave in canal building, a well-developed system of narrow waterways to the north and south of the Pennines, and it was decided to join the two by means of a long tunnel through the Standedge Ridge between Marsden and Diggle. The tunnel would also link up the east and west coasts by a continuous inland waterway. The survey was carried out by Nicholas Brown, and parliamentary sanction for the Standedge was granted in 1794.

James Brindley arranged the upstream end of the Worsley Canal as a barge basin with a tunnel leading into the coal mine, whereby the barges could be loaded directly from the stopes. By the middle of the nineteenth century the Worsley mine had water transport tunnels with a total distance of about 40 miles.

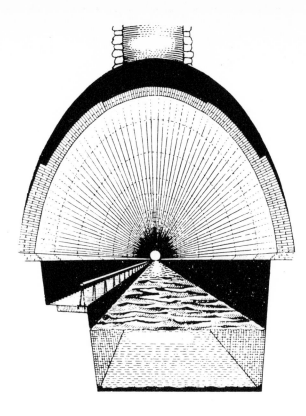

The early English canal tunnels were narrow and admitted barges with a maximum beam of 7 feet, but at the end of the eighteenth century the new canals were given a width of 17 feet to permit a towpath for horses. The sketch shown here gives a view of a tunnel on the Thames and Medway Canal built in 1819.

As determined by the survey, the 9 × 17-foot Standedge would be 5,451 yards long, and hence the longest underground passage ever excavated. Work was started in 1794 by sinking numerous shafts along the line, the deepest being 220 feet. In the hard and difficult grindstone grit encountered, it was necessary to blast with powder. After a great many fatal accidents and tremendous difficulties, the excavation of the Standedge was accomplished and the tunnel was opened for traffic on April 4, 1811. The job had taken 15 years.

Thomas Telford, commonly regarded as the greatest of the nineteenth century English engineers, was also engaged in canal building; his 1,000-foot cast-iron aqueduct spanning the Pontcysyllte River on the Ellesmere Canal, carried on 127-foot masonry piers, is considered the finest work of the early English canal engineers. The canal, including the fantastic aqueduct, was opened for traffic after 12 years, but by that time it had been reduced to an insignificant branch.

Similarly, several other of Telford's largest engineering works proved disappointing from a commercial point of view. His Caledonian Canal, the first ships' canal built in

Britain and conceived as a means of eliminating the dangerous navigation north of Scotland, was a dismal failure. In his report to the Select Committee of the Commons in 1802, Telford estimated the cost of a canal running from Fort Williams to Inverness at £350,000, but upon the passing of the Act authorizing its construction, he increased his estimate to £475,000. The canal as envisaged was to admit a 32-gun frigate and was to be dug to a depth of 20 feet. To carry it over the summit, 29 locks were required. In the end, the cost came to twice the amended estimate, and no 32-gun frigate ever went through the canal since in the difficult ground the depth had to be reduced to 12 feet in the cuts and 15 feet in the lock.

However, Thomas Telford's most magnificent failure, in comparison with which the Caledonian Canal dwindles in importance, was his beautiful design and plans for the Göta Canal, in Sweden, connecting the Baltic with the North Sea. This, incidentally, was the first time a British engineer had been invited to do work abroad; and in 1808 Telford, upon the personal invitation of King Gustav IV and with the blessings of his own government, began the culminating work of his professional life. The canal as planned by him was 120 miles long, 55 miles of which had to be dug to connect a series of natural lakes. The cut was 42 feet at the bottom, a great deal wider than for any existing British canal; the depth was put at 12 feet; and the locks measured 120 × 24 feet. A total of 58 locks raised and dropped the canal to and from a summit elevation of 305 feet above sea level.

Telford, accompanied by two assistants, arrived in Sweden on August 8, 1808, and in 20 days he had fixed the alignment of the canal and the siting of the locks. He was paid £5 a day, and his total consulting fee came to £100. Out of his visit came a set of drawings and maps by his assistants and signed "Thos. Telford, Sept., 1808," which were sent to the local promotor of the canal, Admiral Baltzar von Platen, and subsequently used as the basis of the Canal Act passed by the Estates in 1810. After several requests from von Platen, Telford submitted an estimate of the costs, which he put at 1,597,491 dalers.

The Göta Canal was successfully completed in 1832, after 22 years and a total cost of 10,311,318 dalers, or about six times more than the original estimate. The faulty calculation surely deserves to be a contender for the world record in estimation goofs, but from a technical point of view, Telford as usual had done an admirable job. Nonetheless, the canal was an ill-conceived venture that never paid off. The high costs undermined the country's economy and delayed railroad developments by nearly 30 years. The canal never became a major carrier, nor did it initiate any industrial developments along its long route. After the arrival of the railroads it lost the little importance it had ever had, and eventually turned into a tourist lane through a rural countryside unspoiled by industrial enterprises. Such was Sweden's experience with its first major venture into civil engineering.

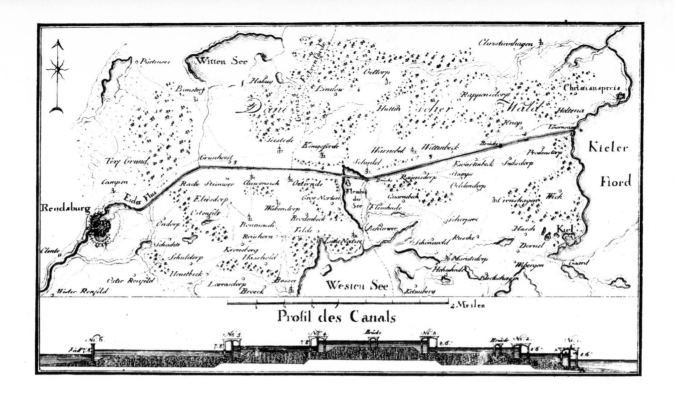

In 1784, King Christian VII of Denmark ordered a so-called ships' canal to be built between Kiel and Rendsburg. From Rendsburg the Eider River was cleared for shipping to the North Sea. The canal and river were open to ships of all nations. This was the beginning of the present Kiel Canal, and apparently it was the first ships' canal joining two seas to be built since the ancient Red Sea Canal.

Here attention has been drawn to the early canals in Britain not necessarily because of their size, since as previously mentioned most of them were narrow, which is evident from the fact that north of the rivers Trent and Mersey the canals admitted barges with a maximum beam of 12½ feet, while the canals south of the rivers admitted boats with a maximum beam of only 7 feet. Nonetheless, these canals constituted the nursery school of British engineering, which put its stamp on developments the world over during the entire nineteenth century.

Canal developments elsewhere

Outside England much more ambitious canal projects were carried out, some to successful conclusion, a great many ending in failure. In Russia, for example, so many local canals were being that the Baltic was joined with the Caspian Sea as early as 1805. By the middle of the century Russia had 50,000 miles of navigable waterways, and the average distance traveled by the barges was 600 miles. This was also the time of improvements of the rivers — the Rhine, Rhône,

Danube, and others were cleared for navigation, sometimes as with the Rhine in such a way that irreparable damage was done to the river and the country alongside it. The river improvements in the upper Rhine increased the current to such an extent that it made upstream haulage nearly impossible, while the scouring of the current silted up the lower reaches; upstream the channel was eroded to below the level of the water table and desiccated the land on both sides. German engineers still had a lot to learn about the behavior of rivers, and the laborsaving method of using the river to do the excavation caused enduring damage that outweighed the temporary benefits gained.

On the Continent, the greatest canal developments took place in northern France and the Low Countries, where canal building had been going on almost continuously since the Middle Ages. Among the works undertaken in Napoleonic times was the famous St. Quentin link which joined the north via the Lys-Scheldt system with the English Channel (at Somme) and with Paris and Le Havre by way of the Oise and the Seine. In France, an inland navigation route connecting the English Channel with the Mediterranean was completed in 1793 with the opening of the Canal du Centre

207

joining Digoin (on the Loire) with Chalon-sur-Saône. The 60-mile St. Quentin Canal with its three tunnels was completed in 1810.

While these and numerous other inland waterways were being dug, there began, at first almost imperceptibly, the construction of canals with altogether different aims. These were the so-called ships' canals, of which Telford's Caledonian Canal was the first British one. With the growth of Ghent as a cotton center after 1800, this inland town was connected to the Scheldt by a canal capable of bringing ocean-going ships to the mills. At about the same time, another ships' canal was opened for the purpose of bypassing the Zuider Zee.

The prototype of this new canal development appears to have been the Eider Canal connecting the Kiel Bay via the Eider Lakes in Schleswig-Holstein with the North Sea. The 25-mile-long and 10-foot-deep Eider Canal was ordered dug by the Danish King Christian VII in 1784, and was completed the following year. Three locks were needed to take the canal over the summit to join the river Eider which was cleared from obstructions and deepened. The purpose of the canal was to bypass the dangerous shoals at Skagen where millions of tons of shipping had been lost, and reduce the distance to the English Channel. The Eider became from the outset an international waterway for ships of all nationalities. Exactly a century after this Danish pioneering effort, it was replaced by the Kiel Canal which follows a level route from Rendsburg, the terminal of the old canal, southwestward to the Elbe estuary. The new canal is provided with locks at the Kiel end and at Brunnsbüttel where the tidal range of the North Sea is 13 feet.

Canal construction also got under way in the United States toward the end of the eighteenth century. The first major development was the Union Canal, which was begun in 1791 and completed in 1827. A long series of small canals followed, such as the Dismal Swamp Canal in North Carolina (left uncompleted), the Hadley Canal and Montague Canal in Massachusetts, the Santee Canal in South Carolina, and numerous others. The first flurry of American canal building culminated with the 364-mile Erie Canal, completed in 1824.

In Canada, work began on improving the St. Lawrence River, necessitating a deep cut to overcome the Lachine Rapids to the west of Montreal. This 5-foot bypass was completed in 1821, while at the same time a 14-mile canal was being constructed to overcome the rapids at Les Cascades, Coteau de Lac, Mill Rapids, and Split Rock, which altogether represented a difference in elevation of 80 feet. By 1847 navigation was possible to Lake Ontario. Two years later, the Welland Canal bypassing the 327-foot drop of Niagara Falls was completed after 25 years of work, whereby Lakes Erie, Huron, and Michigan were connected to the Atlantic for small boats and shallow barges.

The one-mile Sault Ste. Marie Canal joining Huron with Lake Superior—now the busiest canal in the world—was completed by the Northwest Fur Company in 1798. This modest beginning had a lock with a depth of $1\frac{1}{2}$ feet above the sill, and is worthy of mention only because it was the first lock built in America.

In Egypt, too, canal construction got under way at the beginning of the nineteenth century, after an interlude of 2,000 years. Mohammed Ali Pasha, illiterate son of an Albanian fisherman, had made himself ruler of Egypt after slaughtering all opposition, and he now set about to dig the 50-mile Mahmudieh Canal joining the Nile with Alexandria, albeit in his own fashion. He simply put 350,000 fellahin to digging with their bare hands and without preliminary surveys. As a consequence, the Mahmudieh Canal zigzags across the plains, short straight sections being joined with sharp bends. As for the labor force, 20,000 died before the canal and the barrages at Rosetta and Damietta were finished. Brutal and stupid it may have been, but it was this combined irrigation and navigation scheme that made possible the planting of sugar cane and cotton in the Nile delta.

The Suez Canal—a study in evil

Space does not permit a detailed account of the building of the Suez Canal, the greatest man-made waterway brought to conclusion during the period under consideration. If in the following interest is centered on the financial swindles, political machinations, and remorseless exploitation of

Mohammed Ali Pasha, the viceroy of Egypt, ordered 350,000 corvée laborers to dig an 50-mile-long canal from the Rosetta branch of the Nile to Alexandria in 1800. The canal was dug without previous survey and without tools. The slave laborers used their hands for digging.

Ferdinand de Lesseps as he looked at the time of the opening of the Suez Canal in 1869. By then he no longer had any real power, having been relieved of management responsibility five years earlier upon the international scandal arising from his use of slave labor.

labor, it is done for the purpose of highlighting some of the unsavory features of nineteenth century engineering developments. Lesseps, promotor of the Suez Canal, was admittedly the unchallenged champion in unscrupulous dealings, but numerous lesser lights in all countries tried their best to emulate him, at the cost of degrading engineering as well as human dignity.

The chain of events leading up to the digging of the Suez Canal was set in motion in 1844 when a band of bearded and velvet-robed French crackpots descended on Cairo with the intent of rejuvenating the ancient lands around the Mediterranean by means of vast public works. They called themselves Saint-Simonians, and the band, led by their "Pope" and Jesus incarnate Prosper Enfantin, had just been released from a French gaol where they had been confined for preaching free love and in general making public nuisances of themselves.

In his double capacity of Jesus incarnate and graduate of the École Polytechnique, Enfantin obtained permission from Ali Pasha to survey a canal route across the Isthmus of Suez. The survey, including levels, was completed in two weeks, but the Pasha refused to go along any further, and after wasting two years, frequently in company with the young French consul in Cairo—Ferdinand de Lesseps—Prosper Enfantin and his disciples returned to Paris where he quickly amassed a fortune from railroad speculations and formed the "Société d'Études pour le Canal de Suez" in 1846. When

two years later Lesseps was sacked from the diplomatic service, he joined the society and became immersed in its vast canal plans.

Lesseps, unemployed and on his uppers, was repairing his mother-in-law's house when in 1854 he received a letter from a friend from his early Cairo days. Ali, the old brute, had died and been succeeded on the viceregal throne by S'aid Pasha, a bloated monster of a man weighing 300 pounds, whose greatest joy was to have people whipped to death before his eyes. Lesseps wrote a syrupy letter of congratulation to the new ruler and was invited to Cairo where he arrived November 17. Lesseps was then 49 years old. A few days later, S'aid Pasha publicly announced his intention to dig a canal through the Isthmus and that "our friend Monsieur Ferdinand de Lesseps" alone would have the right to develop the enterprise.

This sudden decision caused an uproar in diplomatic, financial, and other quarters. The Saint-Simonians did not take kindly to having their plans snatched from them. Lesseps invited an international technical committee to study the project, and the 13 experts reported that "the execution of the work is easy and its success is assured." That, at any rate, was the conclusion of the published report, but whether the experts actually wrote it is another matter. The cost was estimated at 200 million francs.

Nonetheless, the fraud helped launch an international finance scheme, although it did not achieve its success. Investors outside France took a dim view of the project, and in the end Lesseps could only place 218,494 shares at 500 francs apiece, out of 400,000 shares outstanding. The Ottoman Empire took 96,000 shares and the rest, or 85,506 shares, Lesseps forced on S'aid Pasha. What started out as an international financing scheme ended up with French investors and S'aid Pasha owning 70 percent of the equity.

Under the 1856 concession whereby the Suez Canal Company obtained the right to build and operate the canal for 99 years, S'aid Pasha undertook to supply corvée (or conscript) labor to the extent required, and also deeded 60,000 hectares of land and granted numerous privileges to the company. On April 25, 1859, the first levies of corvée labor had been rounded up and Lesseps drove a mattox into the sand on a desolate stretch of the Mediterranean coast and declared the work on the canal to have begun. He named the place Port Said in honor of his patron.

To the protestations of the British Government against the wholesale use of slave labor Lesseps replied that Her Majesty's Government would do better to concern itself with abolishing slavery in America and Russia. Furthermore, he added, in Egypt slavery is the normal thing, and hundreds of boys and girls from 10 to 13 work on the cut with remarkable zeal and enthusiasm. He might also have added that the thousands of fellahin put to work had to dig into the sand with their bare hands for lack of tools; indeed, in Lake Menzala south of Porth Said, the fellahin removed the liquid

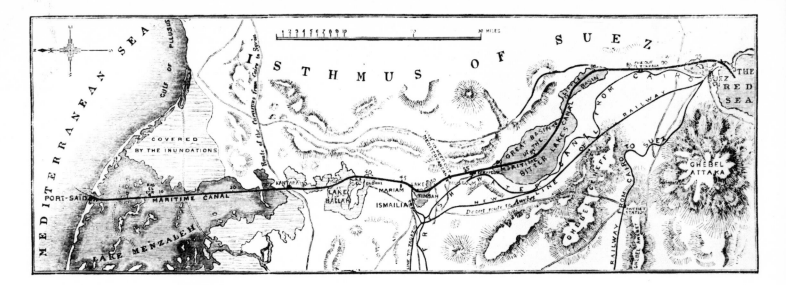

This contemporary map of the Suez Canal is based on a more detailed French original but gives the essential features of the isthmus. In addition to the maritime canal, marked every fifth mile, it shows the soft water canal from Cairo to Suez and the new Cairo–Suez railroad. Another projected line that was never built is also given, as well as the ancient caravan routes. (London Illustrated News, 1869.)

mud by scooping it up with their hands and pressing the water out of it against their chests. When they had obtained a large ball in this manner, they carried it up to the bank. There the sun dried it into a firm mass.

If Lesseps had had his way, the entire canal would have been dug in this fashion, but, fortunately, in January, 1863, S'aid Pasha finally collapsed under his own weight and died. He was succeeded by his son Ismail who like civilized people everywhere was disgusted with Lessep's remorseless use of slave labor—which by this time had reached 25,000—and stopped the work. By then some 8.5 million cubic meters had been excavated by hand.

Some time previously the narrow canal had reached Lake Timsah, about 35 miles from Port Said. Lesseps made the best of the occasion. He struck a Napoleonic pose and commanded the waters of the Mediterranean to enter the lake. The event was trumpeted around the world, and Emperor Napoleon III made Lesseps a Commander of the Legion of Honor. This, without a doubt, was his greatest hour.

The work was too important and had advanced too far to be stopped by unilateral action by Ismail. The two parties to the dispute agreed to have it arbitrated by Napoleon. The Emperor ruled that the canal company was entitled to a payment of 38 million francs as a compensation for the abolition of the corvée labor and 30 million francs for the return of the 60,000 hectares that Lesseps had wheedled out of S'aid Pasha, plus 16 million francs for a fresh water canal

that had been dug from the Nile to Suez, or altogether 84 million francs.

The abolition of the corvée was the most important decision because it saved the enterprise. Although retained as the nominal head of the company, Lesseps ceased in actual fact to have any direct influence on the work which was now turned over to more competent men. The remaining excavations were divided among four contractors who brought in steam excavators and hired 8,000 men in England, France, Italy, and Greece to complete the job. In the next five years they excavated 60 million cubic meters at a cost of 2.5 francs per cubic meter.

But all was not well. In 1865 cholera broke out, and hundreds of men died in Ismalia. Then the money gave out, and the company was forced to sell its El Wadi estate to Egypt to tide it over while it raised a 100 million-franc loan in France. Later, it obtained a further 30 million francs from the Egyptian government by relinquishing all its remaining privileges, such as customs and fishing rights, as well as its workshops, other buildings, and hospitals built on the Isthmus. In the end, the cost came to 453,645,000 francs, as against an estimated 200 million francs.

The inaugural journey through the canal began in Port Said on November 16, 1869, with a flotilla of 68 ships from various nations led by the French "Aigle" with Empress Eugénie aboard. The Empress was sick with fear that her ship would get stuck in the sand and spent most of her time in her cabin sobbing over the lost glory of France. When

210

nothing untoward happened, she recovered sufficiently to hang the sash of the Grand Cross of the Legion of Honor on Lesseps in the new palace at Ismalia built by Ismail Pasha for the occasion. The event was witnessed by 5,000 invited guests who then spent a miserable night in the cold desert for lack of accommodations. On November 19 the inaugural flotilla reached Suez, and and the canal was opened for traffic.

The Suez Canal was 100 miles long and had when opened a bottom width of 72 feet and a depth of 26 feet. At water level the width varied from 196 feet—at the deep cuts—to 327 feet; the slope of the banks was 1:2. Between Port Said and Lake Timsah the canal was provided every six miles with "gares" or sidings, where ships were moored to let others pass. The canal was at this time not a very efficient traffic lane, and during the first year of operation only 489 ships, totaling 437,000 tons, used it. This traffic produced a revenue of only 4.4 million francs, which was altogether inadequate for its operation, not to mention paying a dividend to the stockholders. Indeed, it was not until after 1885, subsequent to the British occupation of Egypt, that the canal was made commercially viable when widened and deepened.

But by that time Lesseps had nothing to do with the Suez Canal. Instead, he was submerged in trouble in the jungles of Panama where after two years of work 40,000 men had died of "vomito negro" or yellow fever. Worse, it subsequently developed upon the collapse of his "Compagnie Universelle du Canal Interocéanique" that he had been engaged in swindles, bribery, blackmail, libel, and had frittered away something like £100 million. Lesseps was sentenced to five years in prison, a term which he never had to serve. He sank into a senile apathy that lasted to his death in 1894 at the age of 89 years. The mills of the gods grind slowly but thoroughly.

Contemporary view of the mechanical dredging at the southern end of the Suez Canal immediately before the opening in November, 1869. At that time the Suez Canal was from 196 to 327 feet wide at water level and 26 feet deep.

The Railway Age

"A clever man he was, a wonder, to rise from common digging in the Newcastle pits to engineering—observe how highly."

JAMES WHISTLER

It was greed rather than inventiveness and enterprise that spurred on the coming of the railways. The canal companies in Britain in absolute control of the most efficient transport system in the country treated their customers with disdain. They insisted on deciding what goods could be shipped, and they charged what they pleased for a service that by 1820 had become so sloppy that it took longer to move a cargo from Manchester to Liverpool than from Liverpool to New York.

Frustrated millowners in Manchester, forced to close their mills for days on end when American cotton destined for them got stuck on the docks in Liverpool, began to cast around for some other means of transport, and a few of them investigated the newfangled tramways already in operation in some collieries. As early as 1801, Richard Trevithic, a mine foreman from Cornwall, had tested a steam carriage on a road at Camborne in Cornwall. Another one of his locomotives ran on flanged rails in the Wylam colliery at Tyneside in 1805. They had proved to be practical and promising experiments, and although Trevithic ceased to have anything further to do with the early tramways, others such as John Blenkinsop and Timothy Hackworth kept on improving both locomotives and track.

At the Killingworth colliery, the owner Sir Charles Liddell and a foreman by the name of George Stephenson watched with interest the work of the early pioneers, and when the latter was invited to try his hand at building a steam locomotive to be used on the existing wagonway, he succeeded in turning out a satisfactory engine which he named Blucher. This engine made its first test run over the short tramway on July 25, 1814. By 1820, a number of Stephenson locomotives were in operation on 14 colliery tramways.

In the manner of all leading engineers and inventors of his generation, George Stephenson got a humble start in life. He was employed as a cowhand until the age of fourteen when he became an engine fireman; three years later while still unable to read and write, he was promoted to tending a pump engine. By 1812 he had become enginewright at the High Pit and spent his spare time cleaning and repairing watches. Two years later, at the age of thirty-two, he built his first successful locomotive.

In 1821, an Act was passed "for making a railway from the river Tess at Stockton to the Willow Park Colliery," using horse traction, but two years later upon the advice of the newly appointed engineer of the railway, George Stephenson, the Act was amended to permit the employment of steam traction. Previously, the rails used for the tramways had been made of cast iron, but now John Birkenshaw had patented a method of rolling double-headed wrought iron rails. They were adopted for the new railway despite their high price—twice as high as that for cast iron rails.

Choosing gage for the first steam-traction railway was no particular problem for George Stephenson. He simply took the traditional gage of the colliery tramways which was 4 feet $8\frac{1}{2}$ inches. In 1846, the ancient gage was made standard by an Act of Parliament, and so it has remained ever since in the majority of countries the world over, with a few exceptions. Russia uses a 5-foot gage, and South Africa has a 3-foot 6-inch one. The Irish railways have a 5-foot 3-inch gage, and the Indian a 5-foot 6-inch one. No other railway has followed the example of the Great Western in using a 7-foot $1\frac{1}{4}$-inch gage as put in by Brunel.

After three years of work, the Stockton and Burlington Railway was opened by two steam locomotives, "Locomotive" and "Experiment," coupled in tandem and pulling a set of 21 coal wagons temporarily converted to passenger use. The line was $8\frac{1}{2}$ miles long, and the journey took 65 minutes.

This, then, was the situation when the Manchester merchants drew up a memorandum complaining bitterly about the great difficulties and obstructions attendant upon the movement of goods from the port to Manchester. A prospectus was issued offering "the establishment of a cheap mode of transport for merchandise as well as passengers." The only man experienced in the new mode of transit was appointed engineer in 1824, and the following year a bill was passed—after hard fights with the canal and property owners along the proposed route.

The building of the 30-mile Liverpool–Manchester line, the first public railway in the world, was beset with severe birthpangs, most of them stemming from the obstinate character of the engineer. George Stephenson was a native genius as an engine constructor, but he knew nothing about civil engineering, and, as subsequently proved, he lacked business and administrative sense. At the time, many engineers, including Stephenson himself, had their doubts about the ability of iron-wheeled carriages being able to move on iron rails, owing to the slight friction between tire and rail. But these early doubts were soon found to be exaggerated. Gradients, however, were something else again; they had to be avoided at all costs. For this reason Stephenson adopted from the outset his "straight-

The Liverpool—Manchester line became unnecessarily costly because of George Stephenson's insistence on choosing the shortest route and accepting all obstacles in the way. His engineering concept also severely encroached upon the countryside, as exemplified by the Olive Mount cutting where 800,000 cubic yards of rock had to be removed.

George Stephenson at the height of his power and influence.

through" concept, i.e., choosing the shortest distance between two points and accepting all obstacles to a level line. As applied to the Liverpool–Manchester line, the gradient never exceeded 1:880 from the portal of the Liverpool tunnel to Manchester. To attain this practically level grade, tremendous excavations were necessary, at a cost of nearly £200,000. The Olive Mount cutting alone required the shifting of 800,000 cubic yards of soil. Then Stephenson insisted on crossing the large bog known as Chat Moss, in the face of opposition by George Rennie and Josiah Jessop, called in as consultants by the company. In the end, Stephenson actually succeeded in laying his line across the bog, but at the price of high costs and long delays. The Edge Hill tunnel at the Liverpool end also involved great difficulties, owing to soft clay encountered during the excavations.

The technical consequences of the Stephenson concept became evident in the need for 63 bridges, including the huge Sankey viaduct across the Sankey Canal on nine arches lifting the line 70 feet above water level. This piece of construction in brick and ashlar came to £45,000.

Construction of the 30-mile line—estimated to cost £800,000—was managed by Stephenson and his nineteen-year-old son Robert. Stephenson employed the labor and supplied the material needed. In 1828, when the company found itself in financial difficulties and requested a loan of £10,000 from a special Exchequer postwar development fund, Thomas Telford was brought into the picture to report on the work. Telford's inquiry revealed an astonishing mess. There were no contracts or standard norms for paying for the work performed. The price for excavations was set by Stephenson on the spur of the moment and varied as much as 100 percent for similar work. Although

213

The Sankey viaduct across the Sankey Brook Canal was another costly structure on the Liverpool—Manchester line. The nine arches built of brick and masonry lifted the railroad 70 feet above the water level and cost £45,000. But it was well built, is still in service, and is capable of carrying the far heavier loads exerted by modern trains.

Telford was impressed by the work done, he refused to approve the loan, largely because the directors of the company had not yet decided on the traction power to be used, whether horses or engines. Indeed, some directors were in favor of rope haulage, using a stationary engine.

To resolve this dilemma, the Board arranged a public trial to determine whether a locomotive could be used. The historic contest was held on October 28, 1829, over a level 1½ mile of track at Rainhill. Four engines took part the "Rocket" designed and built by Robert Stephenson, "Sans Pareil" by Timothy Hackworth, "Perseverance" by Burstell, and "Novelty" by John Ericsson. Stephenson's Rocket won the contest by reaching an average speed of 13 miles and a maximum speed of 21½ miles an hour. The output of the Rocket was 12 hp, and the fuel consumption came to 17 to 20 pounds of coke per hp per hour. The outcome of the Rainhill trials convinced the directors that the steam locomotive could be used for traction on this and subsequent railways.

The Liverpool–Manchester Railway, which in the end cost £1,200,000 and came close to bankrupting the company, was officially opened in miserable weather, with squalls and thunder lashing the Lancashire country. On September 15, 1830, eight trains carrying 700 guests, including the Duke of Wellington and the Prime Minister, pulled out from Liverpool. While making a brief stop at the Parkside station, the Rocket accidently ran over the Right Honorable William Huskisson, M.P. for Liverpool;

he died of his injuries the following night. It was an ominous beginning, particularly since Huskisson had been a firm believer in the future of the railways and had led the fight against the hostile canal and property interests that were doing everything in their power to stop the project.

The poor auguries notwithstanding, the Liverpool–Manchester Railway was a success from the outset. Not only did the Manchester millowners get their cotton without undue delay; they also made a handsome return on their investment. By the end of 1830, about 70,000 passengers had paid to be carried the 30-mile stretch in one hour. By 1835, net receipts had risen to £80,000, or 25 percent more than had been expected.

Three years after the opening of the first public railway, Parliament passed the enabling acts for the first two trunk lines, the 80-mile Grand Junction line connecting Manchester with Birmingham, and the 112-mile London–Birmingham line ending at Chalk Farm to the north of London, although subsequently extended to Euston Place. The building of these two lines made engineering history, not so much because of their length and the difficulties encountered and overcome, but because of the two schools of railway engineering that were brought to fruition on them.

The Grand Junction was engineered by Joseph Locke, besides George Stephenson foremost among the early railway pioneers. Locke had obtained his training under Stephenson who was originally commissioned to build the

Grand Junction. He had gone about it in his usual fashion, keeping grade to a maximum of 1:330 and including a number of tunnels in his project. The estimate was considered too high, and the directors had young Locke survey another route which avoided tunnels in favor of cuttings and followed, as did the early canals, the contours of the country, accepting a maximum grade of 1:180. With this survey Locke demonstrated his method of saving money by spending time and effort on careful, elegantly executed traverses, going around rather than through obstacles. Like Telford, he believed in writing his specifications in clear unambiguous language permitting contractors to put in competitive tenders without undue risks.

The Grand Junction also gave the start to a third pioneering giant, Thomas Brassey, who contracted for the part of the line which included the Penkridge Viaduct, south of Stafford. Thus began the astounding career of the greatest railway contractor of all time, who before his retirement had built 1,700 miles of the early British railways, in addition to something like 3,000 miles in practically all parts of the world. Brassey and Locke became a team that left an indelible imprint on the map of England and continental Europe.

Except for a leaky aqueduct carrying the line over the Bentley Canal, the Grand Junction, mainly due to Locke's scrupulous engineering, was finished on schedule in 1837 after two years of work. The track was laid on sleepers with double-headed rail, a forerunner of the bullhead rail.

The Birmingham–London line, on the other hand, had an entirely different construction history. Robert Stephenson was in charge of the work which was divided into four divisions, each under a competent engineer. In each division work was contracted out for 6 miles of line, with special tenders for viaducts and tunnels. In all, there were 29 separate contracts; and from 12,000 to 20,000 men were put to work under grievous difficulties, particularly when driving the eight tunnels on the line. The deep cuttings, too, such as the one between Camden Town and Euston, caused trouble, and the retaining walls had to be provided with an invert to prevent them from slipping because of the heavy loads acting on them.

The tunnel through the Kilsby Ridge between Denbigh Hall and Rugby was the most troublesome one. It was only 2,400 yards long, and a canal tunnel had been driven through it before, so that the geology was known in advance. Robert Stephenson ran a line of test borings over it, but managed to miss a 400-yard pocket of quicksand lying below a blanket of clay. The contractor John Howell encountered a great deal of water and when, in addition, he also struck the quicksand, the shock was too great for him and he died. Since his sons refused to go on with the work, Robert Stephenson had to take over the job him-

On the cuttings along the London—Birmingham Railway some contractors used a variation of the horse whim for pulling the heavily loaded wheelbarrows up the steep timber ramps. The navvy's job was to steer the heavy barrow up the ramp. The muck was dumped on an elevated loading platform and then shoveled into horse-drawn carts.

BUILDING RETAINING WALLS NEAR PARK STREET CAMDEN TOWN 1836

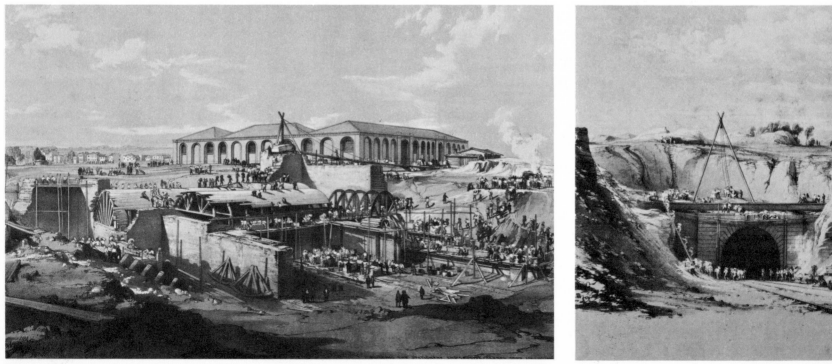

The construction of the London—Birmingham Railway brought about what appears to be the first picture reporting of a construction development. Artist John C. Bourne followed the work from beginning to end and made numerous sketches on the sites. They were later transferred to lithographic stones and included in a book published in 1839 entitled Drawings of the London and Birmingham Railway. *Some of these lithographs have been reproduced on this spread.*

Left: *The deep cutting made in 1836 near Parry Street in Camden Town gave a great deal of trouble owing to the instability of the London clay. Since nothing was known about the behavior of soil under load, not to mention soil mechanics, the engineer had to proceed by trial and error. When a conventional retaining wall proved inadequate, it was lined with a cast-iron invert; and when that failed, iron struts were put across the invert. These struts are still in place.*

Bottom left: *The cutting leading to the engine house in Camden Town also gave considerable trouble. The picture shows the state of construction in April, 1837.*

Bottom center: *The drawing made on June 6, 1836, shows the final stage in the construction of one of the portals of the Watford Tunnel on the London—Birmingham Line. The tunnel was driven through gravel and a bad spill caused great loss of life.*

Above: *In the foreground are two of the thirteen steam-driven pumps used to drain the workings in the Kilsby Ridge Tunnel. Note the long wooded transmission shaft from the steam engine to the pumps. In the background are two conventional horse whims employed to hoist up the muck through the main shaft. It had a diameter of 60 feet and was retained for ventilation.*

Below right: *The picture, dated June 18, 1837, shows the state of the large fill on the Woolver Valley crossed by the London—Liverpool line. The drawing, incidentally, bears witness to the brutal encroachments made by the early railroad engineers on the gentle English countryside. It is easy to understand why the property owners and the educated middle class fought the railroads so frenetically.*

self. He sank a number of additional shafts, installed 13 steam-driven pumps with a capacity of 1,300 gallons per minute, and kept 1,300 men at work on three shifts for 18 months before he had drained the workings enough to continue the advance. It took 30 months to complete the tunnel at a cost of £300,000, as against a contracted price of £99,000.

Other sections also proved too difficult for the contractors, and the work had to be performed under the management of Robert Stephenson or his division engineers. Therefore, the costs kept mounting until in the end they added up to £5.5 million as against an estimated £2.4 million. Whereas the Grand Junction engineered by Locke was built at £18,846 per mile, the Stephenson London–Birmingham line came to £50,000 per mile. The railroad was officially opened on September 17, 1836, 14 months after the Grand Junction.

Another early controversial railroad was the Great Western between Bristol and London. The 3,218-yard Box Tunnel on the line between Chippenham and Bath met particularly violent opposition, but the design of the road-bed and numerous technical oddities introduced by the engineer, the twenty-eight-year-old Isambard Kingdom Brunel, also kept the controversy alive in the parliamentary committees.

Nonetheless, after long battles the Great Western Bill was passed on August 31, 1835. The line was 120 miles long, and its cost was estimated at £2,805,330. The ruling gradient was 1:1320 from Paddington to Didcot where it sharpened to 1:600. The approach to the Box Tunnel had a gradient of 1:100. After the Bill had been passed without mention of the gage, Brunel sprang his secret on the directors: he proposed to make it 7 feet $\frac{1}{4}$ inch. He also had his own ideas on how the rail should be designed. It was given a section of an inverted U and laid on longitudinal balks instead of on transverse sleepers, but in order to keep gage the rails were tied together by crossties at long intervals. The crossties were bolted to piling driven along the roadbed. Brunel also had an idea of how a locomotive should be designed; it should have short strokes and enormous driving wheels. A few of the monsters were actually built, but when tried out, they had difficulty staying on the rails and had to be scrapped as useless.

Work on the Great Western started at both ends in September, 1835. It was completed in nine stages; the one from Maidenhead to Twyford was opened for traffic on July 1, 1839. This is the section that has the elegant bridge crossing the Thames by two shallow spans 128 feet in length and rising only $24\frac{1}{2}$ feet. By December 17, 1840, the western end of the line was operating to Bath, and the eastern as far as Wooton Bassett. There remained only the most difficult stretch of them all, between Chippenham and Bath, including the Box Tunnel.

Work on the Box Tunnel had started in November,

1836. A number of shafts were sunk to grade, and tunneling began through oolite, fuller's earth, marl, and lias. Serious trouble was encountered, and a year later the entire works were flooded. Brunel's assistant, William Glennie, was responsible for the work, and by putting on 4,000 men and 300 horses he succeeded in saving the headings and completing the tunnel in the summer of 1841. By that time the line had been extended to the tunnel portals, and on June 30 Great Western could be opened for traffic, the odd roadbed having been torn up and replaced by a conventional sleeper bed. The broad gage was kept, but eventually that, too, was changed to standard gage, although not until 1892.

After Brunel's original engineering extravaganzas were deleted, there remained enough sound engineering to make

Kilsby Ridge was the most troublesome of the numerous tunnels on the London—Birmingham line. Here a pocket of quicksand and water delayed the work; and it took 1,300 men working 30 months, at a cost of £300,000, to complete the tunnel.

the Great Western one of the finest railroads in the world.

By the time these first British trunk lines were put into operation, France had built 341 miles of railroads, of of which the St. Étienne–Lyon line, opened in 1830, used stream traction. In Germany, the first railroad, connecting Nuremberg with Fürth, was put in traffic in 1835. Austria, Russia, Italy, and Switzerland got their first railroads in the 1840s, all or nearly all of them using British-built locomotives.

In the United States, John Stevens, "the father of the railroads," obtained the first railroad charter from the State of New Jersey on February 6, 1815, but he did not get around to forming a company until 15 years later. Instead, the Delaware and Hudson became the first American railroad built for locomotive traction. The section from Carbondale to Honesdale was opened for traffic in 1829, after the first test run with a locomotive in America had been carried out on August 6. As in England, there had been railroads, or more accurately gravity tramways, previously in operation for local transport of quarrystone and coal, and the Delaware and Hudson was also a coal carrier.

The first railroad in the United States planned from the outset as a carrier of passengers and freight was the Baltimore and Ohio, which obtained its charter from the State of Maryland in 1827. Construction of the line began on July 4, 1828, and on May 24, 1830, 13 miles was opened for traffic to Ellicott's Mills. Hence, the B & O can claim to be the first steam railroad to carry passengers and goods on its tracks.

In New York State, the Mohawk and Hudson Railroad Company obtained its charter April 17, 1826, but legal squabbles delayed the construction of the line between the Hudson and Mohawk Rivers until July 18, 1830. Actual traffic operations between Albany and Schenectady began August 10, 1831. This, then, was the modest beginnings of the New York Central system.

Thus within one year the nuclei of the two finest railroad systems on both sides of the Atlantic came into being. By 1840, the United States had 2,799 miles of railroads, as against 2,000 in Britain. By the end of the century, American track had been extended to 193,346 miles and kept growing rapidly, while the British had reached 21,855 miles and were approaching the ultimate limit. After 1920, the decline had definitely set in on both sides of the Atlantic. The railroad boom had run its course.

The men working on the railroads

The 90 years of railroad expansion saw more dirt shifted, more iron and steel laid, more bridges built, longer and larger tunnels excavated than ever before, indeed through-out the entire history of civilization. In retrospect, particularly when viewed through the dust raised by snorting bulldozers and excavators, it seems incredible that this tremendous bulk of work was accomplished entirely by human muscle. Where did they come from, these laboring giants who were capable of lifting 20 tons of dirt a day to a height of 6 feet, and then spending the remaining 12 hours in boozing, fighting, gambling, and whoring—when the opportunity for the latter presented itself.

In all countries they derived from a common source, from the crofts and rural hovels where potatoes had produced an oversupply of manpower, healthy and strong of limb and wind. After Waterloo there had been no occasion to slaughter them on the battlefields; the rural hovels had been saved from the epidemics which still occasionally raged and decimated the city populations. There were also during these 90 years important changes in legislation which in some countries put an end to the villenage system, in others to indenture contracts which served the same end, namely to keep labor in servitude on the land. In America during these years there was an increasing flood of immigrants drawn from the same sources, and the Scandinavians among them were attracted to laboring on the railroads. "Give me five hundred Swedes and a carload of snuff and I will build you a railroad clean through hell" was the way a western railroad superintendent referred to this particular source of cheap, complacent, and physically strong labor.

These, then, were the men who built the railroads. In countries where the work was done by contractors, the men had no occasion to regret leaving forever the peonage of their fathers. They were better paid than rural labor, better fed, and, after raising hell, they ran no risk of being caned by a farmer or other legal master. They were supplied with tools, and the more intelligent contractors employed such laborsaving devices as horse whims or mules for moving the excavated dirt in major cuttings. Brassey, the pioneer among English railway contractors, appears from all accounts to have been an ideal employer to his rough, mostly Irish labor. Beyond being a great organizer of vast construction works, he had a gift for managing men, whom he treated in what resembled Roman equestrian fashion: they were well fed, honestly paid for what they accomplished, and what they did to themselves or others in their spare time was no concern of his.

But conditions were altogether different in other countries where because there were no private contractors, the government had to manage the construction of the state trunk roads. Admittedly, there was a considerable difference in the manner in which this was done; the wholly admirable way in which the Cavour government of Sardinia built the railways connecting up the Fréjus Tunnel, the first long mountain tunnel ever built, can be compared, for example, with the manner in which the Swedish government managed to get its trunk lines constructed. There is a recognizable pattern in all this: the farther north the rails were extended, the beastlier were the methods used to carry out the work. The arctic railroads in Sweden came into being in a manner that would have made an overseer of the porterage work on the Cheops pyramid snort with disgust. The men had to fend for themselves when they arrived in subzero cold in the wilderness. They had to bring their own digging tools and pay for the metal worn off sledges and drill steels drawn from government stores. They had to build their own huts, buy and cook their own food, sleep in their working clothes for weeks on end. They warmed the dynamite with their own bodies, and when they got hurt, and that happened with distressing frequency, they had to get themselves to a doctor as best they could.

Upon the end of the First World War a few managed to escape from their subhuman mode of life and get jobs on the Murmansk Railway being built by the American Expeditionary Corps. There they were put up in heated shacks, fed three times a day with meat without having to cook for themselves, got new boots and working clothes, and were paid five times as much as they had been accustomed to. Forty years later they were still talking about their brief glimpse of Paradise to their grandchildren and to roving folklore investigators.

The Alpine tunnels

After this brief social note, a mere dip into the ocean of hair-raising detail accumulated from Victorian construction sites, let us return to the general survey of the railway age. The bulk of the grading work may have been done everywhere with pick and shovel, and some hundred million tons of steel rails carried on the shoulders of men, but the construction of a railroad often included, as well, the excavation of long mountain tunnels and the building of numerous bridges and viaducts. Spanning a river or tunneling under a mountain required engineering of a much higher order than shifting dirt, and inspired new inventions and mechanized methods of construction, in order to surmount the high costs and difficulties involved.

The numerous canal tunnels in England and on the Continent had been hacked out of the soft rock; only in exceptional cases was blasting with gunpowder applied. The methods of excavating these tunnels were originally borrowed from mining, but since even the modest English tunnels were considerablly bigger than the normal mining drifts, a number of new methods had been developed to cope with the larger areas. Thus there was an English method of advance and a Belgian one; whereas Germany, France, and Austria, and subsequently also America, had their own special systems of tunneling a large area.

With the need for larger tunnel areas (an old-time single-

track railroad tunnel was 16 × 22 feet and a double-track 28 × 22 feet), these cut-and-tried methods were further improved upon, and new ones developed. To speed up the advance, rock blasting with gunpowder became general practice. In wet workings the contractors began to install steam-driven pumps, as for example in the Kilsby Ridge tunnel.

Although Britain held the record for long railway tunnels, such as the 3-mile Woodhead finished in 1845 and the Standedge (3 miles 62 yards) completed in 1849, railroad construction on the Continent was from the outset accompanied by tunneling on a large scale. By the middle of the century France had 126 tunnels totaling 135,000 feet in length; Austria had 60 tunnels 43,300 feet in length; and Italy 32,000 feet. In Switzerland, Thomas Brassey advanced the 8,198-foot Hauenstein Tunnel on the Basel–Otten Line in 1853 to 1858. The United States had at the time 29 railroad tunnels and 16 large water tunnels, all of them on the Croton aqueduct.

Although the methods of excavation and timbering employed in advancing these tunnels differed in each country, as previously mentioned, they had one feature in common: the excavation was accomplished from numerous shafts sunk to tunnel grade along the projected axis on the ground. The shafts were used for hoisting the excavated rock and served as ventilation chimneys while the work was in progress. Some of the construction shafts were retained to remove the smoke and steam generated by the locomotives using the tunnel.

However, by mid-century the railroad promoters were confronted with the challenge of linking up the railroads to the north and south of the Alps; in America, too, and elsewhere there were mountain barriers separating lines built up on both sides of them. But it was the Alps that presented the greatest problem and induced a long, bitter debate that split the learned world into two camps and eventually spilled over on the floors of the national assemblies. The prestigious British engineers consulted by Continental promoters took one look at the formidable mountain peaks and declared that a railroad across the passes was impossible. Continental professors of physics and engineering believed otherwise, and suggested that it would be possible to extend a line up to a summit elevation of 9,000 feet, and, if necessary, build a short tunnel under the existing road passes. They were bitterly opposed by other equally learned men who wanted to place the tunnels much farther down the slopes and accept the consequences of the long galleries required.

At the time, both schools of thought preached sheer nonsense. To bring a railroad up to the elevations suggested by the summit school was impossible for two reasons: (1) at such elevations in the Alps snow lasts for up to nine months of the year, and in winter it drifts up to a height of 50 feet or more. A summit line could not then,

This drill carriage mounting eight rock-drilling machines was used in advancing the pilot heading in the Fréjus Tunnel in 1863 to 1870. The rock drills were designed by Germain Sommeiller of Savoy, chief engineer of the tunnel. It was the first time in history that machines were successfully applied in underground developments. A similar viable application of rock drilling machines took place at the same time in the United States, in the Hoosac Tunnel. By the time the Hoosac Tunnel was completed in 1873, the time- and labor-consuming work of extracting hard rock had become mechanized on both sides of the Atlantic. Subsequent developments followed the American line, and few if any innovations in mechanized construction have originated in Europe since the turn of the century.

and cannot now, be kept open through the winter. (2) Even with the absence of snow, bringing trains up to such high elevations would be such a slow and fuel-consuming operation that it would wreck the transport economy of the railroad.

The alternative of a base or intermediate tunnel propagated by the opposite camp was equally foolish and unrealistic, at the time. Such a long tunnel could not be advanced since it was impossible to sink the numerous shafts needed to remove the muck and ventilate the workings. Moreover, according to the geologists, the height of the rock cover above such tunnels would raise the heat of the rock to such temperatures that the blood of men would boil. The professors of medicine added their arguments: there was no method known to science of ventilating a tunnel adequately beyond a few hundred yards from the portals.

Beneath the din of professorial combat and ignored at the time, some commonsense ideas germinated which eventually became technically viable. In Geneva, for example, an engineer by the name of Joseph Alby was commissioned in 1840 to prepare plans for tunneling Mont Blanc. His scheme, which in many respects resembles the

motor tunnel under the massif completed 125 years later, ended with the comment that the tunnel could no doubt be built "when human labor was replaced by workers with muscles of steel and lungs breathing fire." Alby saw clearly that machines would be needed if long mountain tunnels were ever to be realized.

Similarly, G. F. Médail, a former shepherd of Savoy, outlined a railway connecting Turin, capital of Piedmont, with Chambéry, capital of Savoy, both important provinces of the Kingdom of Sardinia. But since Savoy was separated from Piedmont by the Cottic Alps, Médail proposed in 1838 that Col de Fréjus be pierced by a tunnel linking up the lines coming to an end at both sides of the massif. Just how this 40,138-foot tunnel was to be built did not worry the nontechnical Signor Médail. If the King of Sardinia wanted to prevent Savoy from being drawn into the French orbit, the transalpine province simply had to be tied to the rest of the realm by a railroad, and there was no alternative to the one suggested by Médail.

Although the Médail proposal received offical attention and was properly filed away in a pigeonhole where it gathered dust until long after Médail's death, the political communication issue so clearly recognized by him refused to be swept aside. When Cavour became concerned with the difficulties of keeping Savoy tied to a united Italy, technical opinion—as represented by I. K. Brunel, H. Mauss, and others—was ready to accept Médail's scheme, and a commission appointed by the Kingdom of Sardinia roamed far and wide, looking for machines and methods for advancing the long tunnel under the Fréjus massif.

The machine the commission searched for—"the workman with muscles of steel and lungs breathing fire"—did not exist. But an English engineer, T. Bartlett, had invented a steam-driven rock drill that was tried out by the commission in 1855 with some success. Other inventions were also tested with less success. The commission visited America and found a direct-action machine which had been patented by a Joseph Fowle of Philadelphia in 1851, but had yet to be tried out in operation.

In the end, a satisfactory air-powered rock drill designed by Germain Sommeiller, chief engineer of the Fréjus Tunnel, and incorporating the best features of both the Bartlett and Fowle machines, was developed and proved its mettle in the Fréjus Tunnel. Although the full-face area of the tunnel was about 650 square feet, the critical part of the excavation was to break out an 11×8-foot pilot gallery, both to release the tension of the rock and to obtain points of attack for enlarging the tunnel to full area. When the Fréjus work started on August 18, 1857—and for four years thereafter—the pilot gallery was driven by manual drilling, for an average daily advance of 9 inches. In January, 1861, the first machines were introduced, and an average daily gain of $1\frac{1}{2}$ feet was achieved. The following year somewhat more than 3 feet were gained, and in 1864 the improved machines produced a record advance of $9\frac{3}{4}$ feet. By 1870, the last year of the mining work, the daily advance when at its best reached nearly 15 feet.

Thus, during the decade 1860 to 1870, and thanks largely to Germain Sommeiller, excavation work in hard rock became mechanized with dramatic results. At about the same time, similar developments took place in America where the 24,416-foot Hoosac Tunnel, on the border of Massachusetts and New York, was being advanced. In the Hoosac the miners drove a 94-square-foot pilot heading at a rate of 1 foot per day, using manual drilling and gunpowder for blasting. When in 1866 rock drilling machines were introduced, the rate of advance increased to $1\frac{1}{2}$ feet per day, and gradually kept improving with better machines for drilling and nitroglycerin for blasting, until by 1873 a monthly advance of 162 feet, or $5\frac{1}{2}$ feet a day, was attained. During the 23 years of the Hoosac advance, American engineers also succeeded in developing reliable mechanical air compressors, whereas in the Fréjus, hydraulic compressors were used.

The Fréjus and Hoosac set contractors on both sides of the Atlantic on the track of mechanization. The huge input of labor previously required on underground developments was cut in half, and so was the time required to complete a tunnel. By 1870, a new safe and powerful explosive—the dynamite invented by Alfred Nobel—had been made available, which, together with the rapidly improving rock drilling machines and air compressors, established a pattern of production that became standard more or less the world over. It should be noted, however, that the break with the past applied only to the drilling and shooting of a round; the rest of the work—"mucking out a round"—was conducted in the old back-breaking manner of men wielding shovels and pushing loaded wagons.

Thus, in starting up the St. Gotthard Tunnel on September 13, 1872, the contractor Louis Favre of Genoa was in a position to apply the experience and the mechanized facilities developed in both Fréjus and Hoosac, as well as dynamite. The Gotthard Tunnel was the longest railroad tunnel attempted and was regarded at the time as "the work of the century." The double-track tunnel had a width of 25 feet and height of 26 feet, and extended for a length of 9 miles 452 yards (14,900 meters) from the hamlet of Göschenen in the north to Airolo in the south. Like Fréjus it was a "political" tunnel, an expression of the economic and military might of the rising German Reich. Its more immediate communication aim was to connect the Ruhr with Milan in the shortest and most efficient way.

The Gotthard Tunnel was driven from both ends, using the Belgian method; i.e., an 8×8-foot pilot gallery was advanced along the crown of the tunnel, after which the top half was broken out to full width and the arch lining put in. Under the protection of the arch, the remainder of the tunnel area was excavated and lined. The rate of advance of the pilot gallery was from 10 to $14\frac{1}{2}$ feet per 24 hours at each end, when conditions were favorable. Unfortunately, conditions were seldom favorable during the 10 years required to excavate and line the tunnel. The chief trouble encountered was water which jetted out of the rock at the rate of 3,000 gallons per minute and sometimes with the force of a firehose. The dynamite flowed out of the holes as a yellow sludge, and it was necessary to enclose the explosive in tin tubes in order to be able to use it at all.

Working conditions in the tunnel became lethal, and 25 deaths and hundreds of casualties were reported each year. Rock dust, dynamite fumes, the exhalations of men and animals, plus increasing heat from the rock raised the temperature at times to 122°F and caused a fatal sickness called "miner's anemia." A man became incapacitated after a few months, and if he persisted he died or became an invalid for life. Thirty horses and mules died each month. In the end, 310 lives were lost and 877 men incapacitated. The contractor died, after seven years, and one by one the directors of the railroad company succumbed to the strain of the mounting death toll and costs.

Hence, after the initial successful pioneering effort, the

222

next mechanized drive through the Alps turned into a major tragedy. At the subsequent technical inquest, it was decided that the primary reason for the failures was the lack of adequate power. Favre had 750-hp water turbines installed at the northern portal and 1,870-hp at the southern; but lack of water in the winter prevented the use of the full capacity, and men as well as animals gasped in the workings. The choice of method was also a grave error, since in sections of unstable and plastic rock the abutments sank and the vault collapsed. The absence of elementary hygiene was the cause of the afflictions incapacitating the miners.

The Gotthard debacle had several major consequences. It delayed by about 10 years the plan for a third long tunnel through the Alps, since the Bundesrat, haunted by the ugly memories of St. Gotthard, kept procrastinating approving the Simplon project. Technically, it led to the development of hydraulic machines as a replacement for pneumatic rock drills. Since 80 percent or more of the primary power used to compress air was wasted in heat, and since water in winter declines to a minimum inadequate to bear such waste of energy, it would be better, it was argued, to compress water instead, and use it directly as a prime mover of hydraulic machines. Finally, the Bundesrat decided to appoint an international commission to report, inter alia, on the sanitary and hygienic requirements for underground work.

Ultimately, after an interlude of 17 years, permission was granted in 1898 to go ahead with the long delayed Simplon project. The tunnel has its northern portal at Brig and runs under Monte Leone for a distance of 12 miles 1,438 feet, including pilot adits at both ends. Simplon was then—and still is—the largest tunnel ever advanced, and because of its length and, not least, the height of the rock cover under Monte Leone and the pressure troubles anticipated, the development included two 16×18-foot single-track tunnels 55 feet apart and joined by cross headings at intervals of 656 feet. The two parallel galleries were driven simultaneously by means of an 8×10-foot pilot heading along the floor in the Austrian manner, and the eastern one was broken out to full area and lined. The second pilot gallery was not expanded to full size until 1920. The pilots were driven by means of four Brandt hydraulic rock drills in each heading, worked by water at a pressure of 1,700 psi. Each machine drove a heavy steel with a 70-millimeter three-pronged crown into the rock under a pressure of 15 tons and turned it 4 to 6 times per minute. In this manner a hole was milled out of the rock at the rate of $1\frac{1}{2}$ to $2\frac{1}{2}$ feet per hour. A round consisted of seven to ten 4-foot holes, and the total dynamite charge was 40 kilograms. A round produced on the average about 350 cubic feet of rock and an advance of $3\frac{1}{2}$ feet.

As in St. Gotthard, the miners struck numerous springs, some with a temperature of 131°F, which flooded the workings and impeded the advance. But the worst trouble in the Simplon drive was encountered after three years of work at 14,596 feet from the southern portal. Here the miners struck a 137-foot-long zone of limestone streaked with gypsum and shot through with springs. The water turned the rock into a doughy mess which exerted a tremendous pressure on the gallery. Timbers up to 2 feet in diameter snapped like matchsticks, and for a while it looked as though the heading would have to be abandoned. In the end, the zone was bridged by means of steel frames bolted together, an operation that required six months and cost one million francs.

Simplon was the last long mountain tunnel to be driven without access to electric power. Electricity was used to illuminate the workings; and when opened for traffic in 1906, the tunnel was equipped for electric traction, thereby simplifying the ventilation problem. But throughout the seven years of the mining phase, power for operating the rock drilling machines derived from Pelton wheels working under a head of 172 feet on the northern side and 577 feet on the southern. The four northern turbines developed altogether 2,239 hp, which proved inadequate for continuing the advance beyond 34,000 feet.

The total cost of the finished tunnel and the unfinished pilot gallery came to 78 million francs, or 3,625 francs per linear yard, as against a contract price of 69.5 million francs. Under the contract, the firm Brandt, Brandon & Company and the Winterthur banks backing the enterprise had lost 8.5 million francs, but since the completion was delayed 700 days beyond the $5\frac{1}{2}$ years stipulated, and the contractor was to be fined 5,000 francs a day, the accumulated fines amounted to 3.5 million francs. Hence, the company's total loss came to 12 million francs, or just about the same as for St. Gotthard.

However, the outcome was altogether different. Whereas the St. Gotthard Railway Company was legally bound to insist on its contractual obligations, even if the contractor was thereby driven into bankruptcy, the Swiss railroads had become nationalized during the Simplon advance, and for political reasons the national administration could not insist on its pound of flesh. The balance was adjusted, and it was in the end the Swiss taxpayers who had to make up the losses suffered in building the world's longest tunnel. The rugged enterprise that had characterized Victorian construction had by no means come to an end, but its crudest features were about to be eliminated.

Building with iron

A characteristic feature of nineteenth century engineering was its increasing reliance on iron for construction, and toward the end of the century also on mild steel. Indeed,

The first iron bridge in history was built in 1776 to 1779 by Abraham Darby III. The bridge crossing the Severn at Coalbrookdale has a span of 100½ feet, rising to a height of 50 feet. It is built of cast-iron ribs hinged at the springings and the crown. The drawing is contemporary with the bridge and was published in London Magazine in 1784. Iron Bridge is now a national monument.

the use of iron is, historically speaking, the greatest contribution of the British pioneers to civil engineering. Iron in building was, of course, no innovation; it had been used in Greek temples, Roman bridges, Byzantine churches, and Gothic cathedrals, but always sparingly, for clamps and reinforcement rods. Iron was scarce and too costly to be squandered on construction.

The technological developments that lowered the price of iron and permitted it to be used in engineering will be sketched in a following chapter, whereas here we shall be concerned only with some of the early iron structures of historical interest. As so frequently happens in technological history, the first iron bridge was erected by a man possessing no experience in bridge building or engineering. It was designed and erected in 1776 to 1779 by Abraham Darby III, owner of an ironworks, and his partner Reynolds, and it spans the river Severn near Coalbrookdale. The bridge—now a national monument—consists of five semicircular 12 × 6-inch cast-iron ribs with a span of 100½ feet rising 50 feet. Each rib is hinged at the springing

and in the crown. The rib castings are 70 feet long, and were poured in open sand molds directly from a blast furnace. No bolts or rivets were used in the assembly; the components were simply mortised into one another and secured by wedges. The total weight of the Iron Bridge, as it is now called, has been estimated at 378 tons.

Tom Paine, author of *The Rights of Man,* designed another cast-iron bridge spanning the river Wear at Sunderland. It had a length of 236 feet and rose 100 feet above the water level. The bridge was built in 1796, by which time Payne had become mixed up with the French Revolution and forced to leave England. Thomas Telford also tried his hand with the new material and designed a small cast-iron bridge across the Severn at Buildwas, but this one dwindles in importance compared to the Menai Bridge designed by him while the 130-foot Buildwas was being constructed.

Telford's Menai Bridge crossing the Straights between the Caernarvon shore and the island of Anglesey on the west coast of Wales was a stupendous achievement and

224

the largest iron stucture hitherto erected. The suspension bridge, which linked Telford's London-Holyhead road, had a span of 579 feet and a clearance of 100 feet to conform with Admiralty requirements. The suspension members consisted of malleable iron chains made up of 10-foot-long and $3\frac{1}{2} \times 1$-inch links with an eye at each end. Five such bars were bundled into a set of chains, and four sets into a group, and four groups into a suspension chain cable. The 30-foot deck was carried by 1-inch rods placed on 5-foot centers, but since the deck was inadequately stiffened when the bridge was completed in 1826, it was exposed to the periodic oscillations that wrecked so many of the early suspension bridges, and Telford's Menai Bridge did not escape a similar fate. It was damaged during a storm in 1829, and then repaired and provided with a heavier and stiffer road deck which lasted to 1893 when it was replaced by a steel deck. The original chains remained in service until 1940.

A number of long suspension bridges were also built on the Continent, among them the 896-foot Grand Pont crossing the Seine at Fribourg in Switzerland. This bridge, built in 1822 to 1823, was the boldest structure up until that time and had suspension cables made up of 1,000 iron wires bound together with iron wrappings. The Grand Pont was entirely successful and remained in service for a hundred years until its demolition in 1923. Otherwise, the early Continental suspension bridges suffered the same fate as so many English and American ones; inadequate bracing caused them to collapse in the first violent storms that hit them.

Not surprisingly, it was in America that the initial difficulties with suspension bridges were overcome, because this type of bridge had many appealing features in a country with broad rivers separating rapidly growing centers of population. It was economical in material and labor and did not require the skills needed for building masonry bridges. Judge James Finley of Pennsylvania erected the first stiffened bridge on the catenary principle in 1796, and the following decade saw the construction of numerous suspension bridges, among them the 244-foot one spanning the Merrimac River in Massachusetts, completed in 1809 and still standing. The original chains were replaced by stranded wire in 1909. This length of service appears to be a world record for suspension bridges.

But the great American suspension bridges from the 4,200-foot Golden Gate in the·west to the 4,260-foot Verrazano Narrows in the east derive from a patent granted to John A. Roebling in 1841, whereby the cables could be spun, or erected *in situ,* by carrying wires (usually 0.19 inch in gage) back and forth across the span, and thereafter bunching them into a uniformly tensed cable held together by binding wire. The Roebling method was applied with great success in 1855 on the Grand Trunk Bridge crossing the Niagara Gorge. It had a span of 820 feet and was a double-deck structure carrying a single-track railroad on the top deck and a highway below. This was the first suspension bridge that proved rigid enough to bear railroad traffic. The two decks were carried by four 10-inch cables built up from parallel wrought-iron wires, and the deck structure was stiffened by 18-foot girders and stayed with wire rope to prevent oscillations. The Grand Trunk Bridge lasted until 1896 when it was dismantled.

John Roebling's masterpiece is of course the famous Brooklyn Bridge begun by him in 1867 and finished by his son in 1887. Here the span is 1,545 feet, and the river clearance 113 feet. The Brooklyn structure was built in the same manner as the Niagara bridge except that the wires for the cables were of galvanized steel with a tensile strength of 71.5 tons per square inch.

However, the bridges previously discussed, whether of the arch or suspension type, were not well suited to the requirements of the railway age. The innumerable canal and river crossings consequent upon the expansion of the railroad net, at first in Great Britain and subsequently in all other countries, required a type of bridge that could be built rapidly at minimum cost. The original impetus to the development of the steel girder bridge was the objection of the Admiralty to placing piers in the Menai Straits as requested by Robert Stephenson, in order to carry the Chester and Holyhead Railway across the Straits in 1845. Stopped by the Admiralty, Stephenson flirted with the idea of converting the existing Telford bridge to railroad traffic, but in the end he hit upon a structure consisting of two huge rectangular box girders made of wrought-iron plates similar to those used in shipbuilding. Each rectangular box was to be large enough to permit a train to pass through it. The original idea was to have the two box girders suspended in chains from tall towers at each end of the span placed close to and parallel with the Telford suspension bridge.

The way in which the idea was further developed serves as an excellent illustration of the empirical character of mid-century English engineering. The French and other Continental engineers would have attacked the problem along theoretical lines, but whether they would have arrived at a better solution seems doubtful, in view of the limited knowledge of the new structural material. Instead, Stephenson enlisted the aid of William Fairbairn, a mechanical engineer and an expert on materials testing, and the mathematician Eaton Hodgkinson to interpret the results. Together the three carried out a number of tests: 12 on circular tubes, 7 on elliptical sections, and 14 on rectangular sections. They tried different sizes of plates of varying gage; and all designs were tested to destruction by placing loads mid-span.

From these experiments undertaken in Fairbairn's shipyard at Millwall, it was found that the rectangular section was the strongest, and six tests were conducted on a

1:6 model 75 feet long, from which it was determined that by adding 20 percent material weight, the carrying capacity of the structure would be increased 2½ times. As a result of the tests, girders were made with a maximum length of 459 feet and a 14 foot 9 inch × 30-foot section. To stiffen the box, the top and bottom were given cellular construction. The box girders were fixed to the tower built on the Britannia Rock in the middle of the Straits, but were carried on expansion bearings at the shore towers, a necessary precaution because it has been found that the structure expands as much as 4 feet in warm weather.

The huge box girders of the Britannia Bridge were constructed on the site, those for the land span on timber stagings, and the 459-foot water span weighing 1,288 tons along the shore. The plates were joined by rivets driven by hydraulic riveting machines. After assembly, the river span was floated on pontoons at high tide and positioned so that as the tide receded, it landed gently in the recesses in the two masonry towers. It was then raised to the final position by means of hydraulic jacks at the rate of 2 inches per minute. The Britannia Bridge, whether by design or accident, was turned into a continuous girder—or as it is known today, continuous span—because after assembly Stephenson decided to stress the girder by jacking up the shore end, while the tower end was riveted to the junction piece embedded in the tower, after which the other end was lowered to its permanent bearings.

The Britannia Bridge was opened for traffic March 4, 1850; and it is still in service, carrying an infinitely heavier load. Considering that hitherto the longest wrought-iron bridge span had been only 31½ feet, Britannia was a tremendous advance in the construction of iron bridges. It was the forerunner of the hundreds of thousands of girder bridges subsequently built all over the world. Britannia

Britannia Bridge crossing the Menai Straits is the first railroad bridge built with iron. The bridge is of historical importance also because its construction was preceded by a series of systematic tests for determining the strength of the material and the shape and dimensions of the structure. The two 14 foot 9 inch × 30-foot box girders still carry the trains to Holyhead.

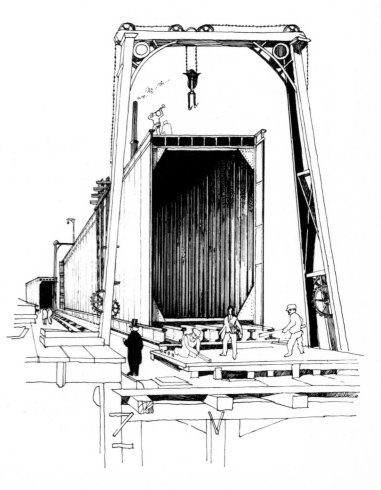

Bridge is an ugly structure which was not improved by the naïve attempt to design the towers like a Norman keep, with battlements and cruciform archers' slits.

While Stephenson struggled with the tests leading to this bridge, indeed some years earlier, Thomas Wallis Pratt had obtained a United States patent for an iron lattice truss, developed from the wooden trusses used for bridge building in America begun some 20 years before. When around 1850 Karl Culmann, subsequently professor at the newly founded Eidgenosisches Polytechnicum in Zurich, ran into some early truss bridges on his American travels, he became enthusiastic about their possibilities and introduced the new bridge type to Europe. As previously mentioned, he developed a graphic method of stress analysis whereby the design of the latticed truss was greatly simplified.

Owing to the work of Culmann, Ritter, Cremona et al., the next truss bridge of historical importance was the Forth Bridge carrying the North British Railway line from a point 6 miles west of South Queensferry to Garvie Island and North Queensferry. The Forth Bridge was designed by B. Baker as a cantilever, and 54,000 tons of "Siemens-Martin" (open-hearth) steel was used in its construction. The two major spans are 1,710 feet long and made up of two cantilever arms, joined by a 350-foot span in the middle.

Not only was the Forth Bridge the longest steel bridge erected up until that time (1882–1890), but also many of the engineering techniques subsequently used in bridge building were developed during its design and construction. There was no longer any guesswork in such designs. The Board of Trade specified that the stresses should be not more than a quarter of the ultimate strength of the material, and that the structure should stand up to a wind pressure of 56 pounds per square foot. The designers stipulated that a steel with a tensile strength of 34 to 37 tons per square inch should be used for the compression members, and a 30- to 33-ton steel for the tension members.

The Trenton viaduct over the Delaware River is one of the early examples of the low-cost truss bridges developed to meet the special needs of the railroads. The type emanates from a timber lattice truss invented and patented in 1820 by Ithiel Town, a New Haven architect. It was accommodated to iron by Thomas Willis Pratt in 1844 and was introduced to Europe by Carl Culmann, professor of engineering at the Polytechnicum in Zurich. It became known as the "Pratt truss," and has been employed for medium-span bridges on railroads the world over.

All rivet holes were drilled and the edges of each plate planed to remove metal affected by stressing.

The steel fabrication shop of the contractor, William Arrol, covered an area of 50 acres at the South Queensferry end of the bridge. The heaviest compression members were 12 feet in diameter and fabricated from $1\frac{1}{4}$-inch plates 16 feet long and $4\frac{1}{2}$ feet wide. Some 42 miles of steel plates were bent into tubes, each tube 400 feet long and provided with internal stiffeners. After fabrication, the tubes were dismounted, and the elements re-erected in the bridge structure and riveted into place by hydraulic riveters. No staging was used in building the bridge, since erection began at the piers and the cantilever arms were extended simultaneously so that the structure was always kept in balance. The suspended spans in the middle were also built from the cantilever beams and joined in the middle.

This final operation illustrates the gains that had been made in statics and materials testing since the Britannia Bridge was built 30 years earlier. The plates overlapping each other at the middle joint were drilled in the shop; and the bolt holes were calculated to come fair at an air temperature of 60°F, at the time of erection. But when the closure was attempted, the temperature was only 55°F, and a chilly-northerly wind was blowing, so that the holes did not meet. By lighting fires of wood shavings and oily waste over a distance of some 50 feet on each side of the middle span, the steel was made to expand so that the holes came fair and the bolts securing the two halves of the suspended span could be inserted and drawn. Structural engineering in steel had reached adulthood.

The Firth of Forth Bridge, completed in 1890, was the graduate exercise in steel construction.

New Materials, New Sources of Power

"It is industry that gives birth to and develops in mankind new needs and gives them at the same time the means to satisfy them. . ."

MARC SEGUIN

Historically speaking, the most important contribution of the early British civil engineers was their use of iron in construction, although they themselves did not take part in the technological revolution that finally broke the bottleneck in iron production. When the Smeatonians needed iron in ever-increasing volumes, it was merely a matter of ordering it, since owing to the work of an altogether different set of early pioneers, the iron mills of their time were in a position to supply it at an acceptable price.

Put in its simplest terms, the metallurgical revolution meant severing once and for all the bondage to charcoal as a fuel and replacing it with mineral coal. Charcoal requires access to forests and a rural proletariat of the lowest order, both of which vanish with the advance of civilization. The supply of iron and other metal ores was seldom exhausted in the ancient mining regions, but the inroads made on the local timberlands for charcoaling undermined the balance of nature; the forests ceased to thrive and vanished entirely, whereupon metal mining and smelting came to a stop.

This happened everywhere—in the Taurus and Caucasus, Cyprus, Elba, Attica, Spain, and England. Indeed, by the early half of the eighteenth century, only such heavily forested countries on the marches of the civilized world as Russia, Sweden, and the United States were able to maintain an iron industry of any scale. But then, after a century or more of confused attempts to use coal as a blast-furnace fuel, Abraham Darby of Coalbrookdale in Shropshire succeeded in 1710 where others had failed. He coked the coal and thereby managed to remove the sulfur which previously had adulterated the pig. The new fuel that was available in limitless supply immediately reduced the cost of pig iron; of equal importance, the coke could carry a heavier burden of ore than charcoal, permitting the capacity of the furnaces to be increased practically indefinitely.

Fifty years later, Richard Reynolds, a Quaker like the Darbys and married into the family, succeeded in developing a process whereby coke could be used in converting pig into malleable iron. (The idea of refining pig iron using coke as a fuel in a reverberatory furnace was originally suggested by the brothers Thomas and George Cranage, who received a patent for the process in 1766.) Thus, during the latter half of the eighteenth century, the impediments to volume production of both pig and malleable iron had been overcome by the efforts of this remarkable Quaker family. Iron for most purposes was now available in ample volume and at a low price. The rails for the many

The invention in 1783 of the coal-fired reverberatory puddling furnace by Peter Onions of Merthyr Tydfil enabled wrought iron to be produced in volume at low cost by the refining of pig iron made in blast furnaces using coke as a fuel and reducing agent. This invention broke the bottleneck in iron smelting.

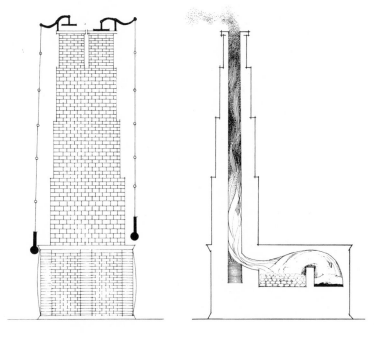

tramroads in the colleries were first made of cast iron and later of malleable iron turned out in the new reverberatory furnace fired with coke.

But although much cheaper and available in large quantities, the Coalbrookdale iron was inferior to the charcoal iron produced abroad and imported in increasing volumes. It still contained too much carbon and its remaining traces of sulfur made it "red short," i.e., it did not behave well when hammered hot. This serious defect was eliminated by the puddling process invented in 1783 by Peter Onions of Merthyr Tydfil. His puddling furnace permitted stirring the iron while in a plastic state, thus removing the excess carbon and sulfur.

Finally, one year later—on February 13, 1784, to be exact—a patent was granted to Henry Cort which summed up all the previous attempts to industrialize iron production; indeed, the Cort patent covered what today would be termed an integrated mill. There was nothing new about the details, but what Cort described in his patent is what may be called a continuous process. He either melted the pig iron or ladled it directly from the blast furnace into a reverberatory furnace fired with raw mineral coal. Through apertures in the furnace the plastic iron was stirred intermittently, and when sufficient carbon had been burned away, the bloom was lifted out from the hearth and shingled under a power hammer into bars, rods, or flats, which were then immediately put through a *rolling mill* and given final shape.

In this manner Cort managed to produce 15 tons in 12 hours, using the same power and labor previously required to turn out 1 ton by the conventional hammer process. As usually happens with great technological breakthroughs, Henry Cort got nothing but misery out of his

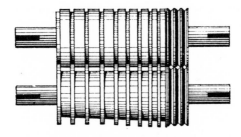

In 1783, Henry Cort introduced a continuous method of iron production. Instead of forging wrought-iron bars, he rolled them in a mill with grooved rolls. His continuous method, including the rolling mill, increased the output fifteenfold with the same input of labor and energy.

great vision. The Crown appropriated his experimental works and patents, which were thereafter free for all to use. When Cort died broken-hearted in 1800, Merthyr Tydfil was booming under the impact of the new iron technology, permitting one works alone to turn out 70 tons of bars per week. By 1806, the Cyfartha works—largest in the world—had expanded to include six furnaces and two rolling mills worked by 1,500 men; in 1819, the output was 11,000 tons of pig iron and 12,000 tons of bars, equivalent to the entire British output during the decade 1740 to 1750. Within one decade Britain had leaped to the forefront as the incomparably greatest iron producer in the world, and traditional suppliers sank back into a state of insignificance and fell victims to an economic crisis from which some of them never recovered.

Thus, owing to the Darbys and Henry Cort, the naval paymaster turned ironmaster at the age of thirty-four, British engineers found themselves in possession of a construction material for the innumerable new structures required by the new age for which they acted as midwives. The railroads alone needed iron on a scale beyond the wildest dreams of traditional ironmasters, but iron was also essential for tens of thousands of bridges, railroad stations, mills and factory buildings, as well as increasingly heavy machines.

Some of the early applications have been discussed in a previous chapter, but, in addition, cast iron became used for columns, beams, window mullions, roof work, etc., in building. Columns in the classical orders and other fancy applications for buildings and bridges could be purchased from foundries; such standard items could be ordered from catalogues up until the First World War. The vogue of cast-iron ornamentation culminated at the middle of the century in a large number of atrocities the world over, some of which, unfortunately, are still standing. For about 50 years cast iron was used for many odd purposes, and in numerous instances the fashionable material failed miserably. These structural failures led to the appointment in 1848 of a Royal Commission which reported a year later on "The Application of Iron to Railway Structures." Yet, before the official report, valuable work in testing the properties of iron had been carried out by William Fairbairn in collaboration with the mathematician, Eaton Hodgkinson. Their combined work eventually turned materials testing into a scientific procedure. Their historical tests on the trusses of the Britannia Bridge have been discussed briefly in a previous chapter.

However, regardless of the applications and misapplications of iron by the first generation of civil engineers and architects, the British output of iron continued to grow. By 1850, British mills turned out $2\frac{1}{2}$ million tons of iron a year, practically all of it puddled wrought iron; whereas the output of steel was much more modest—or 60,000 tons—produced by cementation. The difficult crucible process made steel a very costly material used only for cutting tools and cutlery.

The systematic testing of iron revealed a number of curious features, some of them contrary to the preconceived notions of engineers as to the strength of the new material. The malleable iron turned out in such tremendous tonnages by the British mills was an excellent material in many ways; but it was a poor substitute for the traditional construction materials—stone, timber, and brick. If iron was to have a future as a construction material, it had to be given rather different properties from those possessed by the bloom manipulated by sweaty furnacemen in the

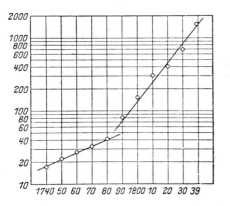

The logarithmic diagram sums up one reason for the British technological and economic hegemony during the nineteenth century. It gives the output of pig iron from 1740 when coke began to be used as a fuel for blast furnaces. In 1760, there were 17 blast furnaces using coke in England; 15 years later there were 81. The output grew from 18,000 tons per year in 1740 to almost 1.8 million tons a century later.

The round blast furnace built by Gibbon at Corbyn's Hall in Staffordshire in 1832 had a height of 50 feet as against 35 feet for the largest furnaces previously used. The round section permitted the continuous development of the furnaces to practically unlimited height and output. The sketch is a copy of the original drawing of the Gibbon furnace.

ironworks of Merthyr. But no one at the time knew exactly what these properties were.

Mr. William Kelly certainly had no idea what he was doing when in 1847 he began some experiments in his ironworks at Eddyville in Kentucky. Kelly made a living by making wrought-iron sugar kettles for the local farmers; and like many ironmasters the world over, he obtained his wrought iron by burning out the excess carbon of the pig iron in a finery furnace. And, like numerous forgemen the world over he had noticed how the iron turned fluid and formed drops near the tuyeres. But he differed from the rest of his colleagues in his deduction that the excess carbon in the pig iron, which made it hard and brittle, might be blown out by air alone. After all, the carbon already in the pig would supply the fuel to consume itself and leave a residue of soft iron.

Naturally, Kelly suffered the fate of nearly all pioneers. The ironmasters to whom he described his new discovery rejected as a poor joke the idea of making wrought iron "without fuel"; even his wife thought him mad. Kelly had to proceed with his experiments in a secluded spot in the Kentucky woods, where between 1851 and 1856 he built seven steel converters in deepest secrecy. When in 1856 he learned that an Englishman had taken out a patent for a similar process, he also applied for one and obtained an American patent June 23, 1857. That year he was also declared bankrupt.

The name of the Englishman whose patent was rejected in America was Henry Bessemer, who in his historical paper presented to the British Association for the Advancement of Science at its meeting in Cheltenham on August 11, 1856 attributed his success to not being burdened with preconceived ideas of iron metallurgy. Although they were similar in method and conception, there was a considerable difference between the Kelly and the Bessemer processes of "making malleable iron without fuel." Kelly's aim was to get a better finery product without liquefying the metal; whereas Bessemer insisted that "to manufacture malleable iron and steel without fuel," it was necessary to keep the metal completely liquid at high temperature.

It would lead too far afield to trace even in rough outline the horrible disappointments that were to dog the initial commercial applications following Bessemer's laboratory successes. Suffice it to say that the Bessemer process —the first of the mass production methods of making a mild steel suitable for construction—was turned into a stabilized production process, not by Bessemer himself, or by the British ironmasters who had bought the rights, but by a couple of anonymous Swedish ironmasters appointed by the College of Mines; the latter paid for the continuance of the trials when the Swedish licensee was no longer able to carry on alone. In Sweden, in a secluded spot in the woods resembling the one where Kelly had come to finan-

231

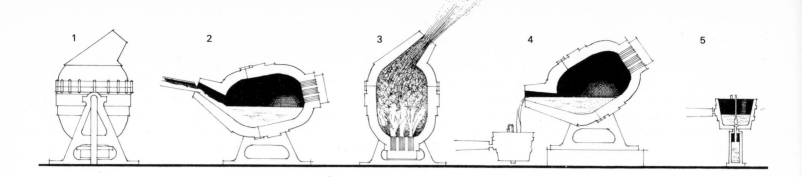

The first steel converter built by Bessemer for his successful experiments was a stationary furnace, which he replaced in 1860 by a tilting type that has remained largely unchanged ever since. Although limited to pig iron unadulterated by phosphorus and relatively uneconomical, the Bessemer process laid the foundation of the modern steel industry. The mild steel produced by the process was superior to puddle iron in wearing quality and was therefore immediately adopted for railroad construction. The sketches show five phases of the Bessemer process: (1) the converter in upright position, (2) the tilted converter receiving its charge of fluid pig iron, (3) the converter in upright position during the blow (when the excess carbon in the pig is consumed in a shower of sparks), (4) the converter tilted after the blow to discharge the steel into a ladle, and (5) casting the steel in a mold to obtain an ingot.

cial grief, the reason for the commercial Bessemer failures was revealed: the British ironmasters had used an iron smelted from phosphorous ores, whereas Bessemer's own initial success was due to a fluke of using pig iron free from adulterants. The Swedish trials also set a limit to his process: only high-grade pig iron free from phosphorous adulterants could be used.

Although the Bessemer process was rapidly adopted everywhere—in France in 1858, in Germany by Krupp in 1862, in Austria in 1863, in the United States in 1864—nobody was aware at the time of the potentialities of the product turned out by the converters. It was not malleable iron nor was it the kind of steel that cutlery manufacturers were accustomed to. It was a different material, a mild steel harder than wrought iron, softer than crucible steel. This was indeed a steel suited for the building of skyscrapers, bridges, railroads, machines, and motorcars, although no one realized it at the time.

Nor was anyone particularly interested in a patent filed by Frederick Siemens in 1858 for his invention of a method of heat generation which when subsequently applied to blast-furnace operation, increased the output by 20 percent, and in 1861 was employed also in open-hearth furnaces for melting glass and steel. A few years later Pierre Martin, operating a family ironworks at Sireul in France, added some scrap to dilute a pig-iron bath, whereby the basis was laid for the Siemens-Martin process. Owing to its superior economic and technical features, this soon outdistanced the Bessemer process and became the favorite of the rapidly expanding steel industry, which nearly doubled its output per decade up until 1910. To complete

this summary, it should be added that by 1875 a third process using the basic converter, invented by Sidney Gilchrest Thomas, had been made available, whereby the vast deposits of phosphorous ores, hitherto completely worthless, could be opened up and used for steel production.

To the civil engineers working in these heady times, the processes used by the steel mills were of no particular interest. To them the important thing was that they had obtained a new construction material that could be applied without technical inhibitions. The inevitable happened: the hubris that made the Sumerian builders pile brick upon brick until the ziggurats threatened to collapse under their own weight, and the architects of the pharaohs pile rock upon rock to "the useless structures of the pyramids," now found expression in even more useless engineering exercises.

The lunatic fringe of the engineering profession was seized by fixed ideas of towers, the higher the better. There were countless designs for steel towers, one higher than the next. Most of them moldered on paper, but some of the megalith structures were actually built.

The outstanding example of this engineering megalomania is, of course, the Eiffel Tower built for the Paris Exhibition in 1889. This ugly and stupid structure was brought to a height of 875 feet on a concrete base extending to 50 feet below ground. The tower was constructed of 7,300 tons of wrought iron. The really interesting thing about the Eiffel Tower is that it was built with obsolete material at a time when world production of mild steel exceeded 12 million tons, and that of France alone was 670,000 tons; in sum, it is not even a monument to the brave new world of steel to which the future belonged.

A much more intelligent conception of the potentialities of the new structural material was shown in Chicago and led to the evolution of the skyscraper. The first tall building erected in Chicago was the 10-story Home Insurance Building begun in 1883. Confronted with the problem of fireproofing a block of offices that also had to be given maximum light in every room, architect William Le Baron Jenney decided to construct an iron skeleton encased in brickwork. The vertical members were made of cast iron, and the first six floors were carried by wrought-iron beams; but above that level the beams were of Bessemer steel. This appears to have been the first time that steel was employed to any major extent in a building. Seven years later steel was used throughout in the Rand-McNally Building, also in Chicago; and in 1892 the 21-story Masonic Building was completed.

The original kiln built by William Aspdin at Northfleet in Kent. The height was 36 feet, and the diameter of the base 17 feet. The capacity of the kiln has been given as 80 barrels.

In the Masonic Building the walls consisted of thin and light screens inserted as panels between the columns and beams. These screening panels could just as well have been made of glass, but such an idea was too radical for the times, and instead there followed an interlude lasting nearly 50 years during which architects devoted their energies and talents to shaping the masonry screens into strange appliqués in Gothic, Renaissance, or other historical styles that they happened to fancy. It took some additional time before the esthetic possibilities of steel skeleton construction were brought to full fruition by such outstanding masters as Mies van der Rohe, Walter Gropius and others.

Building in concrete

By that time, i.e., after the Second World War, another structural material of noble ancestry had also been mastered, and with further insight into its potentialities become respectable. Concrete was the most common building material used by Roman imperial builders, in the form of a grout or mortar poured over broken stone or brick. Concrete was indeed used for bridge foundations in the early nineteenth century, and the French in particular relied on it to provide mass in seawalls and jetties. In fact, many of the refinements in the design of concrete structures derive from French pioneers. But for a number of reasons civil engineers did not take to concrete and relied up until the present century on masonry for applications, such as tunnel linings, for which concrete would have served just as well or better—as subsequently proved.

Concrete is a mixture of cement, sand, broken rock, and water which after being poured into a mold will set and become capable of carrying a load. A good cement, like the natural Italian pozzolana, will also set under water, whereas the usual lime mortar used by masons sets only in air.

The modern manufactured cement is commonly regarded as being derived from the numerous experiments conducted by John Smeaton in his effort to find a suitable mortar for bonding the masonry of the Eddystone Lighthouse. He found that clay mixed with limestone fired into brick and then milled into a powder had the desirable properties, but in the end he relied on the ancient Italian pozzolana for his historical lighthouse.

After these early experiments nothing further happened for the next 50 years until on October 21, 1824, a patent "for making a cement or artificial stone" was granted to the bricklayer Joseph Aspdin of Leeds. Aspdin described in his patent how he took dust derived from the crushed limestone used for repairing roads and mixed it with clay in water. He then exposed the wet mixture to the sun or heated it with steam or fire until the water evaporated. The dry cake thus obtained he broke into lumps which he

233

calcined in a furnace similar to a lime kiln. The calcined product was ground into a fine powder and was then in a fit state for making cement or artificial stone. Since Portland stone was the finest limestone used at the time by English masons, Aspdin called his concoction "portland cement," a name that has stuck to it ever since.

Aspdin's portland cement was used by Marc Isambard Brunel in 1828 for making the mortar for the masonry lining of the Thames Tunnel, despite its high cost of 20 shillings a barrel as against 12 shillings for the natural so-called "Roman cements" made by calcining a natural blend of limestone and clay. Although the Aspdin method got the early British cement industry started, modern cements derive from the discovery by I. C. Johnson in 1845 that an overburned and spoiled clinker produced a far superior product. What had previously been thrown away as worthless clinker because of the accidental exposure to high temperature was actually the best output of the bottle kiln. Instead of the lime and clay mixture being fired to 2000 to 2400°F, a temperature of 2550 to 2650°F was required to produce the right clinker from which a strong cement could be ground.

Cement manufacture also needed an altogether different kind of kiln from the conventional one used for calcining lime, which was terribly wasteful of fuel and left a lot of clinker inadequately fired. The continuous rotary cement kiln was invented by Crampton in 1877, but several improvements were necessary to make it workable, and the first rotary unit was not built until 10 years later at Arlesley, near Hitchin. It did not come up to expectations, however, and the credit for the first successful rotary cement kiln goes to D. O. Sayer, owner of a cement works at Copley in Pennsylvania. Further American improvements established the continuous rotary kiln as a reliable production vehicle in the early 1890s.

But the fact that Portland cement could now be made in well-nigh unlimited volume at acceptable cost did not necessarily mean that concrete came to be used in construction. Indeed, all but a few engineers stayed clear of it, and for good reason; in many countries public authorities refused to accept the material after some early structures built of it had collapsed. Its heterogeneous character prevented the application of conventional methods of stress calculation, and it was not until well into the present century that confidence in the new construction methods became buttressed by compulsory standards specifications for the design and construction of reinforced-concrete structures.

The operative word is *reinforced*. Concrete has a high resistance to compressive forces, but possesses hardly any resistance to tensile stresses. In this respect it resembles masonry. This weakness of the material was overcome by Joseph Monier, a Paris gardener, who embedded in his concrete tubs for orange trees an iron net, for which he obtained a patent in 1867. Although the initial response was not encouraging, Monier kept on filing concrete patents, for containers, floors, beams, pipes, bridges, even railroad sleepers. From such elements he proceeded to buildings and water tanks, and in 1875 he even managed to build a small reinforced-concrete bridge.

Monier was concerned with shape and general strength, but the theoretical calculations for more accurate dimensioning of the reinforced concrete were beyond his ability. However, his inventions aroused the interest of a German firm of building contractors who commissioned M. Koenen to conduct some theoretical studies of the behavior of reinforced-concrete structures. His work "Das System Monier," published in 1887, stimulated the interest of German professors of civil engineering and of the Frenchman Edmond Coignet who wrote a paper in 1894 on a method of calculation closely related to modern theories. But nobody understood it at the time and it was largely ignored. It remained for Koenen to express in simple terms the elementary fact that in a reinforced-concrete member the steel serves the function of absorbing the tensile stresses, while the concrete alone absorbs the compression stresses.

Although eventually the theory of concrete construction became mastered, due to the work of theoretical investigators as well as building practitioners, particulary François Hennebique who licensed his designs to numerous countries, there were still impediments in the way of a general adoption of the new construction material. One, and perhaps the most important from the point of view of architects, was esthetic. The dull gray and uniform texture was not pretty to look at, and when—much later—some daring pioneers began to build churches in concrete, they were aptly named "Seelen Silos" (soul silos) in Switzerland. But concrete was also a labor-consuming material. It was made by men who had to turn over the heavy aggregate by hand, and unlike mortar it had to be used immediately because it set within an hour of mixing. It also required "falsework" or board shutterings for the pouring, which entailed a wasteful use of timber and carpentry to build the molds. The pouring was also labor-consuming, since great care had to be taken to rod the mixture firmly into place and cover the reinforcement bars with dense concrete.

Stone crushers and mechanical mixers were needed before the new material could be used on a large scale, and although concreting machines were, in actual fact, used over a century ago—at least in Germany—their full development required electrified sites. Now, of course, concrete is supplied to all but the very largest construction and building sites from special concreting plants manufacturing a finished product delivered in rotary hoppers. Major dam and underground developments are served by automatic rock crushing and mixing plants erected on the sites.

The immediate consequence of the increasing confidence

The Gmündertobel Bridge crossing the Sitter River in Switzerland is one of the early bridges built of reinforced concrete. It has a 262-foot span and was erected in 1908 from a design by E. Mörsch. The first concrete railroad bridges were built in Transylvania the following year.

and official sanction of reinforced concrete and the early mechanized means of handling it was its application to bridge building. Among the early concrete bridges was the one spanning the river Isar at Grünwald with 230-foot arches. It was built in 1903 and 1904. The second major concrete bridge was the Gmündertobel, crossing the river Sitter at Teufen in Switzerland, with a span of 262 feet, built in 1908. Both these bridges were designed by E. Mörsch. In Italy, the Risorgimento Bridge over the Tiber in Rome was built 1910 and 1911 by the Italian subsidiary of the Hennibique Company. The bridge has a span of 328 feet rising only 33 feet at the crown, and was the first structure in which the boldness of the design revealed the potentialities of the monolithic building method. The Langwies viaduct, built 1912 and 1913, carrying the Coire-Arosa Railroad across a deep gorge in Switzerland, is commonly regarded as the first successful railroad bridge; previous attempts to build concrete railroad bridges had been failures, owing to the vibrations set up by passing trains which tended to break up the adhesion between the steel and the concrete. Not only was the Langwies highly satisfactory from a structural point of view; it was also slender and beautiful, and demonstrated what a gifted designer could do with reinforced concrete.

Deferred mechanization

Building and construction have been the most backward of the major industries in the application of mechanized facilities and methods. Indeed, it seems from available, albeit spotty, evidence, as though in building at least there actually was a reversion from a modest use of simple machines on medieval sites to the exclusive utilization of human labor for porterage work on sites throughout the nineteenth century and well into the present. The overwhelming majority of the buildings now standing in the western world have at one time or another been carried on the backs of men and women. It is no accident that the first efficient building cranes were not introduced until the second half of the twentieth century, against the strenuous opposition of the building trades workers who insisted on their old privilege of breaking their backs carrying hod.

In regard to construction, it has already been noted in a previous chapter that most existing canals and railroads were built by human muscle, and all the millions of tons of rail were borne on the shoulders of the men who hammered the rail into place.

Needless to say, this bestial use of human labor was at all times unnecessary. With some elementary planning, draft animals could have been used to advantage on the canal and railroad cuttings, the rail could have been rolled instead of carried, a horse whim could have been used for raising a pile driver instead of a dozen men being employed to tug on ropes, water could have been drained in other ways than bailing by hand or having a couple of men work a primitive diaphragm pump. But the labor-wasting methods persisted until well into the present century, and on many sites into the 1930s. The mechanization of the construction industry did not begin in earnest until after 1945.

The reason for the long delay was the same everywhere: as long as there was a rural proletariat available as recruiting ground for construction labor, there was no need to consider investments in laborsaving machines. Furthermore, with an abundant supply of local labor, small contractors with little or no capital were able to bid on work. It was merely a matter of hiring laboring men who brought their own tools with them. Similar conditions applied to government relief work in most countries during the postwar depressions when a foreman risked being jailed for using more efficient tools than pick and shovel.

Conditions were different in America where for more than a century there had been more work to be done than labor to perform it—a situation unknown in Europe until the 1950s. The chronic American labor shortage created a situation conducive to the introduction of machines, which is the reason why virtually everything that has happened in the way of the mechanization of working sites derives from America.

With the rapid postwar shrinkage of the rural population in Europe too, accompanied by competition for labor in other sectors of the economy, the former seemingly inexhaustible supply of crude construction labor has dwindled. Rising wages have put the small undercapitalized contractors out of business or forced them to merge into large firms capable of financing the machines needed to rationalize the industry. Reversed conditions of supply and demand caused the unions to become aware of the futility of fighting the machines. Thus it came to pass—within a span of merely a few years—that a mile of four-lane motor road could be built by only 25 men plus dieselized facilities, whereas up to, say, 1930, a mile of railroad required 500 men wielding pick and shovel.

These notes refer to the general trend. There were almost from the outset of western engineering many instances of outstanding engineers and early contractors utilizing machines to overcome difficulties met with in construction. One such instance has already been recognized: Fontana's use of horse-driven capstans to raise the 327-ton obelisk in Rome. Steam power was used for pumping at an early date—in the Tronquay Tunnel on the St. Quentin Canal in 1803. Brunel employed steam-driven pumps to drain the Thames Tunnel in 1827, and less than 10 years later

steam engines were installed to recover the flooded Kilsby Ridge headings. Horse-driven capstans were commonly used over the shafts to hoist the spoil from canal and railroad tunnel excavations; Telford used horse power to haul the heavy chains of the Menai suspension bridge across the Straits in 1825. The Severn Tunnel was drained by steam-driven pumps with a capacity of 11,000 gallons per minute.

While steam power was largely applied to drainage pumping, it was also used with notable success on some major canal developments, in the Suez Canal after 1865 and the Manchester Ships' Canal in the 1880s. This $35\frac{1}{2}$-mile, 28-foot-deep and 172-foot-wide canal necessitated the removal of $53\frac{1}{2}$ million cubic yards of soil; its excavation was the largest mechanized operation recorded up to that time. The contractor A. O. Schenk used 97 Ruston-Dunbar steam excavators, 173 locomotives, and 6,300 trucks on the canal. To operate them he had to build 228 miles of railroad track, on the canal bottom, on the banks, and along the sides. A Ruston-Dunbar "Steam-Navvy" was operated by two men and was capable of removing 600 to 1,000 cubic yards in a 10-hour working day.

The cut made by a "Steam-Navvy" was of course equal to the operating radius of the jib, and in order to excavate the full width of the canal bottom, a number of machines worked in echelon, each one confined to its own track and digging a trench in front of it. To remove the dirt, a second track parallel to the excavator track was needed.

This brief reference to the mechanized excavation of the Manchester Canal suggests another reason for relying on human labor for construction. A steam engine is not very suitable on a construction site. For lack of anything better, it can be applied at the top of a shaft to drive pumps and on such gigantic excavation jobs as the Suez and Manchester Canals. But for innumerable minor railroad and road cuttings, foundation work, and so on, a steam plant is more of a nuisance than a laborsaving facility.

This is the reason why the main line of development did not follow steam but rather compressed air as a power medium for driving construction machines. And the primary mover for driving the compressors delivering high-pressure air was seldom a steam engine but rather in most cases a water turbine.

Although compressed air had interested such early investigators of physical phenomena as Denis Papin– not to mention Ctesibius and Hero in ancient Alexandria—the evidence points to Lord Cochran as being the first to conceive the idea of using the medium for work in water. In 1830, he obtained a patent for the "plenum process," as it subsequently came to be called.

Compressed air found its first important application in caisson work. Before 1850 there was only one method of placing a bridge foundation in water, namely, the cofferdam described by Vitruvius. A cofferdam was built by

The Grand Trunk Bridge over the Niagara Falls was a combined railroad and road bridge designed and erected by John A. Roebling in 1855. It had a span of 840 feet, and was the first bridge built according to Roebling's patented method of spinning the wires making up the suspension cable on the site and then bunching them together with binding wire. Twelve years later, Roebling designed and began the construction of the Brooklyn Bridge, his world-famous masterpiece of bridge engineering.

The Langwies viaduct on the Coire-Arosa line in Switzerland was the first really successful railroad bridge built in concrete. With its 328-foot span it was also the boldest structure erected with the new material up until 1912. It still remains one of the most beautiful concrete bridges ever built.

The Verrazano-Narrows Bridge between Brooklyn and Staten Island is the longest suspension bridge hitherto erected. Its central span is 4,260 feet and the side spans are 1,215 feet. The height of the towers above mean water level is 690 feet. The bridge was designed by the Swiss-born engineer Ammann when he was nearly 80 years old. Previously, he had built some of the longest suspension bridges in the United States, among them the Golden Gate Bridge.

driving a double row of piling in water so as to obtain a boxlike enclosure, open at the top and bottom. To make the shell watertight the space between the pilings was filled with clay, whereupon the water inside the dam was bailed or pumped out until the bottom was reached. If the water was not too deep and the bottom not too rocky or soft, it was then possible to remove the silt from the bottom of the dam until bedrock was reached. The rock was smoothed off, and on the solid and level foundation, masonry was laid and bonded with metal clamps, and the courses were carried up to a predetermined elevation above the surface of the water. With the pier in place, the cofferdam was destroyed.

This ancient method was abandoned in England in 1851 in favor of a compressed-air caisson sunk to 61 feet for the piers of the Rochester Bridge over the Midway. I. K. Brunel used it for the Wye Bridge at Chepstow (1850–1856) and on the Royal Albert Bridge at Saltash (1853–1856). A much improved method, invented by Fleur Saint-Denis, and essentially the same as now used, was applied in 1859 for building the pier foundations of the Kehl Bridge crossing the Rhine. In America, caisson work was done on a grand scale in order to reach bedrock for placing the piers of the St. Louis Bridge more than a hundred feet below the surface of the Mississippi.

As used in these early sites, the caisson consisted of a huge iron cylinder that was towed out in the river and sunk to the bottom at a predetermined position. The ensuing operations were simple in principle but difficult to execute. The water having been pumped out, the caisson was sunk through the bottom deposits by digging out the muck inside the cylinder. As the diggers worked themselves down, the caisson kept sinking by its own weight until eventually bedrock was reached.

This feature of caisson work did not differ from that performed inside a cofferdam, except that as the soft ground was removed, water kept flowing in under the edge of the cylinder. To prevent this leakage the caisson was provided with an airtight floor placed at a height leaving room for a working chamber below it. This chamber was supplied with compressed air of sufficient pressure to balance the hydrostatic pressure acting on the open, or working face of the caisson. Enclosed in this pressure chamber men working with spades, and later also with sand pumps, removed silt, sand, or gravel down to bedrock. As the caisson sank with the removal of the material inside it, the hydrostatic pressure kept mounting and had to be balanced by air pressure until, as in the St. Louis caisson, the men worked under a pressure of 45 psi a hundred feet below the surface of the river.

To get men and material in and out of the pressurized working chamber, the basic feature of Lord Cochran's patent was applied. Above the roof of the working chamber was an air lock to prevent the air from getting lost. In this

"Labor market" is now an abstract term used by economists and bureaucrats, but a hundred years ago it was still a reality, like any market square where buyers and sellers met to bargain. In Paris, the building trades workmen assembled at 6 o'clock every morning with their tools and were met by Parisian contractors in need of labor. Since medieval days the market was held on Place de Hôtel de Ville (formerly Place de Grieve).

lock the air pressure could be varied between the highest working pressure and atmospheric pressure. When material was removed, the lock was put under working pressure and a hatch opened to the working chamber. Buckets of material were hoisted up into the lock, whereupon the lower hatch was closed. The air was let out of the lock, and when atmospheric pressure was reached, an upper hatch was opened and the buckets were hoisted to the surface. Men entered and left the pressure chamber in the same manner.

Modern regulations stipulate that a man working under pressure must be decompressed at the rate of one pound per minute. The St. Louis diggers, who were able to work only half an hour at a stretch a hundred feet down, should have spent 45 minutes in the lock entering and leaving the pressure chamber; instead, they spent two minutes getting decompressed. The consequences of ignoring the inherent dangers of caisson work were soon evident: all the diggers suffered from the "bends"—an agonizingly painful disease caused by nitrogen bubbles being liberated in the tissues—and 14 of them died.

To protect the St. Louis sandhogs the contractor issued bands consisting of alternate scales of silver and zinc to be worn around the waist, ankles, and arms and under the feet; and it was thought that the galvanic action between

239

the two metals would protect the men from the bends. Numerous caisson workers were to die a painful death during the next 20 years until, finally, the cause of the bends was discovered and precautions were taken to prevent it. But galvanic belts continued to be worn by quarrymen and tunnel workers who had no occasion to suffer from the bends but who had their bodies racked by rheumatism and their muscles jarred from the recoil of hand-held pneumatic rock drills.

The technically appealing but dangerous plenum process was also applied to subaqueous tunneling, the first time in an attempt to advance a railroad tunnel under the Hudson River in 1880. After 360 feet had been advanced under the river, the 35-psi air pressure used to hold back the water at the face blew a hole through the soft silt over the heading. The diggers fleeing from the rush of water through the hole took refuge in a faulty air lock, and the entire crew of 20 men drowned.

It was another of the many unnecessary tragedies that always seem to accompany the introduction of a new technology. In this particular case, there existed—but apparently unknown to the promoter of the Hudson tunnel—a superior method of subaqueous tunneling invented by the British engineer Henry Greathead five years earlier, in 1875. The Greathead method employed a steel shield provided with an air lock. The miners were protected by a steel canopy while working under an air pressure balancing the hydrostatic pressure, and as they excavated the face by spade or sand pump, the shield was advanced 2 feet at a time by hydraulic jacks. In the space thus gained, a cast-iron lining was erected at the rear of the shield and bolted to the lining already in place.

The Greathead shield was first used on the City and Southwark subway in London during the second half of the 1880s, and subsequently also in all the tunnels under the Hudson River bringing railroad and motor traffic to Manhattan. It was during the second abortive attempt to tunnel the Hudson in 1889 that the first hospital lock was introduced in a successful effort to save men from the bends. Before the hospital lock, 25 percent of the work force had died from the disease.

After the initial application of compressed air to caisson work, the efforts of inventors and major contractors became directed toward developing air-powered rock drilling machines for work in the long mountain tunnels that had to be built in some way or other—no one at this time knew quite how—in order to join the expanding railway nets. For this type of work compressed air machines were thought to be particularly well suited, because having worked the machines, the spent air could be used to ventilate the workings. The successful development of rock drilling machines has been discussed in a previous chapter, and here it need only be noted that by 1870 the early bulky units had been slimmed down by half, and 30 years later

American rock drills possessed practically all the mechanical features found in modern machines.

Up to the end of the nineteenth century it was chiefly the long Alpine tunnels and similar mountain developments in the United States that inspired the advance of mechanization. Indirectly they also contributed to the early development of electric power generation because the hydraulic power required to drive the compressors supplying air to rock drills, shop tools, and locomotives necessitated the installation of water turbines, of the Pelton and Francis types, and the extensive hydraulic structures that went with them. The major Alpine tunnels, such as St. Gotthard and Simplon, had a large power station at each portal, the construction of which required two years or longer. The turbines were coupled to mechanical compressors or, as in Simplon, to hydraulic pumps, but on the two Simplon sites one wheel drove an electric generator which delivered energy for lighting the underground workings.

When therefore electric alternators and the three-phase system of power transmission became available at about the turn of the century, permitting the development of distant water power sources, the long concern with hydraulic structures made the transition simple; it was merely a matter of replacing the air compressors with electric alternators. Thus, in October, 1906, a few months after the opening of the Simplon Tunnel, when mining work began on the Lötschberg on the Berner Alpen Line, the portal sites were fully electrified.

By "fully electrified" is meant in this instance that the air compressors, pumps, ventilation fans, repair shops, etc., were driven by electric motors deriving their energy from a local transformer fed by a 15-kilovolt power line from two existing power stations on the slopes of the Berner Alps. Instead of having to develop his own hydraulic power stations outside the portals, the Lötschberg contractor purchased the power needed to run his mechanical plant, thus establishing a pattern that has been followed ever since.

But it should be noted that the heaviest drudgery in construction as well as in mining—the backbreaking labor of loading the excavated material—was still performed by hand. Indeed, it was not until the 1930s that a mechanical loader suited for underground work became available.

The primary energy accessible for construction machines during the first decades of the twentieth century shared in one sense the awkward feature of the steam engine of being tied to the source by a navel cord that restricted the operating radius. The compressed air tools were connected to the compressors by a pipeline; the electric motors to a power station by copper conductors. Although much more flexible than steam, both pneumatic and electric power have limits that are exceeded in numerous types of modern construction, such as highway work, grading and compacting of airports, large ditching schemes, and so on—work that moves rapidly over large distances and vast areas. Overcoming the final impediments to full mechanization required the development of the internal-combustion engine into a rugged, reliable, and economical prime mover that could roam freely over a site–that was *portable* in the true sense of the word.

At the time of Simplon there were motorized carriages puttering along narrow macadamized roads, driven by capricious gasoline engines. There was also a new invention by Rudolf Diesel, a clumsy, unreliable, and as yet completely useless internal-combustion engine that relied on internal pressure to vaporize and explode heavy oil. It took several decades before these two types of engines could be put to useful work on a construction site. Via long circuitous routes they were gradually improved, the gasoline engine in motorcars and trucks, and eventually also in farm tractors, until at first almost imperceptibly the largest truck engines were employed in excavators and later also in such newly developed machines as graders, dump trucks, and loaders.

The diesel engine began its useful career at sea and withstood for a quarter of a century all attempts to slim down its heavy bulk to make it suitable for road vehicles. To turn the diesel engine into a viable prime mover, at first for military vehicles, required the impetus of a world war. When hostilities ended in 1945, the engines had become economically viable for peacetime tasks, for clearing up the mess left by the dieselized armored columns and, from mid-century, for the vast amount of new construction generated by the accelerated postwar economy. The diesel engine permitted the construction industry to become thoroughly mechanized and its methods rationalized. With the mastery of steel-reinforced concrete and unlimited power at its disposal, civil engineering could turn to tasks of a scale and complexity that would have impressed even the god-kings of the ancient world.

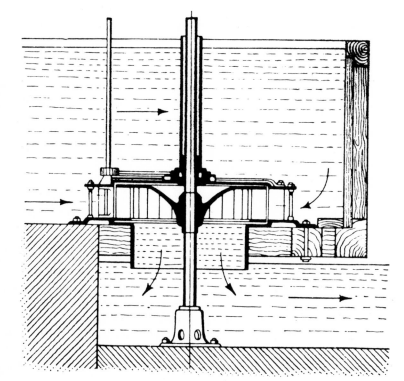

The water turbine was developed in France during the first half of the nineteenth century, and one type was immensely improved by James B. Francis who turned the inflow reaction turbine into a very efficient prime mover. His turbine, designed in 1851, has retained its main features up to the present. The runner with its fixed vanes operates immersed in water which is deflected against them by adjustable guide vanes. The sketch shows an original Francis turbine which today would be enclosed in a volute chamber and would discharge into a draft tube to improve the head.

Useful Pyramids

"There is almost nothing, however fantastic, that a team of engineers and administrators cannot do today. Impossible things can be done, are being done in this industry."

DAVID E. LILIENTHAL

Britain's century, the most amazing in the annals of men, had come to an end. Bobbing far beyond distant horizons, frigates of the British navy, later its gunboats and dreadnaughts, had maintained the Queen's peace and in the silent and unobstrusive manner of sea power protected the liberal ideas that had released the unbound energies and creative gifts of the island people. It had been a century that had been kind to the gifted and to the ruthless. Peasants' sons had grown rich, artisan apprentices had become scientists and engineers. It had been a century of individual efforts; when a man was permitted, to an extent limited only by his own ability, to accomplish anything he desired and in his own sphere attain wealth and status formerly possible only for those in the service of kings.

For the great anonymous mass it had been, as always, sheer hell alleviated only by a plentiful supply of cheap liquor. Uprooted from their natural environment and immediate dependence on the vagaries of nature, at first thousands and subsequently millions of people had lost their dignity in degrading millwork and a nasty existence in jerry-built slums. But the former peasants turned millworkers were not the only slum dwellers. Their new masters also lived in slums of their own making, because freed from the esthetic discipline enforced by the feudal courts, they sank into a trough of romantic lies and false clichés. Everything they touched turned into a slum world of ugly pretensions, the likes of which the world had never seen since the worst days of the Roman Empire.

But on balance, the inheritance left was on the asset side of the ledger, despite the speculations of a couple of German émigrés seeking refuge in the library of the British Museum from the ugly realities around them. Particularly in science and engineering, of which the two knew virtually nothing, the inheritance left by the nineteenth century pioneers held promises altogether different from their dismal prognostications.

A previous chapter has discussed how by the turn of the century all the factors bearing on the development of twentieth century engineering were already at hand. Steel was being made in ever-increasing tonnages, cement was being turned out in continuous rotary furnaces, electric power was available, and the internal combustion engine had been improved. The physical laws governing the strength of most structures were also understood and being taught to a growing number of students in technical universities the world over.

The genesis of dam building

To what ends were the new materials, engines, and advanced knowledge being applied? To the same ends as before: railroads were being extended, tunnels advanced, mills and bridges built. But while this bulk of useful but undistinguished work was being done, the efforts of civil engineers were gradually being directed to a new type of structure that eventually came to characterize the new century, just as the railroads had put their stamp on the previous one, and the Gothic cathedrals on the terminating centuries of the Middle Ages. The structure was the *dam*.

Dam building was assuredly no new invention. There is really nothing new in civil engineering; in this it differs from other engineering disciplines. However, the Romans do not seem to have engaged in dam building; and after the gigantic efforts of the Assyrian kings, little or nothing was done to impound water behind man-made structures until the Spaniards erected the 135-foot-tall Alicante Dam in 1579 to 1594 for the purpose of storing irrigation water. A number of similar Spanish masonry dams followed during the next 200 years, although the Alicante Dam remained the highest one in the world up to about a hundred years ago, after the 168-foot Puentes Dam, built in 1790, collapsed, causing a flood in which 600 persons drowned.

Industrialized Britain was the first country in the West to experience the need for impounding water to supply her rapidly growing urban population. Most of the early nineteenth century dams consisted of barrages made of compacted earth with a central core of puddled clay. These attempts at dam building were accompanied by a series of disasters caused by leakage through the earth embank-

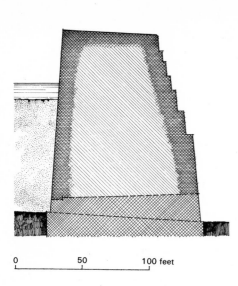

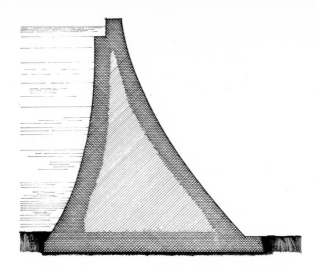

These two early dams in Europe can be regarded as the prototypes of subsequent dam construction. To the left is the 135-foot Alicante Dam in Spain, erected in 1579 to 1594 for storing irrigation water. It is a gravity dam built of stone lined with ashlar. To the right is the Furens Dam (Barrage du Gouffre d'Enfer) built in 1861 to 1866 for impounding drinking water for St. Étienne. It is constructed with ashlar and has an arched crest of 164 feet and a height of 325 feet, and can be regarded as an early example of a scientifically designed dam.

ments undermining the structures; such was the case with the Holmfirth Dam in Yorkshire, which collapsed in 1852. Twelve years later, the Dale Dike gave way, and 244 persons were swept to their death by the raging waters.

In 1852, one of the early American dams was completed, the one across South Fork, 12 miles above Johnstown in Pennsylvania, to impound water for the Pennsylvania Canal. The dam remained in service until May 31, 1889 when, after a period of heavy rains, it suddenly collapsed. A wall of water originally 70 feet high swept down the broadening valley, and when it reached Johnstown it was only 20 feet. But that was enough to destroy the town and seven more, with a loss of 2,200 lives. In the lengthening series of dam catastrophes the Johnstown flood still holds the record of lives lost, but before its centenary it is likely to have been exceeded many times over.

Dam building, in sum, followed the same pattern as the early attempts at building churches in stone and extending the height of Gothic vaulting. In France, however, engineers took a dim view of the British and American predilection for earth dams, and insisted that extending the height of such a dam beyond 60 feet was inviting disaster. Since in France, owing to topographical conditions, dams for river control and for impounding water supply usually had to be built in narrow gorges and to a height of 150 to 200 feet, French interest became centered on the theory governing the construction of tall masonry dams. In 1853, de Sazilly demonstrated that besides considering tilting and slipping it was also necessary to take into account the internal stresses of the dam structure. Delocre suggested in 1866 that a dam blocking a narrow gorge should preferably have a convex plan to utilize the vault effect.

Both these mid-century French engineers recommended a cross section of dams having the maximum compression stress on the upstream face with the reservoir empty and on the downstream face when full. The first large dam to be designed according to this principle was the Furens Dam, built by Delocre and de Graeff in 1861 to 1866 to impound drinking water for Saint-Étienne. It had an arched crest of 164 feet and a height of 325 feet. The dam was carefully built of ashlar, and its profile became known as the "barrage français." However, in view of subsequent developments, it should be noted that by 1843 the archetypal Zola Dam had been constructed for impounding water for Aix-en-Provence. The Zola was for many years the only dam relying on the arch for its stability. It was 120 feet high, and the radius of the upstream face was 158 feet.

The principles developed by these French engineers in dam design are still of basic importance, although the calculation of the internal stresses has been further elaborated by such men as W. J. M. Rankin and H. Ritter. The latter developed a method of analyzing the stresses acting on arch dams.

Power dams

Thus, from the wisdom gained from numerous dam failures the world over, and the theoretical investigations of engineers and mathematicians, all the problems—or nearly all—involved in dam construction were mastered by the turn of the century. But the dams investigated served only the purpose of impounding drinking water supply, river control, and irrigation. The greatest dam builder in the world, the U.S. Bureau of Reclamation, founded in 1902, reflects in its name this limited use of dams. The waterpower stations built at the time and for some years afterward were modest-capacity run-of-the-stream affairs, financed and operated by local companies content to utilize the varying flow of the rivers. Most electric power was gen-

243

The Vajont dam, situated in a narrow canyon to the east of the Piave river in the Dolomites, formed a part of the most sophisticated hydraulic development carried out in postwar Europe. As designed by the SADE power company in Venice, it constituted a power ladder from the Pieve di Cadore dam at an elevation of 683.0 meters to the Soverzene power station at 329.0 meters and 30 kilometers downstream. Known as the Piave-Boite-Mae-Vajont scheme, it involved the building of five arch dams, nine power stations, and two siphon lines bringing the waters of tributaries across the Piave to a 25-kilometer-long pressure tunnel through the left escarpment. The total length of the circular 4.2 to 5.0-meter galleries is 45 kilometers. In the picture is shown the 4-meter aqueduct crossing the Vajont canyon on a 52-meter bridge.

The Vajont dam—highest in the world when completed in 1960—has a height of 261.6 meters and a crest length of 169 meters. It is 22.7 meters thick at the base and 3.4 meters at the crest; 353,000 cubic meters of concrete were used in its construction. The reservoir behind the dam was 6 kilometers long and held 150 million cubic meters of water.

On October 9, 1963, a landslide with an area of 2.0 × 1.6 kilometers and containing 240 million cubic meters of rock spilled into the reservoir, creating a 100-meter-high wave when it splashed over the crest, subjecting the dam to a force of 4 million tons. But the dam stood—to the everlasting credit of its designer Carlo Semenza and the builder Mario Pancini.

A POSTMORTEM OF THE VAJONT CATASTROPHE

Following a slide into the reservoir in 1960 a number of precautions were taken, among them sinking test holes 90 m into the ground and laying out a grid of geodetic stations.

1961–1963. No evidence of creep was observed in the test holes, but geodetic stations showed movements of up to 30 cm per week. In the summer of 1963, the creep was 1 cm per week.

1963, *September 18.* Some stations moved 1 cm per day.

September 28. Heavy rains began to fall and runoff raised the reservoir level to 690 m or 30 m below the crest of the dam.

October 1. Cattle grazing on the left slope of the dam abandoned the area and refused to be driven back.

October 2—7. Geodetic stations moved at a rate of 20 cm a day.

October 8. Stations kept moving uniformly at a rate of 40 cm a day. Geological experts expected a collapse by the middle of November.

October 9. Italy's top experts were assembled at the dam. The power company's senior people were in the control building on the left abutment, government scientists in a hotel across the canyon. A five-member committee followed up and reported on the creep hour by hour. The chief engineer reported station movements of 80 cm that day.

22 h 41 m 40 s the earth moved, as revealed by seismographs throughout Europe. Within 20 seconds the reservoir for a length of 2 km was filled with rock to 175 m above water level.

The compression wave caused by the slide sucked water and rock 260 m into the air and smashed against the right dam abutment, obliterating the hotel. A part of the wave bounced back and swept away the control building, thus eliminating all competent eyewitnesses. As it crossed the dam the wave had a height of 100 m. The two waves joined forces downstream, and a wall of water 70 m high emerged roaring out of the canyon at 10:43 PM, rolled across the Piave to the town of Longarone 2 km, to the west of the dam and destroyed it with a loss of 1,970 lives.

The compression wave and the decompression following the water wave caused most of the damage to the underground structures. In the powerhouse, heavy I beams were twisted into corkscrew shape until they snapped. Heavy steel doors were lifted off their hinges, hurled 12 m, and turned into scrap.

10:55 PM. The waters that had reached 2 km upstream and downstream Longarone had receded and all was quiet in the valley. Only one witness, L. Broili, who lived 200 m up the slope from the right dam abutment, survived the catastrophe by escaping from his house before it collapsed.

The direct cause of the slide was pitching slide planes filled with clay several hundred meters below the test holes. The heavy rains raised the ground water level to the critical seams, and the waterlogged rocks lost their shear strength and began to move. The Vajont collapse pointed up a previously ignored hazard presented by the environment being changed by the impounded water, leading in the end to adjacent rock masses losing their coherence.

Vajont stands intact but abandoned, a monument to superb engineering, but an inadequate insight into the forces at work in the earth's crust.

erated by thermal plants which rapidly grew in size in metropolitan centers, particuarly those with ready access to coal.

Gradually, almost imperceptibly at first, the water impounded for river control, water supply, irrigation, and navigation came to be used also for power generation. In countries lacking local sources of coal, such as Sweden and Switzerland, rapid industrialization stimulated the utilization of their rivers to produce the power needed for industry and railroad transportation. As the hydropower plants grew in capacity, the erratic natural flow could no longer be tolerated; nor could the electrified manufacturing plants supplied by them jeopardize their output because of seasonal power shortages due to lack of water. Thus the need for costly dam building was enforced on the one hand by the increasing size of the plants, and on the other by the power consumers' insistence on "prime" power, i.e., a guaranteed block of power throughout the year.

As one station after another became established along a river valley, frequently by different companies, other problems developed. Obviously, it was the most economical sites that were developed first, which often meant the lower and middle reaches of a river. As plants came to be built farther upstream, a modern variation of the ancient rivalry flared up: an upstream plant could rob a downstream one of water; the upstream plant with its early access to water would be in a better position to generate power at times of the day when it was most needed, whereas a downstream one with inadequate storage facilities would receive the tailwaters of its rival at such times, for example, in the middle of the night, when there were few buyers of the energy generated by its machines.

This instance of friction is only one of the many arising from piecemeal development of a river valley. In the multipurpose schemes where the damming of a river serves other ends besides power production, then irrigation, navigation, and even malaria control are at cross-purposes with power generation. Letting water through the gates of the dam instead of through the turbine of the powerhouse infers a waste of potential revenue needed to pay the interest charges on the immense capital invested in the dam. The value of water stored behind a tall dam becomes evident from the fact that one cubic foot of water released in a second from the 730-foot-tall Hoover Dam represents nearly 80 hp in power. In one hour, this modest flow of water would produce 50 kilowatt-hours of energy.

Thus, numerous factors combined during the 1920s to bring governments into power production, since in most countries there was not enough private capital available to finance the large developments needed. In the United States, the huge dam projects surveyed during the decade —such as the Boulder Dam and the Tennessee dams—were so comprehensive in scope and served so many different purposes that only the federal government was able to cope with them.

Elsewhere, central governments had become involved much earlier in river developments, including power generation; but in the United States, the year 1930 was the watershed between privately financed piecemeal exploitation of convenient sites and federal development of huge water resources beyond the means of the power companies. Although here as elsewhere politics played its part, ideology had no bearing on the decision to build the Boulder Dam, for example, or the Grand Coulee which the power industry regarded at the time as completely worthless. For that matter, what private enterprise could have undertaken the comprehensive recovery of the exhausted physical and human resources of the Tennessee Valley?

With the federal government involved in the development of the virgin western waters, the pyramid building of the ancient pharaohs dwindled in scale and significance. The Hoover Dam was built with 3.4 mlllion cubic yards of concrete; the Grand Coulee required the excavation of 22 million cubic yards of earth and rock; into the Fort Peck went 22 million cubic yards of earth, or the mass of six Cheops pyramids. The construction of the dams also made history: the Hoover Dam was built in less than five years; the Grand Coulee in nine years. Considering that both dams were built in remote and exceptionally difficult sites in inhospitable desert regions, the performance borders on the miraculous. The miracle was accomplished by mechanization on a scale never used before, which became a model for postwar developments the world over, although outside the United States it has until recently seldom been possible to equal the gigantic scale.

The great western dams

The Colorado River Canyon was surveyed during the early '20s under indescribable difficulties, resulting in the loss of several lives. On May 29, 1928, Congress authorized the Colorado River Board to find a suitable site for a high dam. Ultimately, the Black Canyon, 25 miles southeast of Las Vegas, was selected for a dam that was to serve the purpose of flood control, irrigation, silt storage, and the production of electric power. As designed by the Denver Office of the Bureau of Reclamation, the Boulder Dam, as it came to be called, was given a height of 727 feet and a crest length of 1,182 feet. With the dam in place, the lake formed upstream extended for 115 miles, with a maximum width of 8 miles, and contained 30.5 million acre-feet of water.

The arch gravity dam—with a radius of 500 feet—was built as a monolith containing 3.2 million cubic yards of concrete, or more than that in all the 50 dams erected by the Bureau since 1902. Curing this tremendous mass of concrete by conventional means would have taken 150

As we look back over the "dam building epoch" of the twentieth century, Boulder Dam—later renamed Hoover Dam—stands out as the most important stage in the technological development as regards design and mechanized construction. The design of the dam, by John Lucien Savage, was the boldest ever, and the concrete monolith that plugged the Black Canyon the largest ever. The construction was fully mechanized and included numerous innovations that were later universally adopted the world over. But not all the novel ideas that emerged from the creative imagination of Henry J. Kaiser, head of the contracting consortium building the dam, were emulated, as when he put circus artists swinging in bosun's chairs to scale and drill the precipitous walls of the canyon. (When similar work was repeated 30 years later in the Vajont Dam in the Dolomites, the Italian contractors had no difficulty finding local mountain people to do it.) The four large diversion tunnels were driven in 320 days, and the construction was finished two years ahead of schedule, two records that have not been beaten.

Left is a simplified view of the dam and the U-shaped power station at its base. Right is shown the hydraulic structure with the supply tunnels and penstocks serving 17 turbine generator sets with a capacity of 1.3 million kilowatts. The plan (right) suggests how the monolith was cast in 120 blocks positioned in a zigzag pattern, while the profile below shows the close resemblance to the Furens Dam (page 243). The entire monolith was packed with 2-inch refrigeration pipes as shown in the lower right-hand corner of the section. The numerous inspection galleries facilitating access to instruments recording the behavior of the structure are also shown. Such galleries have since then become a feature of all concrete dams.

years, or about three-quarters of the time estimated to fill the upstream lake with silt up to the crest of the dam. The concrete mass had therefore to be refrigerated during construction by a method invented by the Bureau but never previously used in practice. The two power plants butting on the dam and flanking the downstream walls of the canyon were given a capacity of one million horsepower when working under a head of 520 feet.

The contract for building the dam and ancillary plant in 2,565 days was awarded the "Six Companies·Incorporated," a group of prominent west coast contracting firms, of which Henry J. Kaiser, Inc., became known to a wider public. The cost came to $48,890,995 and included the following major items:

The advance of four 56-foot circular diversion tunnels involving the excavation of 1,563,000 cubic yards of rock: $13,284,000, or $8.50 per cubic yard.

The excavation of 7 million cubic yards of alluvial to bedrock.

Lining the four diversion tunnels with 312,000 cubic yards of concrete: $3,430,000, or $11 per cubic yard.

The placing of 3.2 million cubic yards of concrete in the dam: $9,180,000, or $2.70 per cubic yard.

The contract also included the construction of a 98-foot cofferdam upstream and a smaller one downstream, two spillways, two outlet works, and a U-shaped powerhouse below the dam. The contract differed from the conventional one in that it only concerned the execution of the work; materials were to be supplied by the federal government.

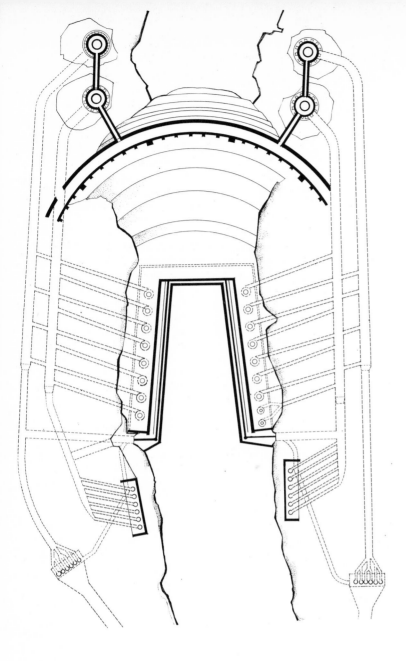

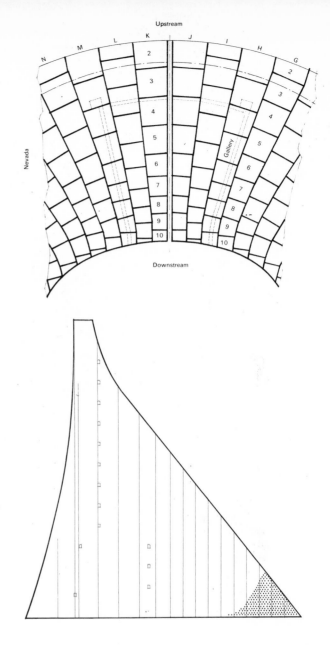

The formal notification to start the work was given on April 30, 1931, but the field superintendent Francis T. Crowe had already pitched his tent on the canyon rim on March 4 and started the work on April 20. By early June he had more than 700 men engaged on camp building and preliminary work. A 200-mile 80-kilovolt power transmission line was built, a camp of 478 houses was created to become Boulder City, a 22-mile railroad and a 10.5-mile highway with five tunnels were constructed—all this accomplished in 200 days. The accommodations offered the workers at Boulder City were the finest in American construction history. Each man was given a private 7 × 10.5-foot room for himself. The eight dormitory buildings were provided with showers and toilets. Family houses, sited on lots with a 50-foot frontage, had two or three rooms. The

fully equipped hospital contained 30 rooms. By the end of the summer, 2,000 men were accommodated in this fashion, and a year later Boulder City had 600 buildings and a population of 4,000.

The Boulder Dam, subsequently renamed the Hoover Dam, was from the outset a record-breaking operation, whether compared with previous or subsequent construction achievements. For example, the four 56-foot diversion tunnels, with a total length of 15,907 feet, involved the excavation of 1,451,369 cubic yards of breccia. The tunneling started in May, 1931, and was completed by April 1, 1932, or in 320 days. Hence the next to the largest tunnels ever mined until then were advanced at the rate of 50 feet per day. The performance was achieved by means of one invention and an innovation which later became standard in postwar tunneling: the multistage drill jumbo—devised by the tunneling foreman Bernard William—mounting up to 30 rock drills operated by as many miners, and the use of a Marion electric excavator with a 3.5-cubic-yard bucket in each heading.

The mining of each heading involved the advance of a 12×12-foot pilot tunnel one round ahead of the 56×40-foot main heading that was drilled up and charged in $4\frac{1}{2}$ hours by a crew of 80 men and shot in one round for an average advance of 16 feet. Owing to the oversize ventilation plant, the heading was cleared of toxic fumes within 5 minutes after shooting the round. Mucking out 1,000 to 1,600 cubic yards of rock required 9 hours. The 1,200 men engaged in the operation were paid $5.60 per hour, while the excavator operators earned twice as much.

The river was diverted through the four tunnels on November 13, 1932, at which time work on the upstream cofferdam got started. It was 480 feet long, 98 feet high, and 750 feet wide at the base. It was sealed with steel piling and puddled clay, and its construction involved the placing of half a million cubic yards of gravel, 151,000 cubic yards of rock, and 3,500 cubic yards of concrete. This work was finished in two months.

By the end of 1932, such main facilities as the aggregate screening and washing plant and the concrete mixing plant were ready. The former had a capacity of 1,000 tons per hour and the mixing plant 7 cubic yards per minute, or 170,000 cubic yards per 26-day month. Since sand and gravel were taken from an alluvial bed 6 miles upstream the dam, the railroad transport was on mainline scale; 60 million ton-miles were produced in the hauling of 8.6 million tons of gravel, 1 million tons of cement, and 3 million tons of muck.

The method of pouring the 3.2 million cubic yards constituting the dam monolith was just as revolutionary as the previous operations. The concrete was poured in air temperatures sometimes exceeding $125°F$, but even when poured at lower temperatures, the curing gave off heat, raising the temperature of the concrete to $40°F$ above the ambient. Calculations showed that the heat that had to be dissipated amounted to 700 British thermal units per degree and cubic yard of concrete poured. In order to do the job at all, the concrete had to be refrigerated. This was done by embedding 2-inch pipes arranged in 1,600-foot loops in the concrete as it was poured. All told, 800,000 feet of pipe loops were embedded in the dam structure and connected to three compressor plants with a total capacity of 15,324 cubic feet per minute. By pumping refrigerated water through the loops, the heat generated by the curing concrete was removed and the temperature lowered to $40°F$.

Concrete pouring began on June 6, 1933, by which time the canyon had been spanned by five 3-inch cableways, each capable of carrying 20 tons. The ready-mixed con-

crete was brought in buckets and directed to any one of the 120 vertical blocks staggered in relation to one another and poured within shutterings to a height of 5 feet at a time. A block was left to cool naturally for six days and then refrigerated to $40°F$. When a block had been approved by government inspectors, a cement grout was splashed over the top of it to a height of 2 inches, and another 5 feet was added.

The spaces between the blocks were concreted and the refrigeration pipes filled with grout last of all to complete the monolith. But the Hoover Dam is not a solid structure since it contains five inspection galleries at different levels, as well as radial galleries and elevator shafts. Thermometers and strain gages in large numbers are embedded in the blocks and the leads brought out in the galleries; from here they are connected to a central control board on which the behavior of the structure is recorded.

The powerhouse at the bottom of the dam has two 650-foot wings, 159 feet wide and 229 feet high, connected by a 400-foot-long central section. Seventeen 8-foot penstocks bring water to turbines working under an average head of 530 feet. The maximum output is 1.8 million horsepower.

The Hoover Dam was completed on March 1, 1936, more than two years ahead of schedule. The total cost came to $155 million, of which $108.8 million was for the civil engineering, $11.3 million for interest charges, and $30 million for the electrical installation. Expressed per kilowatt of installed capacity, the cost came to $115.

Taming the Columbia River

While the Hoover Dam was being built, the Grand Coulee Dam across the Columbia River got started on September 8, 1933. Although it is somewhat lower (550 feet) than the Hoover Dam and impounds less water (9.6 million acre-feet), its great length of 4,173 feet makes it the largest concrete structure in the world, the volume being 10.6 million cubic yards. Its 15 acres of spillways pass a flood of 1 million cubic feet per second, whereby this man-made waterfall becomes three times the size of Niagara in respect to height and five times in the matter of volume. The water backed up behind the dam forms a lake stretching 151 miles toward the Canadian border. The two powerhouses associated with the dam have a generating capacity of 1.9 million kilowatts. A group of seven contracting firms known as Consolidated Builders spent somewhat more than eight years excavating 22 million cubic yards of alluvial material and rock, freezing one shore side, grouting the fissures in the foundation rock, and pouring 10.6 million cubic yards of concrete, until in 1942 this grandiose development "of no more usefulness than the pyramids of Egypt"—to quote the encumbent president of the American Society of Civil Engineers—was completed.

Since professional opinion at the time of building the Grand Coulee was unanimous in regarding this vast and costly development as worthless economically as an Egyptian pyramid, it is tempting to inquire to what extent engineering opinion was right. It does not require much research to discover that this "useless pyramid" paid for itself in 10 years merely in federal, state, and local tax assessments on the aluminum smelters sited near the dam!

But Grand Coulee was not intended primarily as a power producer. Indeed, the original idea, dating from 1918, envisaged the dam only as a storage for water to irrigate the desiccated lands in the Columbia River basin. However, the irrigation project was not realized until 1952 when the first unit of the six 65,000-hp pumps began raising water into a network of 4,500 miles of canals and laterals, crisscrossing 2 million acres of desert land.

The first irrigation water from Grand Coulee began to flow through the huge network on May 22, 1952. To publicize the event, the Bureau of Reclamation created a complete 90-acre farm in one day and deeded it to a worthy war veteran nominated by the local voters. In one eight-hour working day, 80 acres of newly irrigated desert land were turned—by means of 150 tractor-driven plows, harrows, and drills—into 15 acres of alfalfa, 12 acres of red clover, 6 acres of legumes, 31 acres of beans, and 15 acres of corn, while at the same time a team of building trades workers erected a three-bedroom house, barns, and machine-shed. From an engineering point of view, the operation was an amusing divertissement in the shadow of the contracting effort of building the dam and carrying out the canal works, but it brought home to the common man in the American Northwest the magnificent potentials of a mechanized civilization as nothing else could have done.

As mentioned in passing, by the time this veteran's farm was put into being, Grand Coulee had already been paid for. The irrigation scheme, embracing about one million acres, cost $750 million and was financed up to 75 percent by the power sold. The scheme provides a living for 150,000 people raising crops worth $95 million a year, while business and service establishments produce $34 million more. This is the sort of economic accomplishment that can be achieved by grand-scale engineering. Unfortunately, to the untold millions in Asia and Africa they are only achievements a few can read about. Only Americans possess the idealism, will, and competence to accomplish such marvels.

Since the completion of the two great prewar dams in the American West, dam construction has been going on elsewhere on the same and other long rivers. The Columbia has been given eight additional dams and powerhouses, amounting in total to a volume of 36 million cubic yards and a generating capacity of 15 million kilowatts. The Colorado River has been dammed at 10 other sites where-

by the electric capacity has been raised by another 3 million kilowatts. But none of the new structures compared in size with Hoover and Grand Coulee, which still remain the highest and the largest, as well as being individually the greatest power producers of the 2,800 dams in the United States.

These two dams, and some 60 others, were designed by *one* man, John Lucian Savage, during his 21 years as chief engineer in the service of the U.S. Bureau of Reclamation. When he retired in 1944 at the age of sixty-five, his salary had not reached $10,000 a year, which is typical of government appreciation of outstanding scientific and technical contributions on the civil side. But then began his career as a consulting engineer; this lasted up until his death at 88 years of age in 1967; and during this period he designed a score of dams outside the United States, among them the Grand Dixence in Switzerland and the Koyna Dam in India.

TVA and moral-purpose engineering

The great western dams are admirable civil engineering accomplishments in respect to design, and beyond all in the contractor's ability to mobilize and coordinate vast resources of mechanized facilities and skilled manpower. But seen in historical perspective, they represent nothing new. By the time the structures were completed and turned over to their owner, in both cases the U.S. Reclamation Bureau, they were operated, indeed had to be operated, within the narrow limits laid down for the Authority. The flow of the rivers was regulated in accordance with the law, blocks of power were generated and sold to those who could use it, i.e., the power companies. At Grand Coulee, the local dam management also had to be concerned with downstream navigation as defined in its directives. Beyond that, the agency did not concern itself with the wider aspects of its operations, with, as it were, the moral purpose of the large-scale utilization of these magnificent natural resources.

However, it was "moral purpose," as applied to engineering, that was the guiding philosophy for another giant development going on in the 1930s, namely, the rehabilitation of the entire Tennessee Valley. Seen in retrospect, this comprehensive development is beyond doubt the greatest moral achievement by engineering in the history of mankind, and despite its manifest success, it seems likely to remain unique. The example held up to the whole world to behold has not yet been emulated, and it probably never will be.

TVA spent 20 years wrestling with the complex problems involved in the unitary rehabilitation of an area of 89,000 square miles (about equivalent to the size of England and Scotland combined), with a population of 6 million (about the same as Sweden's at that time)—an area that suffered

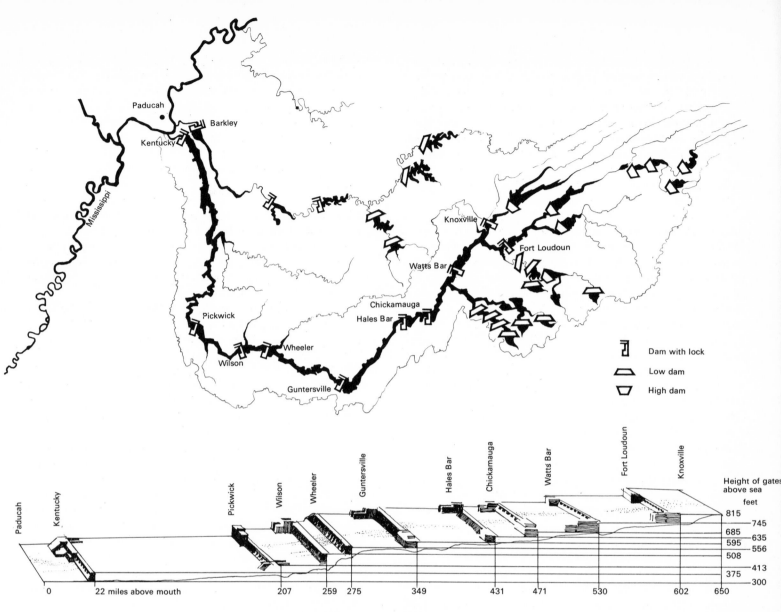

PROFILE OF THE TENNESSEE RIVER

The schematic map shows the area directly affected by the TVA scheme, including the 30 dams constructed in 1933 to 1951 to control the Tennessee River and its tributaries. The river was made navigable from Paducah on the Mississippi to Knoxville, a distance of 620 miles, by means of nine dams provided with locks. The rest of the dams serve the primary purpose of flood control but are also power producers. The total installed generator capacity of the system was originally 9.4 million kilowatts.

from all the evils consequent upon ruthless and thought-less exploitation of its natural resources: Its magnificent stands of hardwood timber had been chopped down and turned into charcoal for metal smelting; the soil had been exhausted by one-crop farming; the streams had been polluted by industrial tailings and human detritus. Heavy rainfalls scoured the slopes and carried away the soil; re-peated floods scourged the bottomlands and took a dis-tressing toll of property and lives. Even the human re-sources of the valley were wasting away; in some areas in the south, malaria ravaged 65 percent of a population liv-ing in squalor and ignorance. Educational facilities were retarded; technical skills were lacking; the electric energy consumed was only 350 kilowatt-hours per person per year.

The Tennessee Valley was by no means unique in its di-lapidation; there were and still are numerous raped areas the world over, where people are forced to live in similar or worse squalor. The Fertile Crescent managed to support a civilization for several thousand years before it became desiccated; in the Tennessee Valley it took less than a hundred years. There are Alpine valleys that have provided a comfortable living since Neolithic days, but have been destroyed in two weeks, or more accurately in the time required to blast a round in a power tunnel, thus severing the rock partition to an underground source of water that had kept the valley alive since before the dawn of European history.

The rehabilitation work in the Tennessee Valley is unique in many respects and not least in the success achieved by the Authority in welding the numerous experts employed into a close-knit team where each member was made aware of the moral purpose of the vast undertaking. The catalogue of TVA's accomplishments reads like an accountant's report of Genesis: 30 million cubic yards of rock and earth was excavated and 113 million cubic yards of concrete poured to erect 30 dams, create 175,000 acres of lakes, build 1,300 miles of highways and 200 miles of railroads; 9.4 million kilowatts of power capacity was installed, 234 million trees were planted; 200,000 men were taught useful skills; malaria was wiped out; 3 million acres of farmland were rejuvenated by new methods of cultivation; a 2-million-ton phosphate plant was built to supply plant nutrients essential for effective farming. The wild life was also restored; and the waters in rivers were nursed back to health and made capable of maintaining fish and higher forms of marine life.

As the work progressed, the income of the inhabitants of the valley rose by 477 percent from 1933 to 1951, or in total from $837 million to $5,500 million. The per capita income rose from $163 to $940 per year. By June 30, 1952, the gross investment in the valley amounted to $1,036 million, or $9 per year and per head of the population residing in the area. The federal revenue from income taxes increased by $300 million, or by six times the annual investments by TVA. The investment in power plants, which in total came to $702 million, in 1952 yielded 4.7 percent after taxes. The flood control measures which cost $2.5 million a year have saved $11 million of damage a year, or 5 percent on the capital invested for the purpose. The formerly ravaged forests produced $200 million worth of timber per year. Private employment in the valley rose from 447,800 in 1933 to 919,400 in 1950.

TVA, the greatest comprehensive scheme of rehabilitation and development ever undertaken, was enacted by Congress in 1933 and was signed by President Franklin D. Roosevelt on May 18 as part of his New Deal program. Like so many of the heroic measures enacted during the First Hundred Days, it was highly unconventional; but unlike many of the others, its unique administrative setup was never seriously challenged. Its principal feature was accentuated by the President when he signed the Act:

"The Tennessee Valley Authority is a corporation clothed with the power of government but possessed of the flexibility and initiative of private enterprise."

With the broad authority given to it, the TVA Board under the inspired chairmanship of David E. Lilienthal was able to pioneer an altogether new, and for the United States revolutionary, approach to dealing with the multitude of problems involved. TVA gathered its own staff of technical experts: aerial and cadastral surveyors, geologists, agronomists, chemists, public health experts, wild life biologists, ornithologists, limnologists, librarians, accountants—not to mention brigades of engineers and architects. The Authority not only planned the infrastructure and designed dams and buildings, powerhouses, parks and recreational facilities; it also carried out the work. To this end it started the largest occupational development scheme ever undertaken outside the Armed Forces. In 1933 alone, 38,000 local men were given aptitude tests preliminary to apprentice training in construction skills. Sharecroppers became skilled craftsmen; and tenant farmers, machinists. The maximum number of people employed by TVA at any one time was in 1942 when it had 40,000 on the payroll, but, as previously stated, altogether 200,000 men and women were taught new skills while in the employ of the Authority.

The manner in which TVA went about its multitudinous tasks reflects the philosophy of its first chairman, David E. Lilienthal, who saw clearly the lethal dilemma presented by modern science, namely, the impossibility, owing to the numerous specialists required, of developing in unity the resources of a region. Simply by the *intensity of their specialization,* disunity would be created instead of order. The "desperate part of the problem" confronting the Board was to make different groups of specialists care about things beyond their own specialties, "because nothing can be accomplished in unity until each technologist has learned to subordinate his expertness to a common purpose."

The Board gradually succeeded in tearing down the barriers between specialist interests, and the herculean work was moved ahead by a team of experts impelled to subordinate themselves and their professional attitudes to a common aim. In the ability of the first TVA Board to resolve "this desperate problem" on which the outcome of the grand design depended, lies its historial importance.

However, as mentioned in passing, the strange thing is that despite its manifest successes technically, socially, economicaly, and even politically, TVA has not been succeeded by any similar projects in the United States. What has followed in the way of giant engineering schemes in the west and elsewhere conforms to the standard pattern of departmentalized control by federal agencies far removed from the sites and supremely ignorant of the local impact of their decisions. A grotesque example of the consequences of decisions made by a centralized authority

251

applies to Grand Coulee. In 1952 the dam was flooded, and the twelve 150,000 kilowatt alternators in the two powerhouses came within seconds of being destroyed. Someone who had been ordered to close an upstream gate opened it instead; and as the downstream gate was already closed, the water rose inside the dam and flooded the galleries and the turbine pits. The man who was previously in charge of the gates had been fired in an economy drive ordered by Washington a few days earlier.

It is futile to speculate on America's failure to pursue the path pioneered by TVA. Politics no doubt played a significant part. The war followed by a long period of economic upswing has not inspired heroic efforts in the public sphere, except of course in space research and nuclear armaments. TVA was conceived and carried out in the trough of the worst depression that America has ever experienced; and there was a large pool of unemployed scientifically trained and highly skilled men from whom TVA could draw its original staff. The desperate social conditions shook the American people and inspired favorable attitudes toward anything that held out promise of alleviating their economic distress. Perhaps the real reason is much more simple: philosopher-managers of the caliber of David E. Lilienthal are in meager supply.

If, then, America has ignored the valuable experience of unitary development pioneered by TVA, it nevertheless has served as a source of inspiration for similar efforts the world over. The Snowy Mountain Scheme in Australia and the High Dam in Egypt, of which more below, the Cassa il Mezzogiorno set up for the rehabilitation of southern Italy and its islands, all owe their existence to TVA. All told, some 40 river valley schemes in Europe, Asia, Africa, and South America have been copied from the American original. Indeed, many of the projects have been planned by former TVA engineers.

Some of these major developments were started by the former colonial governments and aborted with the Balkanization consequent upon the liveration of the colonies. The magnificently conceived Jordan Valley Scheme, planned by members of the TVA staff, which would have irrigated 606,000 acres of desert and produced 600 million kilowatt-hours of energy is one of the more tragic miscarriages. The scheme envisaged, inter alia, bringing seawater from the Mediterranean through tunnels and canals, and allowing the water to disgorge under a head of 1,300 feet into the Jordan rift, where after generating power, it would evaporate under the hot sun.

In Africa, the Kariba Dam on the Zambesi River was actually completed in 1960, after six years of work during which 12,000 Africans were employed by a group of British and French contractors. The result of their labor is a 420-foot-high dam with a crest length of 2,025 feet, containing 1.3 million cubic yards of concrete. The underground powerhouse associated with the dam has six 100,000-kilo-

watt generators and a present capacity of 600,000 kilowatts. An additional powerhouse will raise the generating capacity to 900,000 kilowatts.

The Volta River Scheme inherited from the former colonial power has recently been completed by Italian contractors. The rockfill dam at Akosambo, 66 miles inland from Accra, is 370 feet high and 1,165 feet long, and has a volume of 750,000 cubic yards. The powerhouse has a capacity of 786,000 kilowatts, and the power generated will be wasted making more crude aluminum billets. The development has cost $70 million of which Ghana has contributed half.

That is about what has happened up until now with the many African schemes. Of the numerous Indian plans, the Bhakra Dam on the Sutlej River in the Himalayan foothills was finished in 1962. It is a concrete gravity dam with a height of 755 feet and hence is the second highest dam in the world. The volume of concrete in the dam and powerhouse is 5.6 million cubic yards, and the installed power capacity 1.8 million kilowatts. The dam impounds 7.1 million acre-feet of water, and the associated irrigation scheme includes 4 million acres of land.

The Bhakra Dam, engineered by the International Engineering Company, and owned and operated by the Punjab government, is on the scale of the Grand Coulee and could provide the same opportunities for social and economic rehabilitation as TVA, but there is no information about such "subsidiary" developments.

As a matter of fact, all government projects everywhere in the world lack the unitary features of TVA. Nowhere is the purely technocratic and departmentalized attitude more pronounced than in the U.S.S.R., where at the present a number of gigantic hydropower developments that dwarf anything completed in the West are in the process of development. The 29 Soviet powerhouses now being built have a total capacity of 21 million kilowatts; and 180 more plants with 160 million kilowatts are scheduled for completion by 1980.

Among the largest Russian plants is the low-head Lenin Dam on the Volga, upstream at Kuibeshev, which impounds 58,000 million cubic meters of water and has an installed capacity of 2.3 million kilowatts. The 22d Congress Dam, also a low-head scheme on the Volga, has a capacity of 2.6 million kilowatts. The Saratov development, a third Volga scheme, has a generating capacity of 1.4 million kilowatts, and the Bratsk Dam and powerhouse on the Padunski River 4.5 million kilowatts.

The Soviet high dams are built on a similar gigantic scale. The Inguru Dam in the Georgian Republic is a concrete structure 984 feet high with a crest length of 2,198 feet, containing 3.9 million cubic yards of concrete. The Nureck rockfill dam in the Tajik Republic is also 984 feet high and is 2,395 feet long, and has an installed capacity of 2.7 million kilowatts.

But even these magnificently engineered Soviet power-houses dwindle in comparison with the Krasboyarsk Dam on the Yenise River in Siberia. This concrete gravity dam is 394 feet, and its crest length is 3,428 feet. The volume of concrete is a modest 5 million cubic yards. But there is assuredly nothing modest about its capacity, which is reported at 5 million kilowatts—or nearly 30 percent more than all the stations operated by TVA, and 2½ times the capacity of Grand Coulee, the largest plant in the Western world.

Compared with such figures of Soviet accomplishments in hydraulic engineering, everything done or planned in the West dwindles in size but not in significance. These Russian developments are intended for power generation and nothing else, whereas such major schemes as the Snowy Mountain in Australia and the High Dam in Egypt are much more comprehensive in scope and adhere more closely to the TVA conception of valley rehabilitation. Indeed, the Australian scheme goes beyond everything hitherto accomplished in its large-scale geographic surgery to correct some horrendous blunders of nature.

The Snowy Mountain Scheme

Briefly, the Snowy Mountain Scheme aims at salvaging the runoff waters along the eastern slopes of the coastal Snowy Mountain range, which gathered by the Snowy River, are wasted in the southern ocean, whereas on the western or inland side of the range an annual rainfall of less than 10 inches is wholly inadequate for farming. By means of reservoirs and transmountain tunnels, the waters feeding the Snowy River are impounded and made to flow westward, and when emerging at the western portals, they will be used to generate power under a head of 2,500 feet before being canalized for irrigation purposes. When completed in 1970, the scheme will provide a storage of 1.9 million acre-feet, adequate for the irrigation of 750,000 acres in the inland Murrumbidgee and Murray River valleys, in addition to feeding a generating capacity amounting in total to 3 million kilowatts.

The Snowy River and the Murrumbidgee, the main tributary of the Murray River, have their springs in the neighborhood of Mount Kosciusko, which with an elevation

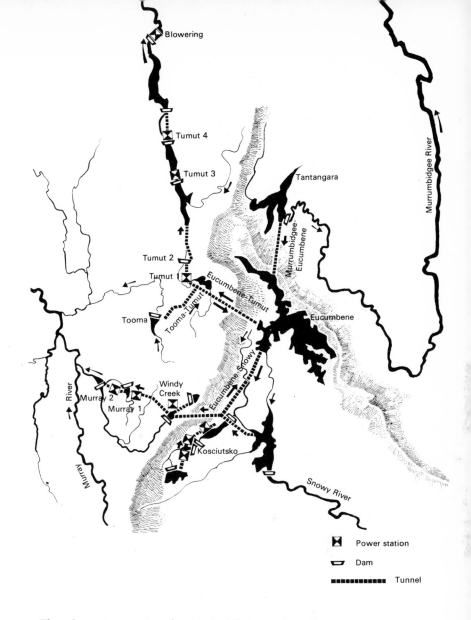

The schematic map gives the principal features of the Snowy Mountain development in Australia. The three major rivers, Snowy, Murrumbidgee, and Murray, have been joined by an extensive system of tunnels piercing the divide. Snowy's water, which previously was lost in the sea, is now stored in a number of large basins and used to irrigate the desert west of the divide. The installed capacity of the power stations is 3 million kilowatts. The profile below indicates the power ladder in the northern part of the development.

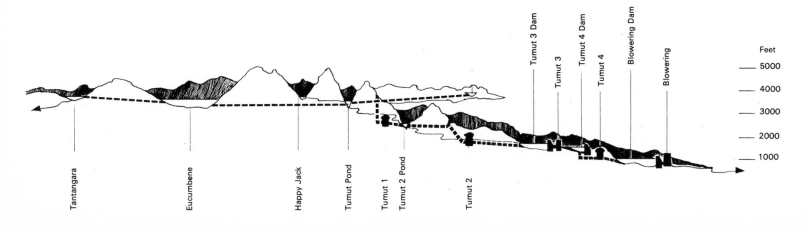

of 7,305 feet is the highest peak in the Australian Alps. The Murray flows 2,000 miles toward the west before turning southwest and discharging into the Great Australian Bight. The Snowy dashes from the mountains for 300 miles eastward to get lost in the Tasman Sea.

The Snowy Mountain Authority was established in 1949 and had to start from scratch in an untouched wilderness that was inaccessible except on horse or on foot. The 300 square miles affected by the development were unmapped, and little was known about the geology. Teams of surveyors, geologists, hydrologists, and engineers were sent out to collect the data needed for the siting and design of the huge hydraulic structures envisaged, in addition to the building of the roads, bridges, camps, power lines, and all the rest of the basic infrastructure needed before construction could get under way.

Eventually, the Snowy Mountain Scheme was turned into two major developments, of which the Snowy-Tumut is finished, after 15 years of work. It involves the diversion of the Eucumbene River, a tributary of the Snowy, through 14 miles of transmountain tunnels connecting with the Tumut River, which is a tributary of the Murrumbidgee. The latter is in turn joined by tunnels to the Eucumbene River. Another tunnel draws water from the Rooma River to the Tumut.

Joining the four tributaries in this fashion requires headwater storage dams on the Eucumbene, Upper Murrumbidgee, and Tomma, all three reservoirs connected by tunnels to the Tumut Pond formed by a dam across the Tumut River. From the latter dam the waters from the four rivers descend through a power ladder with a head of 2,700 feet divided among the four power stations with a total capacity of 1 million kilowatts. The tailwater is collected in the Blowering Reservoir which has a storage capacity of 1.1 million acre-feet feeding the irrigation network of the Murrumbidgee Valley.

The two headwater dams, the Eucumbene Lake on the east side and the Tumut Pond on the west side, are connected by a 14-mile-long transmountain tunnel which provides for a two-way flow. Normally, the water flows from east to west, but when the Tumut and Tooma are in flood, the direction of flow is reversed, and the surplus is stored in the Eucumbene Lake.

The second development includes the diversion of the Snowy River itself by means of a second system of transmountain tunnels to the Murray River. When disgorging through the west portal, Snowy's water falls 2,600 feet through two powerhouses with a total capacity of 1.5 million kilowatts, before discharging 800,000 acre-feet of water into the Murray.

The two developments, Snowy-Tumut and Snowy-Murray, are connected by a 15-mile-long two-way transmountain tunnel joining Lake Eucumbene with the Snowy River at the intake of the 19-mile-long Snowy-Gechi tunnel, en-

abling Snowy's spring flood to be stored in the lake.

When the scheme is completed in 1970, after 20 years of work and a cost of £A400 million, the formerly wasted waters will be salvaged to provide 1.9 million acre-feet for irrigating 750,000 acres of desert land, after supplying a power generating capacity of 3 million kilowatts. The direct return on the investment has been estimated at £A30 million per year, but the indirect economic and social benefits from this amazing development are of course incalculable.

Nasser's High Dam

The Snowy Mountain Scheme is the largest unitary development in the southern hemisphere and is a source of pride to Australians, to whom it represents a tangible expression of their takeoff into the higher reaches of technology and advanced industrial economy. The other major scheme that is expected to be finished at about the same time—the High Dam at Aswan—possesses an altogether different character. It represents a desperate attempt by the young revolutionary Egyptian state to lift itself by its bootstraps out of the country's ancient squalor and constant threat of starvation. For the Egyptian people, the High Dam has become a life-and-death venture that will gain new arable land from the desert fringing the valley and, beyond that, water in regular abundance to make the desert bloom. For the revolutionary "Free Officers" government, it is a political showpiece, something they hit upon when rummaging through government files after seizing power in 1952.

Since the High Dam project possesses nearly all the features of the dilemma suffered by the underdeveloped world —usually lost in the heated debate about aid to the numerous countries set free upon the disintegration of the colonial empires—the events leading up to the realization of this vast project surely warrant a digression from the purely technical developments.

Basically, the High Dam was conceived as a means of keeping the rapidly growing Egyptian population from starving to death. Under "natural" conditions, and using gravity basin irrigation, the Nile in normal flood is capable of feeding 7 million people when administered by a competent government. Then each person would be fed by the annual crop raised on three-quarters of an acre. Not only would he be well fed, but also there would be a surplus, either for vast public works or for allowing landowners in Cairo and Alexandria to splurge in conspicuous consumption.

When Napoleon invaded Egypt in 1799, there were not more than $2\frac{1}{2}$ million people living in the Nile valley, the end result of centuries of misrule, famine, and plagues. At the time when the Suez Canal was opened, the population had recovered to the ancient limit of 7 million people.

...and Coulee blocking
...Columbia River is still
...largest concrete dam
...he world, with a volume
...16.6 million cubic
...ds. It is three times as
...h as Niagara Falls,
...I five times as much
...er flows through its
...es and turbines as over
...falls. The original pur-
...e of the dam was to
...re irrigation water for
...reclamation of about
...illion acres, but with
...installed capacity of
...million kilowatts,
...nd Coulee is also one
...he major power pro-
...ers in the world, only
...ntly surpassed by a
...Soviet dams. The de-
...er was John Lucien
...age, greatest dam
...der in history.

Picture shows one of the Tumut stations in the northern part of the Snowy Mountain development in Australia. Four of the stations along a power ladder with a head of 2,700 feet have an installed capacity of 1 million kilowatts. Their tailwater, equivalent to 1.1 million acre-feet, is used to irrigate the desiccated Murrumbidgee valley.

Then came the British. Lord Cramer, the first British proconsul, declared it the duty of the civilized Englishman "to extend the hand of fellowship and to raise morally and materially the millions of Egyptians from the abject state in which he finds them." And so indeed he did. With the completion of the Aswan Dam in 1902, the water stored behind the dam permitted the introduction of perennial irrigation, whereby five crops could be taken off the land in two years.

When Lord Cramer retired in 1907, Egypt under his administration had become one of the wealthiest agricultural countries in the world. As a consequence, the people had increased to 12 million, living on 8 million crop-acres. The Aswan was heightened in 1912, and again in 1933, and one million more crop-acres were added. But then there were 5 million more people to feed, or only one-half an acre per person. After Egypt became independent in 1936, King Farouk's government added a further 800,000 acres, but again there were an additional 5 million people to feed, and when he was forced to abdicate in 1952, seven persons had to be fed from two crop-acres, compared with three at the turn of the century. Worse, the land now showed signs of exhaustion, in part due to inadequate drainage after many years of continuous perennial irrigation. When the revolutionary government took over, the population in Egypt was 25 million, or three times as much as ordered by nature.

The chief reason for the population increase was humanitarian. Because more children died in Egypt than anywhere else, doctors bent every effort to keeping more of them alive. They fought tuberculosis, schistosoma, trachoma, and other lethal diseases ravaging Egyptians, and when in 1947 a cholera epidemic broke out, the health services of the world came to the rescue and quickly brought the scourge under control.

There is no arguing about such humanitarian work, but the fact remains that the population surplus resulting therefrom has to be kept alive in some way. And the one way hitherto discovered is by the orderly development of industry. But the conventional view has always been that Egypt lacks the prerequisites for industry. Its only natural resource is the Nile, and in the past, good governments have devoted their efforts to getting the best out of it, as permitted by the technology available at the time.

Yet, the best has not been good enough, because of the behavior of the river. Although the Nile usually conducts itself in an orderly fashion, it does not necessarily follow that the annual supply of water carried by it is regular. It varies between excess floods, which after ravaging the valley waste themselves in the sea, to a minimum wholly inadequate for irrigation needs.

This irregularity of supply has always been a worry to the Egyptian water administrations, and scientific studies extending back to the last century have been directed at finding a way of storing water over one year. Thus the idea of the "century storage" arose, implying the control of the water supply at the source, i.e., using the equatorial lakes as water reservoirs. The last scheme, presented as late as 1955, planned to use Lake Victoria as the main storage, with Lake Albert as a regulator for the White Nile, and Lake Tana in Abyssinia for the Blue Nile. The large swamp in southern Sudan known as the Sudd, where half the water of the White Nile gets lost, was to be canalized.

This rational scheme, which could be carried out in 20 years or less, might well have been realized today, had the British managed to retain their former suzerainty along the entire course of the White Nile. But with the split-up of Africa, it has become a political impossibility, because none of the new national states along the Nile would entrust their vital water supply to an upstream independent national state. Hence, the only sensible way of solving the many problems along the Nile valley in the manner of manner of TVA is out.

Although Egyptian and British officialdoms had little faith in the possibility of industrializing Egypt, one private citizen thought differently. Adrian Daninos, a Greek living in Cairo, became an infernal nuisance to successive Egyptian governments by advocating the use of the Aswan Dam as a power producer. In 1912, he insisted on building a nitrogen plant at the power station; in the 1930s, inspired by TVA, he added reclamation, irrigation, and river navigation. In 1947, he presented his grand design to store 13,000 million cubic yards of water behind the dam; build a lock admitting 2,000-ton ships; and construct a 125,000-kilowatt hydropower plant, a 600,000-ton fertilizer plant, and a 100,000-ton steel plant. The entire project, he thought, would come to £E35 million ($101 million).

After publicizing his plan, he visited Aswan to reconnoiter the site in company with an Italian consulting engineer, Luigi Gallioli. Returning, they had a much more exciting tale to tell. They had found a natural basin capable of holding the entire flood of the Nile! On January 12, 1948, they presented their rapidly concocted plan in a paper read before L'Institut d'Egypt. It involved building a barrage farther upstream the Aswan Dam, which would hold 186,000 million cubic yards of water, or two annual floods. It would enable the reclamation of 5 million acres and the conversion of an additional million acres from basin perennial irrigation. The powerhouse would produce 16,000 million kilowatt-hours. Best of all, the whole thing could be had for $210 million.

This, then, was the original idea of the High Dam. Daninos formed an international syndicate and succeeded in interesting a number of distinguished engineers, including Lucien Savage who volunteered to act as his personal consultant. But Daninos had less luck with his own government, for it showed no interest in his visionary plan. Then followed a few years of war and political machinations and "Daninos' 1949 Project," as it was now called, was filed away to gather dust in a government department, unknown to the mass of the Egyptian people and unnoticed by the rest of the world.

When the Free Officers seized power on July 23, 1952, they had no coherent plan of what to do beyond settling old scores with the former administration and its corrupt politicians—these "traitors and saboteurs." They had a vague idea that by a program of land reform and industrialization the nationalistic movement would be able to erase poverty and raise the living standard of the Egyptian people. But they were ignorant of how to go about realizing these worthwhile aims.

Two months after seizing power, they found Daninos' plan and were fired by the simple solution it offered to all of Egypt's age-old problems. Once they had decided to go ahead with the Daninos project, the first step was of course to push Daninos himself out of the way and ignore his existence. The second one was to obtain an engineering study of the project. That proved just as easy as to get rid of the author of the plan. By shrieking long and loud, as required by left-wing protocol, about West Germany's payment of 3,000 million marks to Israel in compensation for the Nazi atrocities, the Free Officers pursuaded the German government to pay the engineering costs.

The German government commissioned the Hochtief and Dortmund Union, a consortium of West German firms, to carry out the study, which began November 22, 1954. Two years later, the Union presented two plans, one of which was approved by an international technical commission appointed by the Egyptian government and consisting of Sir Alexander Gibb and partners; Doctors Karl Terzaghi, C. J. Steele, and Lorenz G. Staub from the United States; Max Prüss of Germany; and André Coyne of France. The chief concern of the panel was to ensure beyond doubt that the rock- and sand-filled dam as proposed by the Germans would not be undermined, and to determine the design of a sealing curtain in the core of the dam. Eventually, the panel decided on a grout curtain extending down to the granite bedrock.

By the end of 1954 Nasser was in complete power, having ousted General Neguib, leader of the revolutionary government. A consortium of British, German, and French firms was formed for the purpose of building the dam. The World Bank, the United States, and Britain offered loans totaling $270 million, and more finance was forthcoming from France and West Germany. Mr. Eugene Black, president of the World Bank, found himself "in substantial agreement" after a meeting with Nasser on February 9, 1955. But Nasser procrastinated signing the agreement, fearing that the offer had been limited to the first construction stage only for the purpose of exerting pressure on his government when it came to financing the concluding stage. He held out for a definite commitment from the western powers on financing the entire project from the outset.

But then Nasser took off on a strange course, and began playing up to the Communist bloc, originally in order to obtain arms for use against Israel. In May, he even recognized Communist China and thereby flaunted not only American opinion but also major American policy, thus arousing the ire of Mr. Dulles, the Secretary of State. Dulles publicly announced the cancelation of the American loan, and since the credits were interdependent, the World Bank and British loans also.

The rest is current history. Nasser retaliated by confiscating the Suez Canal, ostensibly for the purpose of using the canal tolls for building the High Dam. The income derived from that source would not go far, and the action scared off the western construction consortium. Instead of obtaining finance and aid from the West, Nasser got himself into a war with the West and with Israel, too.

The incident, if that is the word for it, is worth noting, because it points up the awkward position of the engineer in his relation with politicians. An engineer is by training and work a mentally disciplined creature dealing with facts as presented by nature; whereas politicians, professing to be concerned with the welfare of hundreds of millions of people, can indulge in small-boy tantrums. Lost in this fracas was the adult engineering fact that the United States loan of $30 million canceled by Dulles, which pulled the rest of the credits with it, was of no particular value at the time. Indeed, the loans could not be used because of the elementary fact that the High Dam could not be built! Before anything could be done in the way of constructing the High Dam, the old Aswan Dam had to be provided with a power station to supply the energy for the operations on the upstream site. And the earliest date for the completion of that hydropower development was 1960, five years hence. A lot of things could happen in five years, including the death of Mr. Dulles and a change of American administration.

As a matter of historical record, when the Egyptian ambassador to Washington became aware of the hostile opinion in the United States following the recognition of Communist China, he hurried back to Cairo and persuaded Nasser to accept the loan on the conditions offered. But then it was too late; Dulles had already made public the American refusal to finance the High Dam.

The vacuum left by the American and British withdrawal naturally sucked in the Soviet government. In October, 1958, Russia offered to finance the High Dam, and in December a contract was signed whereby 400 million rubles ($93 million) was made available as an advance credit for the first stage of the development. The loan was offered at $2\frac{1}{2}$ percent interest and payable in 12 years, beginning in 1964, on condition that only Russian equipment, plant, and engineering methods be used. In January,

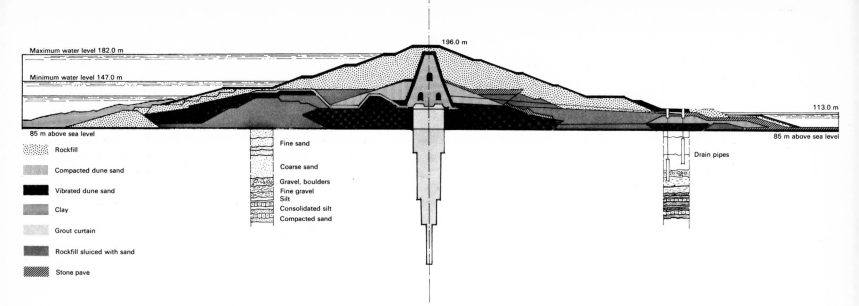

Maximum water level 182.0 m

Minimum water level 147.0 m

196.0 m

113.0 m

85 m above sea level

85 m above sea level

Rockfill

Compacted dune sand

Vibrated dune sand

Clay

Grout curtain

Rockfill sluiced with sand

Stone pave

Fine sand

Coarse sand

Gravel, boulders
Fine gravel
Silt
Consolidated silt
Compacted sand

Drain pipes

The High Dam is built on the model of a moraine ridge, although the masses are placed in a more orderly fashion, in accordance with the best Soviet practice of soil mechanics. Local rock, sand, and clay on the order of 42 million cubic meters have been placed on the undisturbed sand bottom of the Nile. Of this total not less than 15.2 million cubic meters consist of sand sluiced into place — a Russian method opposed by the Western consultants. Rock muck from the diversion tunnels, some mixed with sluiced sand, makes up the rest of the body of the dam. In the center is a clay core, and extending down to the granite base 180 meters below the river bed is an injection curtain 40 meters wide at the top and 5 meters at the bottom.

1960, another agreement was concluded providing a further credit of 900 million rubles ($227 million) for completing the work.

The German plans for the dam were sent to Moscow for review; and when they were returned in June, 1959, numerous changes had been made. Some of the changes were accepted by the international commission; others they opposed, to no effect, because the Russians refused to have anything more to do with the western consultants.

The Russian engineers found it difficult to accept the German designs, and when Nasser at 13:50 o'clock on January 9, 1960, blew the first round on the site, there were in fact no definite building plans for the High Dam. (The final designs for the first stage did not reach Cairo until eight months later, and the ones for the second stage were not approved by Moscow until the fall of 1963.) For several years, white lines on the cliffs on both sides of the gorge had marked the position of the dam; now the whole structure was moved 600 yards upstream. The Russian engineers did not take kindly to the grout curtain in the core, but having made their own tests, they grudgingly agreed with the Egyptian engineers that perhaps it would be

needed. Generally, the Soviet changes were directed at simplifying the construction and thereby reducing the time and cost of building the dam.

The High Dam, eventually built after this troublesome gestation period, is without doubt an admirable exercise in soil mechanics. Roughly speaking, it is a rock and earth fill dam. It is a ridge built across the river resembling that left by a retreating glacier, except that the masses are distributed in a much more orderly and intelligent fashion than the slapdash construction of a moraine ridge. The dam is built altogether of local materials: 28.6 million cubic yards of rock, 20 million cubic yards of sluiced and vibrated sand, 4 million cubic yards of clay, etc.; or a total volume of 55 million cubic yards. It rises gradually from the gravel bottom of the river for a distance of 1,542 feet to a 131-foot-wide crest at 364 feet above the bed, and then slopes 1,542 feet to the bottom downstream the dam. The total crest length is 11,810 feet.

In the center of the transverse section is the grout curtain, descending to a depth of 591 feet to the granite bedrock. The curtain has a width of 131 feet at the top and tapers to 16 feet at bedrock. Above the curtain is the clay core of the dam with a base of 197 feet, tapering to 32 feet at a height of 203 feet. The core has three concreted inspection galleries.

At the maximum storage level of 597.1 feet (above sea level) and 318 feet above the bed, the dam will impound 205,000 million cubic yards of water, of which 39,000 million will be dead storage, leaving 166,000 million cubic yards operational capacity. After evaporation and seepage losses are deducted, the net live storage is 97,000 million cubic yards, of which Egypt will get 72,000 million, and Sudan 25,000 million.

The length of the reservoir—"Lake Nasser"—is 310 miles, extending to the Second Cataract, and the average width is 6 miles. The lake covers an area of 1,860 square miles.

259

The benefits expected from the dam include the reclamation of 1.3 million feddans (1.4 million acres) of new arable land and the conversion of 0.7 million feddans (0.73 million acres) of basin-irrigated land to perennial irrigation. On the basis of seven persons per two acres, the gain in new irrigated land would feed 8 million people, but with a population increase of 0.5 million or more per year, the likelihood is that the benefits gained will have been discounted while the dam was being built.

But then there is the power production, estimated—perhaps somewhat optimistically—at 10,000 million kilowatt-hours per year, or five times more than the consumption when construction began. It is generated at a power station downstream the dam, with an installed capacity of 2.1 million kilowatts.

Since the High Dam is actually a barrage, the water from the storage is brought by a 1,258-yard-long supply canal with a maximum width of 273 yards at the tunnel intake. Six circular 49-foot and 310-yard-long rock tunnels provided with control gates and serving as penstocks bring the water through a downstream power station measuring 961 × 246 × 134 feet before it discharges into a 530-yard-long tailrace canal with a maximum width (at the station) of 301 yards. Twelve Francis turbines, each with a capacity of 180,000 kilowatts when working at a mean head of 190 feet, drive 175,000-kilowatt generators. The power is transmitted to Cairo and the Delta by two 500-kilovolt lines over a distance of 590 miles.

Summing up, the High Dam itself, its storage capacity, and its electric plant are on a considerably larger scale than Grand Coulee, the biggest hitherto achieved in the western world.

Building the dam

Space does not permit more than a brief sketch of the first and most critical stage of the construction of the High Dam—that involving the diversion of the Nile. Before the dam could be built, the diversion canal, the tunnels, and the two cofferdams had to be finished, and regardless of the year chosen, these works had to be in place before the day in July when the Nile rose in high flood at Aswan. A failure would set the work back one year and entail a loss of £E200 million ($560 million) to the Egyptian economy.

The work involved the excavation of nearly 14 million cubic yards of rock in the diversion canal, 9 million cubic yards in the tunnels, the removal of 1.3 million cubic yards in the tailrace canal, plus the dumping of 5.2 million cubic yards of rock and 7.8 million cubic yards of sand in the river for the upstream cofferdams. In addition, 825,000 cubic yards of concrete had to be placed—for the power station, tunnel linings, intakes, and so on. The brutal volume of civil engineering work appears to have been the largest on record.

It was a tall order, and a splendid opportunity to demonstrate the incontestable superiority of state enterprise. The Russian engineers cut two years from the timetable set by the international consultants, and insisted that the first stage be completed in 1964. The Egyptian government concerned with the work went the Russians one better. The diversion canal, it officially announced, would be completed in 1963 and the work would be done with 6,000 men at the most.

One year after the start, no power from the old Aswan Dam had reached the site. Only 70 Russians, 80 Egyptian engineers, and 2,000 workers were at hand. Nonetheless, the Minister of the Interior announced that 95 percent of the work had been completed, whereas in actual fact only 0.9 million cubic yards had been excavated out of a total of 6.5 million scheduled for the first year.

Russians operated the excavators and rock drills and even drove the 25-ton Zis dump trucks. There was a severe shortage of skilled Egyptians, and Ivan Komzin, the Russian site manager, insisted that he would need at least 3,000 experienced Russians to get moving. There was also trouble with the Soviet equipment. The cotton corded tires of the dump trucks, costing $840 apiece, were torn to bits in a day. The 180-ton Ulanskev excavators behaved satisfactorily but had to be returned to Russia for overhaul. The rock drills did not come up to expectation at all. Egyptian engineers who had seen better things sneered at the Russian machines; indeed, the efficiency of the Soviet plant proved upon study to be 23 percent less than that of equivalent Western machines.

So it might have continued, had not the Egyptian government accepted the suggestion of its senior engineers to invite bids from private contractors for the first stage of the job. The Russians, realizing that they were heading toward failure, regretfully agreed. The contract was awarded to the Osman Ahmed Osman firm, later known as Arab Contractors, Ltd., at an initial price of $45 million, later raised to $56 million.

Seen in retrospect, this departure from basic Soviet and leftish ideology saved the operation. The contracting firm appeared on the site in May, 1961. Its site engineer Amin el-Sherif saw immediately that there was insufficient equipment, and what there was, was obsolete by western standards. In spite of Soviet objections, the firm was permitted to order $650,000 worth of Swedish rock drills and, beyond all, 20 Aveling-Barford 35-ton dump trucks, ostensibly to be used by the contractor on later work. When the operations peaked in 1963 and 1964, there were 54 Aveling-Barford dump trucks doing more work than the 200 Soviet Zis trucks. Dunlop was also brought into the picture to deal with the excessive tire wear.

The Osman company also began to hire labor. It paid well (from £E10 to £E30 a month in cash and keep) and was able to recruit a thousand men per month where the government agency had found none. By the end of December, there were 9,000 Egyptian workers on the site. In that month alone, 1,000 tons of dynamite were used to produce 400,000 cubic yards of rock, whereas up until the arrival of the contractor only 300 tons had been consumed.

By 1962, the operations got off the ground. By that time, the Egyptian engineers had also taken the mettle of their Soviet colleagues, and spoke their disdain openly: "Compared to western professionals they are a bunch of amateurs," or "they may be competent at home, here they don't know what they are doing" were some of the comments. Nonetheless, Moscow kept shoveling in equipment and men at a respectable rate, and by the end of 1962 there were 17,000 tons of equipment (worth about $25 million) and 1,500 specialist workers and 450 engineers on the job. The Egyptian force had then grown to 200 graduate engineers, and 9,000 skilled and 12,000 unskilled workers. In December, the Egyptian contractor excavated 550,000 cubic yards of rock—a world record for all categories. The operation had been taken over the hump by Amin el-Sherif, the contractor's site manager. Having done the impossible, he died in December from overwork—like so many other good construction men.

On January 1, 1963, the target date for the diversion of the Nile was set to May 15, 1964. By now operations were in full swing, for the Russians as well as the Egyptians. Aleksandrow, the new Russian site manager, brought in during the course of only two weeks 200 Soviet experts and not less than 11,000 tons of additional equipment, which raised the number of 5-cubic-yard Ulanskev electric excavators to 16, in addition to 74 other excavators, 210 dump trucks, 47 cranes, 107 bulldozers, 2,069 rock drills, 80 compressors, 36 concrete pumps—to mention only a few of the major equipment items among the 40,000 tons of machines put to work on the High Dam.

In July, the contractor with 35,000 men on the site pulled out 654,000 cubic yards of rock, and by the end of the year altogether 13 million cubic yards had been excavated. He started work on the powerhouse in August, and a few months later began dumping 5 million cubic yards of rock into the river for building up the upstream cofferdam; and the Russians began sluicing sand into the dam. In 1964, finally, a further 5.5 million cubic yards was dumped and 8 million cubic yards of sand added for good measure. As the upstream cofferdam rose out of the water and the gap between the arms narrowed, it became necessary to seal the diversion canals with sand cofferdams.

The work was completed well in advance of the target date, the river was diverted without incident, and when in July the high flood came roaring down, the upstream cof-ferdam stood up to it. The first stage of the job was finished; all that now remained was the building of the High Dam itself.

The cost of the High Dam will not be known for many years to come, but the best estimate is £E128 million ($348 million) plus £E90 million ($252 million) for the electrical plant, or a total of £E218 million ($600 million). To this should be added the cost of the reclamation and irrigation developments, new roads and so on, which raises the total to £E425 million ($1,190 million). These investments are expected to increase the national product by £E234 million ($650 million).

The cost of the installed capacity in the power station, including the civil engineering, comes out to $243 per kilowatt, which is 20 percent less than the postwar European mean for hydropower development up until 1960.

However, as in the Tennessee Valley, the incalculable benefits derived from the High Dam will in the long view become more important than the quantified economic ones. Since the work started in 1960, a whole generation of young Egyptian engineers, indeed every graduate in the country, has been gaining experience in applying modern mechanized facilities to civil engineering on a vast scale. Some 35,000 fellahin previously accustomed to working with only the most primitive tools have been taught modern skills in operating and maintaining electrified, dieselized, and air-powered equipment. Vast numbers have been better paid, fed, and quartered than ever before in their lives.

When the big job is done and the bulldozer drivers, excavator operators, miners, and all the rest of the skilled men that have been occupied in building the High Dam return home, life in Egyptian towns and villages can never be the same as before. It is inconceivable that the young engineers who cut their professional teeth on mechanized construction will be content to boss excavations using men carrying baskets of dirt on their heads, nor will they find former bulldozer drivers for that kind of work. The real Egyptian revolution has yet to come.

Despite its odd and frustrating political history, Nasser's High Dam measures up to the high moral aims that David Lilienthal of TVA saw in engineering as a means of improving the lot of the common man and restore to him some measure of human dignity. When viewed against the vast time scale of Egyptian history, it gives a deep sense of satisfaction that the people living in the cradle of civilization after 5,000 years of misery have at long last been given the opportunity to erect a useful pyramid for their own benefit.

Supplement

PALEOLITHIC HOUSES

Although it is reasonable to assume that the Aurignacian and Magdalenian hunters took refuge from the weather in the caves or under the overhangs in the limestone formations lining the rivers in the Dordogne region, there are among the animal paintings in the sacred caves so-called *tectiform* (tentlike) signs that have been interpreted to be houses or, as the word suggests, tentlike huts. But whether the tectiforms represent man-made shelters or possess magic significance is by now only of academic interest. Because it has been definitely established that the Gravettian mammoth hunters in the east, roaming from Czechoslovakia across the Russian steppes into Siberia (where nature does not provide natural caves for shelter from the icy blasts sweeping across the open land) built houses for themselves.

At Vestonice in Czechoslovakia, a group of three paleolithic huts has been unearthed. One of them measures 49 × 30 feet and appears to have been the communal dwelling of a hunting clan. Around one of the five hearths set on a floor paved with limestone grit were found a thousand flakes of worked flint. In another hut, protected in front by a circular 6-foot limestone wall, the oldest man-made structure hitherto discovered, was found a beehive kiln which—judging from thousands of clay pellets, many of them shaped into heads of animals—obviously had been used for firing ceramics.

Of the numerous Russian sites, the Timonovka settlement consisting of six 39 × 10-foot dwellings on the Desna River near Briansk is of particular interest from a technological point of view. These dwellings, sunk 10 feet into the ground and entered by a ramp, had walls lined with timber, while the roofs consisted of logs laid across the trench and covered with earth. The hearth was sited near the entrance and provided with a hood tapering into a chimney extending through the roof. The hood and chimney were shaped of birch bark daubed with clay on both sides to make it fireproof. Such a chimney was not used in Europe until the end of the Middle Ages; before that, kings and peasants alike had to spend their time in smoke-filled halls or huts at times when the barometic pressure prevented the smoke from the hearth from rising through an opening in the roof. Although it usually takes a long time for a good idea to mature, this is the only known case where it required some 20,000 years for a sound invention to be commonly accepted.

THE CANALS OF NINEVEH

The 18 Nineveh canals known by name were constructed in only 13 years, from 703 to 690 B.C. When Sennacherib succeeded his father Sargon, the capital of Assyria was Dur Sharrukin (now Khorsbad) built by Sargon on modern principles of town planning, with avenues crossing streets at right angles and several large squares fronted with palaces and temples. But Sargon's capital did not appeal to his son because it lacked trees and greenery. Dur Sharrukin was a city of stone sited in a desert of stone.

Sennacherib had other ambitions. He was a nature lover who took great delight in parks and gardens, and immediately after assuming power he decided to turn the entire area around the capital into a green belt. Apparently detesting his father's austere and uncompromisingly military administrative center, he began to look around for another site where he could build a new capital more to his liking.

He found one on the shores of the Khosr River a couple of miles upstream from its confluence with the Tigris. Nineveh, the new capital, was built in only two years; at any rate, the palace and the inner curtain wall were ready in 703 B.C. The palace was surrounded by a newly planted park, and around the tall city wall was a broad belt reserved for fruit orchards.

The new parks and orchards had to be irrigated, and to obtain the water needed, the Khosr River was straightened and canalized up to the village of Kisiri 1½ beru (1 beru = 10,692 meters) or about 10 miles upstream. The river water was conducted between dikes, whereby the water level was raised sufficiently to facilitate gravity irrigation of the plains around Nineveh.

The Kisiri Canal satisfied the needs for a few years, but in 702 and again in 700 B.C. the King mounted expeditions to the south of

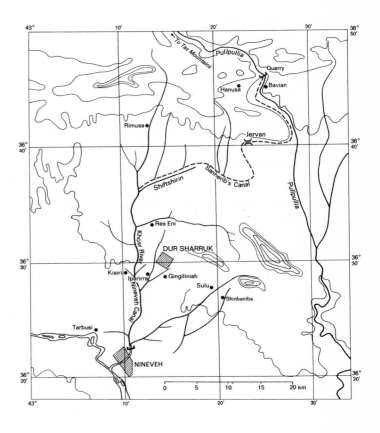

Babylonia where he discovered the marshland flora and fauna and took a liking to them. Upon his return to Nineveh he began to look around for other sources of water. His chronicler tells how the King himself went into the Musri Mountains 12½ miles northeast of Nineveh and how he there found three rivers at the towns of Dur Ishtar, Shibaniba, and Sulu. He had the rivers cleared of obstructions and joined them into the so-called Musri System. The dams north of Nineveh were apparently constructed in conjunction with the new canal system.

With the additional supply Sennacherib obtained enough water to be able to turn a large area into a marsh for planting Babylonian building reeds, as well as luxuriously irrigated fields suited for the planting of sugar cane, mulberry and "wool-growing" bushes—an early mention of cotton. The imported fauna included wild boars, deer, and cranes.

These hydraulic works appear to have been completed in 694 B.C. The next report on Nineveh's water supply is dated 690 B.C. and is recorded in cuneiform characters on panels cut into the rock in the gorge at Bavian, near the border of Armenia. Hence, within only four years an entirely new system including 14 canals, in addition to an outer curtain of 98- to 115-foot-high walls enclosing the city was constructed. The better part of the new water supply was obtained in the Tas Mountains and necessitated the construction of a 148-mile-long canal, contained within dikes, from Bavian to the Khosr River, including the 918-foot-long aqueduct at present Jerwan. The new canals were reported in detail, and they have been traced archeologically or by sophisticated reading of the Bavian inscriptions. They are indicated on the map on page 262. The inscriptions appear to be a transcription of the speech given by the King at the official opening of the canal in 690 B.C. They read in part as follows:

"Now in entrusting that which I have planned to the kings my sons, falsehoods are not befitting. With these few . . . men I dug the canal. By Assur, my great God, I swear that with these men I dug that canal in a year and three months and finished the construction.

"To open the canal I sent an Ashipu priest and a Kalu priest . . . and I presented gifts to Enbilulu, Lord of rivers, and to Eneimbal (Lord who digs canals) . . . The sluice gate like a flail was forced inward and let in the waters in abundance. By the work of the engineer its gate had not been opened when the gods caused the water to dig a hole therein.

"After I inspected the canal and had put it in order to the great gods who go at my side and who uphold my reign, sleek oxen and fat sheep I offered as a pure sacrifice. Those men who dug the canal I clothed with linen and brightly colored garments. Golden rings, daggers of gold, I put upon them . . .

"At the mouth of the canal which I had dug through the midst of mouth Tas, I fashioned six great steles with the images of the great gods, my Lords, upon them; and my royal image in lapin appi salutation (gesture with hand on nose) I set up before them. Every deed of my hands I wrought for the good of Nineveh I had engraved thereon. To the kings my sons I left it for the future . . ."

In March, 1934, one of the steles was discovered, cracked and partly submerged in water, in a narrow gorge formed by the Gomel River in the Bavian Hills. The sculptured stone block measured 16 × 13 × 23 feet and in falling had come to rest on a stone wall. Closer examination revealed that the monument had been placed at the head of the canal. Obliquely across the river was a weir that guided the water into the canal. About 984 feet downstream were the remains of a dam across the river. The width of the dam appears to have been 39 feet, and the upstream end of the canal ended at

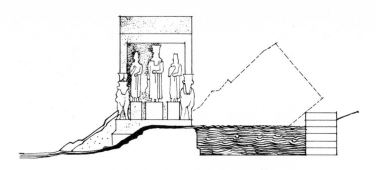

the face of the dam. The water was conducted by a tunnel through the dam to the continuation of the canal on the downstream side. At the west side of the dam was a 1 × ½ × 2-meter pit excavated in the rock, and the investigators believe that the pit served some function in controlling the gates placed at the downstream end of the tunnel.

The width of the canal through the Bavian Hills was 20 feet, whereas the width of the Jerwan aqueduct has been estimated at 49 feet. Whether the canal widened before it reached the aqueduct has not been established, but downstream it was probably this wide to receive the additional water from the four rivers included in the Shiftshirin system.

The relatively well-preserved aqueduct on the Sennacherib Canal is about all that remains of the technically sophisticated hydraulic civilization of the two river valleys, in comparison with which the Greek and Roman contributions seem rather modest. It has really not been until the present century that western civil engineers have surpassed their Mesopotamian colleagues. But it is doubtful whether a modern development on the same scale as the Sennacherib Canal system could be completed in one year and three months.

THE SILOAM TUNNEL

Technically speaking, the Siloam Tunnel scheme involved advancing a 2 × 5-foot gallery for a distance of 359 yards, and the excavation of 413 cubic yards of limestone rock. The difference in elevation between the two points of attack, i.e., the Gihon Well in the north and the Pool of Siloam in the south was 7 feet, which would provide the sharp gradient of 1 :105. From the Gihon Well the direction would be roughly southwest.

The attack was made by two mining teams working toward each other. The existing evidence shows that the southern team set in the point of attack about 15 feet above the proper bottom elevation of the tunnel and had to work downward to correct the grade. As a consequence, the height of the tunnel near the southern portal is about 15 feet and tapers to the roof. The tunnel winds through the Ophel hill in such an irregular manner that it is a marvel that the two teams met at all. The most obvious reason for this peculiar mining performance it that the miners were not guided by a survey, however simple and elementary. At times when the teams found themselves lost, they had to drive raises to the surface to find out where they were in relation to each other. Owing to the inability to keep line, the length of the tunnel came to 560 yards and the volume of rock excavated to 819 cubic yards, or about twice as much as needed. However, some good came out of the sloppy performance because owing to the increased length, the grade became 1 :235, an improvement from the point of view of water conduction.

COPPER TOOLS

About the middle of the Third Millenium B.C., Egypt obtained its supply of copper from Sinai. The Egyptian state sent out an annual expedition which besides miners and smelters also included bureaucrats aligned in proper pecking order and reflecting the monolithic structure of the government. The ores mined consisted of such easily reduced minerals as cuprite and malachite, which were smelted at the mines, whereupon the ingots were carefully recorded and stored under the protection of the Great Seal. The Sinai copper differed from the Oman copper imported to Ur by independent Sumerian merchant adventurers, inasmuch as it lacked such natural adulterants as the tin and nickel which imparted to the Sumerian copper an exceptional hardness; indeed, the smelted product was a natural bronze rather than copper. Until the technology of alloying refined copper with metallic tin was fully developed (ca. 1500 B.C.) the Egyptian artisans had to rely entirely on soft refined copper which for such purposes as weapon and toolmaking was far inferior to the hard material available in Ur. However, by cold hammering the cutting edge the initial hardness of the refined copper (87 Brinell) was raised to 135 Brinell, i.e., the hardness of an annealed medium (0.37 %) carbon steel. A natural or technical bronze, on the other hand, had a hardness of 140 Brinell, or about that of a quenched and tempered 0.37 % carbon steel.

A NOTE ON STONEHENGE

In the Neolithic settlements investigated throughout Europe, it has not been possible to distinguish among the characteristic longhouses any differentiated buildings, serving either ritual purposes or as a residence of a ruler. But from about the middle of the Third Millenium, evidence begins to accumulate, pointing to the existence ot at least ritual buildings. Two such "temples" have been excavated in Denmark, and in Britain also there began to appear about this time circular embanked structures undoubtedly serving ritual needs. One such simple earth structure is Stonehenge on the Salisbury Plain. It consisted originally of a circular ditch with a diameter of 300 feet and a depth of 5 feet with an earth bank enclosing 56 holes or shallow pits, some of them containing cremated bones that have been carbon-dated to 2490 ± 150 B.C. The earthworks are roughly oriented toward the sunrise at summer solstice and were approached from the northeast by a banked trackway.

This ancient ritual site underwent a tremendous transformation some time in the sixteenth century B.C. The entire circular site, which had a diameter of 100.75 feet, was enclosed by 30 upright sarsen stones (hard sandstone) rising $13\frac{1}{2}$ feet from the ground, each one weighing about 26 tons. The uprights carried lintels cut to the curve and joined to each other by tongue and groove. At each end the downside of the sarsen lintel has a mortise hole that fits a corresponding tenon on top of the uprights. Hence each upright has two tenons. By this method of timber joining, the 29 lintels are firmly secured into a circle of stone, each member being joined laterally to abutting lintels and vertically to the uprights. The rock for the construction members was quarried at Avebury in North Wiltshire, some 18 miles distant, and dressed on the site, juding from the chips found.

The outside sarsen circle encloses a concentric ring of so-called bluestones tapering to a height of 11 feet. This kind of bluish gray sandstone is found only in the Preselly Mountains in Pembrokeshire, 150 miles from the site. Inside this ring are five so-called "trilithons"

placed in a horseshoe layout. Each lateral trilithon consists of two 20-foot sarsen uprights carrying a tenoned lintel, while the center one extends 22 feet aboveground and weighs about 40 tons. Inside the horseshoe was an ovoid of bluestones surrounding the "altar stone." The monument was oriented so that the largest trilithon and the altar slab in front of it faced the rising sun at summer solstice.

As mentioned previously, the stones were dressed in a manner unique in Britain but common in the Near East. The slabs were grooved down to the finished surface by hammering with heavy stone mauls, after which the ridges were removed by hammering, and the surface was then finished off. The uprights were given entasis, i.e., were slightly curved in order to correct an illusion of concavity. On one of the trilithons were reliefs of a bronze dagger and axheads obtained by cutting down and removing the entire surface of the stone by half an inch.

The meaning and purpose of this magnificent monument have been debated for the last 800 years. No Roman historian or Saxon chronicler has mentioned it, and the earliest reference to it dates from the twelfth century. In his *Historia Britonum,* written around 1139, Ambrosius Aurelianus states that it was erected as a monument to those slain in the battle with the Hengists in A.D. 470. This explanation satisfied the curiosity of the learned until 1624 when Edmund Bolton suggested that Stonehenge was the tomb of Boudicca. Thirty years later, Inigo Jones made an investigation of the monument at the order of King James I and expressed the view that it might have been built by the Druids but more likely by the Romans. Others thought it had been erected by the Danes. John Aubrey (1626–1697) was the first to take a strong position in favor of the Druids, and his idea has been accepted until quite recently, although contested in 1872 by James Ferguson who insisted that it was a sepulchral monument of the Saxon period.

None of these views are accepted any longer. Aside from the similarity to the grave circles of Mycenae, the shape of the dagger is unmistakably Aegean. For this and other subtle archeological reasons, the erection of the stone monument Stonehenge II has been dated to the height of the Aegean civilization at about 1600 B.C., i.e., at the beginning of the Bronze Age proper. There was at that time a lively trade throughout Europe, particularly in amber but also in metals. The growing need for tin used in bronze manufacture inspired merchant adventurers to prospecting journeys far afield after the meager local supplies were exhausted. The accumulation of Mycenaean objects in southern England and Cornwall suggests that in this area Mycenaean merchants had established trading posts which served as diffusion centers for superior technical skills and new ideas.

Among these new ideas may have been the conception of political authority, of the right of some person to rule over others and impose on them the obligation to labor on a monument designed to perpetuate his authority and to impress on others the hallowed powers granted him by the gods. In case such ideas did not sit well, the strangers also had the instruments to aid an ambitious local bully to impress his will on his neighbors. They supplied him with double axes and short swords of hard bronze against which stone clubs and flint daggers were of no avail.

We may speculate about the details of the transition, but one thing seems clear—the outcome was that in Britain at this time somebody possessed political influence gained by military might to command the local population to labor in distant quarries to break out something like 2,000 tons of monoliths and haul 1,500 tons of them over a distance of 18 miles, and 500 tons 150 miles. Together it becomes a transport performance of about 100,000 ton-miles carried out on rollers. Although it is unlikely that this portage work was done in one continuous operation, but rather carried out in agricultural off-seasons over a period of many years, it was directed by a durable authority capable of enforcing its will over a large area. The very existence of the monument conjures forth a society not unlike the Celtic one as it was encountered by the Romans, with a rude chieftain claiming personal allegiance of a warrior caste capable of impressing freemen peasants and their thralls to work on it. There is, of course, the likelihood that some such society was already in being when the first merchants from the Aegean arrived, since Stonhenge I was after all a ritual place where an ambitious or religiously gifted man could gain power and influence over the others. The rest—bronze weapons, stone monument and all—would then follow as a logical sequence.

EUROPEAN PREHISTORY

"Who was king, who was not king," laments a scribe in Akkad. "Behold it has come to a point when the land is robbed . . . by irresponsible men," writes a bureaucrat in Memphis. The ancient world had fallen on evil times: In Mesopotamia, the Sargon Empire had broken down around 2250 B.C. In Egypt, the death of Pepi II in 2265 B.C. had left the country without a central authority. Both the ancient powers had become weak and unable to resist the attacks of restless barbarians beyond the borders.

These quotations are about the only surviving records of the world-shaking events that left more durable monuments in the ashes of pillaged and burned cities all over the ancient world. In Palestine, the Amorites sacked and killed; in Anatolia, over 300 sites trace the onslaught; the second city of Troy was destroyed, as were Tiryns, Asine, and Lerna in Greece. Everywhere the old civilizations were on trial. The Indo-European people had made their entry on the stage at the end of the Third Millenium B.C. with an orgy of fire and blood and, temporarily at least, the light dimmed over the ancient cultures.

The Hittites settled on the ashes of the former cities on the Anatolian plateau, the Mitanni in Syria, the Kassites in Mesopotamia, while the Aegean islands and Greece were taken over by strangers from the north. Others entered Europe and established themselves on the substratum of peaceful Neolithic peasants. Most authorities agree that the newcomers that made a nuisance ot themselves in the ancient world and put their permanent imprint on Europe came from the plains north of the Black Sea extending to the Ural Mountains in the east. They were a pastoral people versed in farming, although preferring cattle raising, but they also possessed a tremendous technological advantage by having domesticated the horse. They had fine wagons with solid wheels, and they knew how to work metals. They had a social organization of their own, with an elected chief or king, a council of elders, and an assembly of freemen, all of them with definite obligations and privileges that in some way resemble medieval feudalism.

These, then, were the men who were destined eventually to inherit the earth. In the East, the ancient civilizations rode out the first storm; they recovered from their temporary weakness and succeeded in holding the barbarians in check, and when eventually—some 800 years after the newcomers had made their entry—the two met head on in the battle of Kadesh in 1288 B.C., the power equilibrium in the East was maintained. But it was a narrow victory, and the implications of it made a durable impression on Ramses II. By his orders every temple throughout the length of Egypt was filled with inscriptions to commemorate the event that was to give his country a few hundred years more of native rule.

In Europe, however, the newcomers maintained their ascendancy. The old agricultural communities vanished; the longhouses gave way to settlements consisting of small houses for individual families. Fortifications began to appear, and a stratified society headed by a warrior aristocracy emerged. Into this rustic world ruled by petty kings or chieftains came stragglers from the eastern civilizations, either as captives of war or as fugitive craftsmen who had lost their patrons in the universal upheaval, or merchant adventurers and metal prospectors. They opened up the mineral deposits in the Carpathians, the Tyrol, Iberia, and elsewhere where gold and copper, which had been exhausted farther east, here lay easily accessible to those who knew how to look for them. Even factories were established, such as Los Millaries in Spain, and within its protected walls gold and copper were smelted and fabricated into jewelry, vessels, tools, and arms. Trade routes were established, and from the metalworking centers in central Europe copper and later bronze manufactures spread over the entire continent. The amber trade took on importance, and amber was carried over a route that went from southern Denmark via Hamburg, Dresden, and Prague, to the head of the Adriatic, from where it was shipped to the Aegean consumption centers. The reciprocal trade in Mediterranean shells appears to have crossed the Gotthard Pass on its way to northern consumers.

Expansion by chariot

The high technical quality and style attained by the bronze ateliers in south central Europe suggests that the copper mined in the Tyrolian mines was worked by displaced merchants and craftsmen from Syria who, owing to the Amorite troubles, were no longer able to carry on their former trade. So long as they proved capable of producing better and more lethal axes, swords, spears, and daggers, they could always hope to find the protection of some nameless rustic princeling in temporary control of a bog-infested area north of the Alps.

As evidenced by the numerous horse and wagon burials extending in space from Siberia to Sweden and in time from the Third Millenium into the present, it was the taming of the horse and the use of harness that brought about new techniques in warfare and enabled the newcomers from the Russian plains to overrun the ancient civilizations and the bucolic lands to the north thereof. They had learned about the wheel from the Sumerians, and during the initial phase of the onslaught their horse-drawn wagons had solid wheels. But then around 2000 B.C., somebody somewhere (probably in Syria or Anatolia) with access to good bronze tools invented the light spoked wheel which made possible the development of a light cart drawn by a team of fast horses. There is not the slightest doubt but that the *chariot,* as ultimately developed, became a revolutionary engine of war before which the ancient empires crumbled. In 1750 B.C., Egypt fell before the chariotry of the Hyksos, a Semitic confederacy that came to rule the country from the Delta to the Second Cataract, where at Buhen they ceremoniously buried a horse that has earned the distinction of being the earliest horse found in Egypt. When 200 years later the Hyksos were driven out by Pharaoh Ahmose, it was done with the aid of chariotry.

By that time the light spoke-wheeled war chariot had been copied by the barbarians beyond the bounds of the Fertile Crescent. In Mycenae, Armenia, the Altai Mountains, and India, it was common; even in distant Scandinavia, thirteenth century B.C. rock carvings depict chariots with four-spoked wheels. "A chariot maker fells the

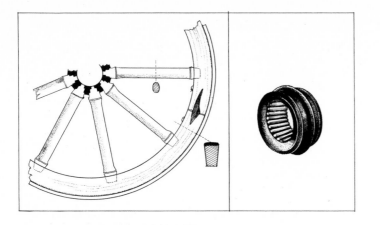

poplar with gleaming iron, to bend a felloe for a splendid chariot," reads a passage in the *Iliad*. "I bend myself before Sindra as a carpenter his felloe of good wood," is how wheel making is mentioned in the Sanskrit *Rig-Veda*. Both are references to the method of heat-bending a piece of suitable wood—in Egypt imported ash—into a single felloe in which four, later six or eight, hornbeam spokes were cunningly morticed and joined to a hub of oak. (Later, in Celtic times, the hub was provided with needle bearings made of boxwood. The iron tires were shrunk on the wheel while red hot and expanded, a method not used in Europe until the nineteenth century.)

In the technically sophisticated ancient East, means were found eventually to overcome this formidable engine of war by new ways of building fortifications, and by the time of the *Iliad* it had also become obsolete in Mycenae. But in Europe its use continued unabated, and chariotry became intimately associated with the military exploits of the Celtic tribes.

Troublemaking Sea People

During the later half of the Second Millenium, Europe, still shrouded in nonliterate darkness, seems to have attained a measure of prosperity owing to the development of its virgin sources of copper and salt. Petty chieftains listened to the heady tales told by travelers of the riches and pleasures to be found in distant lands where the sun never lost its warmth. Pirates began to appear in the Mediterranean; in 1350 B.C., the kings of Cyprus, from where Egypt drew its supplies of copper, complained of raids by some strange people. In 1230 B.C., the strangers—"people of the north coming from all lands"—had joined a Libyan army which was defeated in the western desert by Pharaoh Merenptah and lost 9,000 swords. Who were these "Sea People?" There has been much speculation about their origin, but whatever their domicile, they kept coming. In 1170 B.C., an Egyptian scribe spoke of them in the following terms: "All at once the lands were removed; no one could stand up to their arms, Hatti, Carchemish, Arzava, and Alasiya (Cyprus). They were coming forward toward Egypt while the flames were prepared before them. They laid their hands as far as the circuit of the earth . . ." But again they were defeated, caught apparently in the cross fire of Egyptian ships and archers on the shore. Among the attackers have been recognized Philistines, and Danuna from Cilicia.

But although beaten back by Ramses III, the "Sea People," allied with people in western Anatolia, had by then already wrecked the Hittite empire; Troy had fallen, and so had Mycenae. Everywhere around the eastern Mediterranean the strangers—armed with long swords provided with typical flanged hilts developed in central

Europe a few centuries earlier—slashed their way through the eastern world. And wherever they went, the curtain of darkness descended once more over the Aegean. To the north thereof it had never lifted.

But it seems quite obvious that in this darkness a new technology was brewing, and when the curtain lifts again we are confronted with a fully developed iron technology which was not to be improved upon until the development of the blast furnace at the end of the Middle Ages. What appears to have happened is that with the break-up of the Hittite empire the skill of iron smelting, fully developed in Anatolia at least as early as the thirteenth century B.C. (judging from the diplomatic correspondence between the kings of the Hittites and Assyria) but long kept a secret by the Hittite smiths, was now shared by the neighboring peoples. In Palestine, ironworking Phoenicians and Philistines had settled along the coast; along the Tigris the Assyrians began to arm themselves with weapons of iron; in Greece iron-armed Dorians had swept down from the north, in northern Italy pre-Etruscan armorers had established their ateliers, in the Alban Hills—indeed, in Rome itself atop the Palatine Hill—iron-smelting furnaces fanned by the western winds had been established.

The ironworking heathens in Greece and around the Aegean, in close proximity to the literate East, were bound to be the first to shed their illiteracy, As for the rest, they had to be content with legends, all of them—whether Etruscan, Roman, Irish, or British—tracing their ancestry to the heroes of Troy who had lived somewhere in the distant East in the murky times when the clash of iron swords rang in a new era and the rising hegemony of Europe over the Near East.

Celts and Germans

Now the troublesome Sea People with their strange barbarian names given them by the Egyptian bureaucrats began to take more definite form. The sea raiders, som of them hired as mercenaries by Egyptian army commanders, the hell raisers in the Mediterranean during the long transition period from bronze to iron, fall into a definite slot. For nearly a thousand years they had lived their anonymous semi-nomadic life with their cattle herds north of the Alps, Celtic and Germanic tribes sometimes grazing their herds in the same area, sometimes moving apart, but always fighting and quarreling among themselves. The more lively and zestful Celts kept moving farther west; the plodding Germans stayed behind. Just what the German-speaking tribes did beyond tending their cattle no one knows because their archeology is so drab. But before 1000 B.C., there was probably no marked difference between Celt and German, and except for their languages it is futile to look for fine distinctions between them.

The last pre-Christian millenium, at least north of the Alps, gets its flavor from the Celts, just as the slow-starting Germans had the latter half of the first Christian millenium pretty much to themselves. Both of them were thoroughly savage; the Celts were head-hunters, their Germanic kin kept eating each other's livers and other suitable innards, at least ritually, until less than a thousand years ago.

It may be well to keep in mind these facts of life, or perhaps more accurately such gory ends of life, when we proceed to sketch, all too briefly, the hilarious progress of Europe's new masters. That the ancient Neolithic peace was broken at the end of the Third Millenium is evident from the need for fortifying the settlements at this time. To begin with, they were naïvely constructed earth ramparts with or without a ditch, more suited to keeping out animals than armed men. Gradually, and increasingly so during the latter half of the Second Millenium, the forts became more frequent and more complex; beyond all, they were no longer built as defenses around a village but as strongpoints, to be used for rallying raiding parties or for protecting people and cattle against raiding neighbors. From archeological evidence, life in Europe had become a nonstop Western.

No cultivated area was safe from cattle and thrall rustlers. Every settlement had to find some defensible site in the immediate neighborhood. A promontory extending into a lake was provided with a palisade and ditch across the base; nearly every steep hill was turned into a fort. Wittnauer Horn in Switzerland provided refuge for about 350 persons and their cattle; when dug out, the ammunition dumps of slingstones were found intact.

As the mercenaries returned from their lethal exploits in distant lands, where they had seen better things in the way of fortifications, construction improved. The earth ramparts became interlaced with timber, pretty much in the manner recommended in military engineering handbooks for fieldworks used up until the last war and, for that matter, at Troy II. Sometime around 500 B.C., a Scythian raiding party scaled the timber-laced ramparts of Boskupin in Poland and laid waste to some hundred timber houses behind the defenses.

As the First Millenium proceeded through its various Celtic stages, Hallstatt and La Tène, subdivided by numbers for more accurate dating, the fortifications became increasingly more extensive and labor-demanding. It had become expensive to stay alive. A small insignificant hill fort on the marches of the Celtic world, the one at Abernethy in Pertshire which enclosed a 45 by 135-foot area, required 3,200 linear feet of timber, necessitating the cutting of 640 trees over an area of 60 acres, from which the logs had to be dragged to the site. Consider then the cantonal *oppidum* at Manching in Bavaria which was $1\frac{1}{2}$ miles in diameter and enclosed by timber-laced ramparts 4 miles long. Here nine cross-braced timber courses were used in the ramparts and joined with iron nails. It has been estimated that 300 tons of iron was consumed, surely a record splurge in ancient construction. It was a stupid way of tying together timber; it would have been better to have used juniper or hardwood dowel pins, which would have been much stronger and would have lasted longer. But, at any rate, Manching shows that the Celts certainly knew how to produce iron in ample tonnages, to be able to waste it in that fashion.

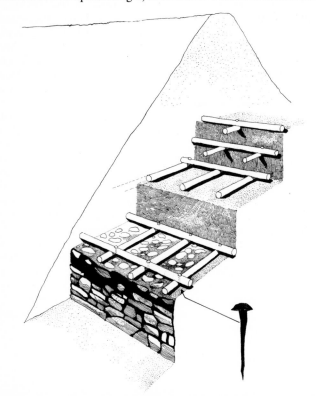

The Manching fortifications were started during middle La Tène times, and this was the *murus gallicus* that Caesar ran up against time and time again in his campaigns in Gaul. The oppidum of Bibrace consisted of 60 acres of built-up area enclosed by 3 miles of murus gallicus, with winding streets framed by timber buildings that do not seem to have differed much if at all from the wall-enclosed towns around the Baltic which began to be established at the end of the thirteenth century A.D. And like them, they are not particularly interesting from a technological point of view.

When the Romans eventually came into deadly contact with their northern neighbors, we get, at long last, some documentary information on the Celts and their manner of life. Like an Irish pub brawl it is best witnessed at a distance, but then it becomes thoroughly enjoyable. If the classical writers can be believed—and archeology and analysis of the ancient Celtic tales generally agree with them— the Celts were beyond doubt a highly gifted people in numerous ways, but altogether lacking constructive sense. We are not here directly involved with their artistic gifts, nor even with their metallurgical prowess—which was formidable—nor their gift of gab which was even more fearsome since it quickly led to quarrels and bloodletting. We are here principally concerned with engineering and have occasion to wonder why the otherwise gifted Celts as well their Germanic neighbors never managed to accomplish anything comparable to what had long been commonplace in the ancient civilizations.

Completely fearless, the Celtic warriors shed their blood and made a thorough nuisance of themselves "around the circuit of the earth," as the Medinet-Habu scribe puts it. They invaded Italy in 400 B.C., sacked Rome in 300, raided Delphi and Syracuse in 368, took Galatea in Asia Minor in 279, and scourged Egypt in 274 B.C. But nothing durable came of this waste of lethal effort. They were from all accounts completely unable to muster the higher forms of mental discipline required to organize and administer their conquests. They did not possess what it takes to become empire builders.

There seems to be reasonable agreement among learned specialists that the feudal society held together by personal bonds of loyalties extending from an elected king or chieftain to barons and freemen in a descending order of liegemen, developed from the rustic order which kept Celtic families, clans, and tribes together in order to maintain a sense of security against a neighboring rustic unit similarly joined in personal relationships. In its developed state in the Middle Ages, it served an elementary need; but even then the impetus toward civilized accomplishments derived from the outside. No feudal baron was instrumental in building Gothic cathedrals or digging canals.

The Celts possessed without doubt the physical means required for the takeoff into a higher order of economy, or civilization as it is also called. They had at their disposal easily worked deposits of gold and metallic copper and the oxides of copper, subsequently also unlimited sources of spathic iron ores. And they had the know-how to mine and smelt and turn the metals into desirable artifacts which from archeological evidence were widely accepted by customers much more sophisticated than the masters of the ateliers where they were made. This trade in metals alone would have produced a surplus that could be used for building palaces, temples, and roads, and for the necessary draining of bogs and much more besides. What became of the Celtic wealth?

Classical writers, although not particularly interested in such mundane details, give an answer to the question. So does archeology. Judging from the large number of transport amphora, of a size equiv-

267

alent to the volume of some 20 modern bottles, quite a lot of the surplus was spent on wine. When a Celtic chieftain gave a party, the ration was three amphorae for each guest. The lower order swilled beer and, as the Roman writer Ammianus Marcellinus describes it, "with wits dulled by continuous drunkenness rush about in aimless revel." A drunken Celtic party must have been something to behold, judging from the impression made on contemporary Romans. At the head of the table sat the mustachioed chieftain shouting, boasting, and belching, his red face gleaming from the grease of boiled pork; and in descending order from the high seat sat his warriors and liegemen getting drunker and more bombastic and querulous with each amphora. A chance remark, an imaginary insult, and the revelry suddenly flared up into a deadly quarrel. The sword of a Celtic warrior rested uneasy in its scabbard.

They did not sow, neither did they reap. Somebody else, their womenfolk, thralls, and captured bondswomen, saw to that. A Celtic freedman saw to his cattle, whether scrawny or sickly did not matter so long as there were many of them. The rest of the time he slept in a drunken stupor, when he did not hunt to relieve his restlessness. His function in life was killing—anything, anyone, anywhere. It was then a Celtic warrior became a man of purpose.

And for that purpose went the remainder of the surplus of the Celtic world—on arms and the accouterments of war, on swords with enameled pommels, on ornamented helmets of bronze, on personal adornments, on chariots of fine wood and wickerwork studded with silver, pulled by stallions with yokes of gold. A fine figure of a man he was as he emerges from the Celtic tales and Homer and Caesar's Gallic Wars. Battle-drunk and screaming, led by their chieftains, the Celtic bucks drove their chariots into the enemy host to the raucous sound of *baritus* and *carnyx*. Some discarded their clothing and charged in ritual nakedness, their arms and chests tattooed in abstract designs. On the return from the fracas the gory heads of the enemies of the day dangled from the wickerwork of their chariots.

All this is grist for the poet's mill. The 27 feats of skill performed on the field of battle by the Celtic charioteers, their balancing acts on the pole, their trick driving and all the rest of the rodeo stunts kept their stay-at-home poets in a creative frenzy. The Celtic tales make heady reading, and their heroes have lived ever since and made people in remote valleys forget for a few brief moments the hunger gnawing at their insides.

But nothing came of it, except possibly some weird sculptures on Romanesque cathedrals nearly a thousand years after the last Celtic chieftain had been thoroughly beaten on the field of battle by the disciplined Roman legions, stripped of his gaudy war accouterments by a corrupt Roman tax farmer, and reduced to the state of a colonial freeman. It was a fascinating early attempt by the people north of the Alps to make history. But it did not succeed, chiefly because they did not possess the mental discipline needed to turn creative gifts and physical courage into statemanship. That, apparently, finds little nourishment in a cow country.

EGYPTIAN SURVEY

One of the intriguing mysteries of the Great Pyramid is the accuracy of the four corners, of which the NW one is only 13″ off 90°, and the opposite one 3′33″, the roughest of the lot. There are few modern buildings where the corners have an accuracy of ±13″, and since

the Egyptian survey priests did not possess any angle-measuring instruments, beyond a mason's square, one may wonder how they went about obtaining this measuring accuracy.

In the body of the text it has been suggested that the Pythagorean theorem was applied. This is not an idle assumption because records of Babylonian mathematics from 1800 B.C. show that the properties of the right-angled triangle were fully understood. An unknown Egyptian mathematician writing about 1850 B.C. discusses mensuration and gives a correct formula for calculating the volume of a truncated square pyramid.

Then we have Herodotus' reference to cadastral surveying (II:109) in Egypt. "This king, he writes (speaking of Sesostris or Ramses II) divided the country among the Egyptians by giving each an equal square parcel of land, and made this his source of revenue, apparently the payment of a yearly tax. And any man who was robbed by the river of a part of his land would come to Sesostris and declare what had befallen him; then the King would send men to look into it and measure the space by which the land was diminished, so that thereafter it should pay in proportion to the tax originally imposed. From this, to my thinking, the Greeks learned the art of measuring land; but the sun-clock and sundial, and the twelve divisions of the day, came to Hellas not from Egypt but from the Babylonians."

Modern Egyptologists have established that Ramses' cadastral surveyors in laying out the tax squares of land used the *double remen*, or the diagonal of the square, and it is logical to assume that they applied the same method when laying out the corners of a building. They would mark a measuring rope, or wire, in the relations 3:4:5, i.e., in accordance with the Pythagorean theorem. The practical application of the theorem to the Great Pyramid would no doubt be as follows: Having established by astronomical observations the direction of a side and measured its length, for example the one from the SW to the NW corner, the latter corner would be marked on the ground. By placing the measuring rope so that the "3-part" coincided with the side and with the "3-mark" firmly held over the NW corner the 4 and 5 parts would be stretched until the rope formed a triangle on the ground. The corner formed by the 4 and 5 sides would be marked and joined by a line to the NW corner. If the distance 3 was made long enough, say up to one-third of the pyramid side, the point set along the NW–NE side would fall far enough from the NW corner to form a 90° angle ± an unknown error, depending on how accurately the three distances had been laid out along the rope, the uneven stretch of the rope, and the manner in which the work had been performed. But such errors could be compensated by taking the mean of many measurements.

From written records, then, it has been established that this method was routine by about 1200 B.C., and from the aforementioned discourse on mensuration, most likely familiar by 1850 B.C. This still

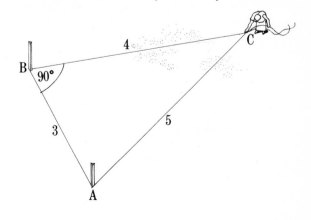

leaves us 700 years from the time the pyramid was built, but considering the conservatism of the Egyptian establishment, it is permissible to assume that the 3:4:5 relations of the right angle were known and applied during the reign of Cheops.

But it is also possible that Cheops' survey priests knew of an alternative method of laying out an accurate right angle without recourse to angle-measuring instruments. Thales of Miletus (624–565 B.C.) learned during his stay in Egypt that an angle on the circumference of a circle subtendend by the diameter is always a right angle. Ever since, this has been known as "Thales' thereom."

If applied to laying out a pyramid corner, the theorem might have produced a more accurate angle since the method did not necessitate measuring the relations 3:4:5 which would introduce an awkward source of error. The work would be accomplished as follows:

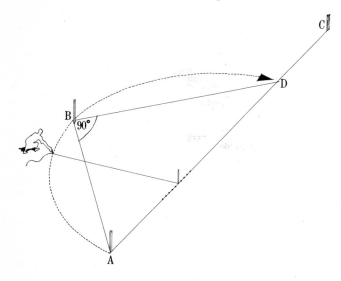

After the first side *AB* had been oriented toward North, the corner *B* was established by measuring the distance *AB*. Then an auxiliary line in any suitable direction, for example *AC*, was laid out from corner *A*. From here on it became a matter of finding, by trial and error, a point somewhere along the line *AC* that would serve as a center for a circle passing through *A* and *B*. The circle would be described on the ground, using a measuring cord or, still better, a copper wire, both of which existed at the middle of the Third Millenium. With the circle established, it was continued beyond point *B* until it intersected the auxiliary line at *D*. By joining *B* and *D* the angle *ABD* had been described on the ground, and owing to the large distances involved, the angle would measure 90° ± an unknown error due to the uneven tension applied to the rope or wire. But by repeating the operation a number of times and using the mean of the several points *D* obtained, the errors could be compensated.

The trouble with this idea is of course that 2,000 years had lapsed when Thales learned about the theorem, and there exists no documentary reference to it anywhere before Thales.

AQUEDUCT RESEARCH

Research into the topography of the Roman aqueducts has always been an exclusive interest, and only a handful of people have been concerned with it during the last 500 years. The very first investigation came about entirely by chance. Pope Pius II accompanied by his court had climbed Monte Affliano (south of Subiaco) on April 7, 1461, and when descending, the illustrious party lost their way among the gorges of the Valle d'Empligione. There they ran across a wealth of ruins described in some detail by Flavio Bióndo, a member of the party. A couple of hundred years later, Lucas Holste, German librarian at the Vatican Library, made the first systematic investigation of the aqueducts, but his maps and descriptions had not been published when he died in 1661.

The first published work dealing exclusively with the Roman aqueducts was written by Raffaele Fabretti of Urbino and is dated 1680. He discovered Aqua Alexandriana and found the springs of Marcia and Claudia. But like so many contemporary scholars, Fabretti devoted most of his energies to polemicizing with his learned colleagues, and he found no time to publish his detailed investigations of the terrain to the west of Tivoli. He was followed by the Spaniard Diego Revillas, abbot and professor of mathematics in Rome, who mapped the courses from Subiaco to Tivoli. Revillas planned an ambitious work in four volumes to be published in 1789 with the financial aid of some English country gentry. But nothing came of it. The material was dispersed; some of it landed in the Vatican Library, and the rest was rescued 150 years later by the British School in Rome.

Then followed Cassio's *Corse dell' Acque antiche* published in 1756 to 1767 when the author was eighty-seven years old. Although his work is uncritical and in part incoherent, his descriptions of Anio Novus and Claudia agree closely with reality, at any rate between Gallicana and Capanelle. He was the first to describe the remains along the west shore of the Anio.

After Cassio no further investigations were made for about a hundred years; and it was not until 1880 that Rudolfo Lanciano published the first critical work, *I commentarii di Frontino interno gli aquedotti*. But Luciano was no aqueduct specialist, and the chief importance of his work was the influence it exerted on the Englishman Thomas Ashby, who came to devote his entire life to aqueduct research.

Thomas Ashby came from an old and rich Quaker family that had been brewers since 1738. His father, in precarious health, was advised by his doctor to move to a more amenable climate and he chose Rome where the family settled in 1890. His son Thomas was then sixteen years old, and when his father soon became a friend of Lanciano, young Thomas came under his influence and decided to devote himself to archeology.

Thomas Ashby's interest in Rome's aqueducts was further enhanced by an event that, seen in retrospect, came to have great importance in interesting the English-speaking world in the subject. A New York water commissioner by the name of Clemens Herschell became fascinated by Frontinus and decided upon retirement to translate his book into English. He made a "pilgrimage," as he called it, to the Benedictine monastery at Monte Cassino where in the library is kept the oldest copy of Frontinus' book (Monte Cassino C 361), dating from the twelfth century. He spent two days at the monastery (November 22 and 23, 1897) and succeeded in getting permission to *photograph* the valuable manuscript.

The detail is worth recording because once again it was an amateur who took the initiative to an important action. Before that day, all printed works of Frontinus, in Latin or translation, had been based on a transcript of C 361 made by Poggio Bracciolini (1380–1450) who had discovered the manuscript in the library of the monastery.

Poggio was without doubt a tremendous scholar, a formidable polemicist with a taste for pornography, who by bribery and theft succeeded in salvaging a vast number of classical manuscripts—including those of Lucretius, Marcellus, and Vitruvius—from oblivion and decay in the monastery archives. But it is not to be expected that such a restlessly active person would be capable of making a faultless transcription of such a highly technical manuscript as C 361 with its wealth of figures.

But now the manuscript, page by page, as photographed by Herschell, became available to interested scholars. Herschell also translated the text into English, but in doing so he cut some corners by using French and German translations. Where they differed he had some Latin professors act as referees. The method did not prove satisfactory, and the first English version of Frontinus was severely criticized when published in 1899. A radically revised translation made by Charles Bennet was published in 1925 in Loeb's Classical Library and is now the standard Frontinus text in English.

Thus, Ashby at the outset of his career had access to the photographic reproductions of Frontinus' manuscript. He was also fortunate in being able to utilize Dr. Esther Boise Van Deman's special researches in Roman building materials and methods. Esther Van Deman was an American archeologist who was the world's foremost authority in this field until her death in 1937. When she and Ashby inventoried the remains in 1925, she was able to distinguish the original masonry work from later repairs and additions. In this way Ashby's own unique knowledge of the topography of the aqueducts was complemented by her detailed insight into Roman building technology.

Thomas Ashby's *The Aqueducts of Rome* seems to have been begun in 1908 and finished in 1928. The 925 pages of manuscript were delivered to the Clarendon Press in April, 1931, for editing and linotyping, but on May 15 that year Ashby was killed when he fell off a train between Cambridge and London. His bulky and apparently untidy manuscript required extensive editing, performed by Professor I. A. Richmond, and a critical reading of the book reveals some errors and/or inaccuracies that the author would probably have noticed if he had lived to correct the proofs himself. Nonetheless, Ashby's are the last words on the subject.

Ashby's most important contribution from a technological point of view was the leveling of the four major aqueducts (Anio Vetus, Marcia, Claudia, and Anio Novus), whereby he succeeded in sorting them out in the difficult Tivoli terrain where they are crowded together and previous investigators had failed to keep them apart. The leveling was done in 1915 by the two Italian engineers, Guido Corbellini and Guglielmo Ducci, working under the supervision of Vincenzo Reina, professor of geodosy in Rome. The survey included the entire route from the intakes to the receiving castelli in Rome. Each *cippus* (right-of-way stone), *puteus* (shaft), *specus* (water channel), *castellum* (watertower), *piscina* (reservoir) found in the terrain was fixed in elevation, and its position plotted on the topographical maps of the Italian General Staff. The survey was published by Professor Reina in a book entitled *Levellezione degli antiche acquedotte romani*. The seven topographical sheets with the aqueducts plotted by Ducci were printed in 1917 by the Instituto Geografico Militare and are included as an appendix in Ashby's own book. Two of them are reproduced in this book on pages 89 and 90.

The seven sheets contain several hundred *caposaldi,* or stations, marked on the ground with wooden stakes painted red. When Ashby and Esther Van Deman went over the ground in 1925, many of the stakes could no longer be found. Now they are all gone. Since the stations were not fixed by bearings to permanent topographical features, and many of the remains recorded by Ashby and his predecessors have been destroyed or have vanished during the past 50 years, the seven maps and Ashby's descriptions are just about all that is left of the four large aqueducts, with the exception of the protected ruins of some viaducts.

CARBON DATING

The C 14 method of archeological dating, invented by Willard F. Libby of the University of Chicago, is based on the aging of carbon, one of the most common elements and the basic constituent of organic life, animal and vegetable. The carbon dioxide released by plants as the tailings of the photosynthesis process consists of the carbon isotope 12. Included in the air that is swept up into the higher reaches of the atmosphere the carbon atoms become exposed to the incessant bombardment of cosmic rays, and some them are inevitably hit, whereby their nuclei absorb the energy and are turned into the carbon isotopes C 13 or C 14, or "heavy carbon."

Plants are unable to distinguish between "heavy carbon" and ordinary carbon and absorb the heavy isotopes together with the light ones. When animals or human beings consume the plants, they will likewise absorb the heavy and radioactive carbon which then becomes part of the cellular structure by the process commonly referred to as growth. Once absorbed, they never vanish. In other words, all living organisms—plants, animals, humans—are radioactive.

Radioactive substances' break down at a fixed rate, different for different elements. The fixed rate is known as the "half-life," i.e., the time required to reduce the radioactivity by one-half. For C 14 Libby determined the half-life to 5,568 years with a theoretical accuracy of 0.54 percent. This means that a tree cut down today will emit twice as much radiation as 5,568 years hence, and, vice versa; a tree cut down and thus having its life process terminated 5,568 years ago will emit half the radiation of a tree alive today.

By interpolating on the half-life scale it is possible to determine the exact time of a prehistoric event ending in death and leaving traces in the form of organic remains. Libby's first dating by his method applied to a piece of birch bark recovered in northern Germany, and from its radioactivity he was able to state that the ice barrier had retreated from this part of Europe 10,800 years ago. Similarly, from a piece of charcoal found in the Lascaux cavern in Dordogne, he was able to state that the campfire that left this particular piece of coal had been lit 15,516 years ago.

It should be possible to apply a similar method for determining the age of inorganic artifacts, such as pottery and pottery shards, which are now dated by rather subjective comparative methods. A pot is exposed to such a tremendous shock when fired that the effect becomes lasting and measurable. Like everything else in the ground, potting materials absorb radiation from radioactive substances, but when the pot is being fired, it is likely that the natural materials of the body will lose their radioactive charge. Eventually, when the pot, or a shard of it, gets buried in the ground it begins again to absorb radiation, and it then becomes a matter of measuring the accumulated charge up until the time when the shard was dug up and stopped absorbing radiation.

Glossary

abutment the part of a buttress, pier, or wall which receives the thrust of an arch, vault, strut, bridge or dam.

actus a Roman linear measure (= 12 *decempedae*) 38.8 yards.

advance in tunneling the distance gained from the portal, per round or unit of time.

apse semicircular projecting part of a building, in a church the eastern end of the choir.

Agricola Gnaeus (A.D. 37–93) Roman general, son-in-law of Tacitus, commander of XX Legion garrisoned at Chester (70–73) and governor of Britannia which became Romanized during his administration.

acropolis (Gr. *akros* highest + *polis* city) name of any fortified height in a city used as a refuge.

ajutage size of water pipe given in *digitus* (finger) used by Roman water administration.

alabaster a fine-textured compact variety of gypsum.

Allaqi one spelling of the name of a wadi in the desert halfway between the first and second cataracts of the Nile where during the New Kingdom the gold mines were worked with political prisoners.

aqueduct a structure above or below ground used for the conveyance of water.

aguefer subterranean water course in rock under high pressure. In the Alps the aquefers bring cold snow water through vertical fissures into the ground where it is heated and then rises in a closed system.

architrave (It. *archi* + *trave* beam) lowest part of an entablature resting on the columns. Derives from a method of building with supporting posts carrying wooden beams, a method of timbering applied to building in stone.

archivolt building member surrounding a curved opening such as an arch, also the molding or other ornament on the wall face of the *voissoirs*.

Argos old city on the Peloponnesus which traded with Egypt, also name of Odysseus' dog.

artesian well (F. *Artois*) a well made by boring into the soil until water is reached which from internal pressure spouts like a fountain, in Europe first sunk in Artois.

articulated (L. *articulare* to divide into joints) in building clearly and logically related parts.

artifact (L. *artis* + *facere* to make) a product of human workmanship as distinguished from a natural object.

asphalt (Gr. *asphaltus*) a black solid bituminous substance consisting of a mixture of hydrocarbons; also a mixture of lime, gravel and sand; in Sumer also with chopped fibres and dung.

attic (Gr. *attikos* pertaining to Attica) a low story or decorative wall above the main orders of a façade.

Aurignacian period early epoch of upper paleolithic culture, ca. 25,000 B.C., of the Cro-Magnons, from Aurignac in Haute-Garonne, France. *Homo aurignacensis* appears to have hunted from the Pyrenees in the west to Czechoslovakia in the east.

azimuth (Ar. *al-sumut* direction) arc of horizon measured clockwise between a fixed point, in astronomy and geodesy south, in navigation north.

baldachin (fr. Bagdad) a rich medieval fabric, in building a structure in the form of a canopy suspended from the ceiling or projecting from the wall, usually over an altar.

ballista (Gr. *ballein* to throw) a Roman military engine in the form of a crossbow for hurling missiles and used in sieges.

barbican (Ar. *barbakh—khanah* a house with piped water) an outer defense work of a gate or bridge.

basilica (Gr. *basilike* king) in Rome an oblong building used as a public hall, with a broad nave flanked by colonnaded aisles, used as a model for early Christian churches.

bastide (Pr. *bastida*) a small fort or fortified house or tower, also small fortified town outside a castle.

bastion (It. *bastione*) a work projecting outward from the main enclosure of a fortification, consisting of two faces meeting in a salient angle.

bay principal compartment of a building or structure, as marked off by buttresses or columns.

bitumen (L. *bitumen*) originally mineral pitch or asphalt.

Boeotia, early republic in Greece.

breccia a rock composed of angular fragments cemented together under water pressure.

bull a papal letter distinguished from other apostolic letters by being sealed with a *bulla*, a round seal of lead bearing on one side St. Peter and St. Paul and on the other the name of the pope who issues the bull. It is written on parchment and in the third person; after giving the name of the pope it starts off with the formula "Bishop, servant of the servants of God."

bulldoze originally used to intimidate Negro voters in Louisiana; also secondary blasting by mudcapping, and to excavate with caterpillar tractor provided with a steel blade.

burghal hidage basis for assessing the number of days Anglo-Saxon peasants had to work on the king's fortifications in a year.

burhbot the Anglo-Saxon law regulating the obligation of laboring for the king.

caisson a box or chamber used for construction under water.

cantilever a projecting beam supported only at one end, also the two beams or trusses projecting from piers toward each other which when joined form a span of a cantilever bridge.

cap in mining the horizontal timber resting on uprights (legs) supporting the back of a working.

capital (L. *capitellum* small head) head or uppermost member of a column or pilaster crowning the shaft and taking the weight of the entablature.

Carrara marble quarries in Tuscana with stone of a finely grained texture, opened up during reign of Augustus.

caponier (Sp. *caponera* cage for fattening capons) a work built crosswise in a ditch to sweep it with fire or to cover a passageway.

casemate (Gr. *chasmata* chasm) a bombproof chamber in which cannon can be placed and fire through embrasures, also used for quartering troops.

castellan (L. *castellanus* occupant of a castle) governor or warder of a castle or fort.

castellum a Roman fort, but also used in the combination *castellum aqua* a water reservoir in the shape of a tower.

catapult (Gr. *kata* down + *pallein* to hurl) engine resembling a large crossbow used by Greeks and Romans to throw stones and spears horizontally.

censor (L. *censere* to value) a magistrate in Rome who took a register of the number and property of the citizens and managed the public finances.

chalcolithic *(chalco* copper + *lithic* stone) a period when copper tools were used together with stone tools.

chariot (L. *carrus* F. *char* car) an ancient two-wheeled vehicle for war commonly drawn by two horses.

chord a straight line intersecting a curve.

chorobates a leveling device consisting of a long board with a groove filled with water used by Roman and medieval engineers for leveling with an accuracy of 1:2000.

citadel (L. *civitas*) a fortress that commands a city, for control and defense.

cippus a post or pillar usually inscribed and set up as a landmark, also to mark the underground conduit of an aqueduct. The distance between two cippi was usually 240 feet.

clan (Gael. *clann* descendants) social group consisting of a number of households the heads of which claim descent from a common ancestor, bear a common surname, and acknowledge a chief, and possess a common totem.

clerestory part of church which rises clear of the roofs of other parts, with windows for lighting the nave.

clinker a hard brick or vitrified and fused stony matter as formed in a furnace from impurities in coal or in a cement kiln.

clustered pier or column composed of several columns connected together.

coffer in building ornamented panel deeply recessed in a vault or dome chiefly to lighten the structure.

cofferdam a watertight enclosure made of piles packed with clay and from which the water has been pumped out to expose the bottom of a river, also temporary dam across a river.

coke the infusible coherent residue when coal is subjected to destructive distillation.

component one of the parts into which a vector or tensor quantity such as a force, momentum or velocity may be resolved.

computate to reckon or count, to make up a bill.

concrete (L. *con* + *crescere* to grow) a compound joined by concretion, specifically artificial stone made of mixing cement and sand with gravel, broken stone or other aggregate.

continuous beam a girder or beam having more than two supports.

contra brevia copies of a rescript.

converter a vessel used in the Bessemer process for blowing off the excess carbon in pig iron thereby converting it into mild steel.

cornice (L. *cornix* crow) horizontal member which crowns a façade, top course of a masonry wall.

corundum native alumina or aluminum oxide Al_2O_3 the hardest mineral except for the diamond and therefore used as an abrasive.

counterscarp the exterior slope of a wall or ditch.

crémaillière an indented or zigzagged line of entrenchment in a system of fortification.

crenelation (L. *crena* a notch) one of the embrasures alternating with merlons in a battlement, an indented pattern.

cubit (L. *cubitum* elbow) the forearm, a measure of length from the elbow to the tip of the extended forefinger; English cubit 18 inches, Egyptian cubit 20.7 inches, Roman 17.5 inches, Greek 18.22 inches, Hebrew 17.58 inches.

cupola (L. *cupa* tub) a roof or ceiling having a rounded form, when large also called a *dome,* also a furnace or kiln.

cuprite a mineral consisting of copper oxide Cu_2O with 89 percent copper, usually found together with metallic copper.

curator Roman administrator, *Curator aquarum* was a member of the water tribunal.

curtain (L. *cortina* court) part of a bastioned front connecting two neighboring bastions, also a similar stretch of plain wall.

Cyclops (Gr. *Kyklops* round-eyed) in Greek mythology one of a race of giants with but one eye which was in the middle of the forehead, believed to have brought together the huge unworked blocks of stone used in the foundations of the Mycenaean and early Greek citadels.

denarius Roman silver coin weighing 1/72 pound in 268 B.C., reduced to 1/96 pound by Nero.

Dialogue de Scaecario a description of the accounting procedure used by the exchequer during the reign of Henry II in the form of a dialogue between the author Richard Fitz Neal and an unknown pupil. The dialogue was written in 1176–1179 when the author had held the office of treasurer since 1158. His father and uncle had held the office before him during the reigns of Henry I and Stephen, and the accounting method is likely to have been developed by one of them.

Diodorus Siculus Greek historian born in Agyrium in Sicily, date unknown. In 56 B.C. he began his history of the world comprising 40 books, visited Egypt in 59 B.C. His works have survived in eight manuscripts, translated into Latin in 1472, into English in 1700, 1804 and 1936.

dip the angle which a stratum or ore body, fissure, or fault makes with a horizontal plane, as measured in a plane normal to the strike.

dolorite a hard and dark rock used as a tool by Egyptian miners and quarrymen.

donjon a massive chief tower in an ancient fortification.

dowel a round pin of wood fitting into a corresponding hole in an abutting piece to hold two pieces together.

dromos passage to a subterranean *tholos* tomb.

druid priest, judge, teacher, conjurer among the Celts.

dynamics branch of mechanics dealing with the motion of bodies.

edile or **aedile** Roman official whose chief duty was to look after the public works and police the city.

ecology a branch of biology which deals with the mutual relations between organisms and their environment.

Elephantine island in the Nile below the first cataract at Aswan, administrative center of the southern nomes and Nubia. The island obtained its name from the elephants that used to roam the area during the early dynasties.

emporium (Gr. *emporion*) a place of trade, a market place, commercial center.

enfilade to rake with gunfire in the direction of the length of a work or line of troops.

entablature in classical architecture the wall resting on the capitals of the columns. It was divided into the *architrave* (part immediately above the columns), *frieze* (the central space), and *cornice* (the upper projecting moldings).

Eridu one of the early cities in Sumeria and a rival of Ur.

escalade (It. *scala* ladder) to attack a fortified place using ladders.

escarp (It. *scarpa*) ground about a fortified place cut away nearly vertically to prevent approach.

Etruscans a technically gifted people living between the Arno and the Tiber, flourished economically and culturally before they were overrun and absorbed by the Romans.

Exchequer (OF. *eschequier*, chessboard) under Norman and Plantagenet kings the department of state charged with the collection and management of the royal revenue, so called because the lords sat around a checkered table when auditing the accounts.

factory originally a trading station where factors, or commercial agents, resided and transacted business for their principals. Many of the cities around the Mediterranean such as Cadiz and Marseilles began as Phoenician factories.

falsework temporary construction usually in timber to support main work during erection.

fellah (Ar. *fallah*) peasant in Egypt, Syria, and other Arab speaking countries.

Fiscus, also **fisc** (L. *fiscus* money basket) Roman public treasury, especially under the Empire.

flint nodule a small round and smooth flint stone which splits well.

gallery (It. *galleria*) in mining a working drift or a tunnel, in fortification any covered passageway.

Gallia or **Gaul** Roman name of areas populated by Celts. Transalpine Gaul (*Gallia transalpina*) comprised present France, Belgium, and Switzerland; Cisalpine Gaul (*Gallia cisalpina*) included the southern slopes of the Alps and northern Italy,

gargoyle (OF. *gargouille* throat) a waterspout often grotesquely carved projecting from the gutter of a Gothic building.

geodesy branch of applied mathematics which determines by measurement the exact position of points on the earth and their relations with each other.

girder one of the main timbers in a framed floor supporting the joists which carry the flooring, also any heavy steel beam.

glacis (OF. *glacier* to slide) the natural or artificial slope from the top of the counterscarp toward the open country.

graffito (It. *graffio* scratching) rudely scratched inscriptions, drawings, etc., found on rocks, walls, vases and other objects. Greek tourists have left thousands of graffiti on Egyptian monuments.

gravette (after Gravette in Dordogne) designation of a type of man preceding the Cro-Magnons. A Gravettian branch hunted mammoth along the corridor between the northern ice barrier and the Würm glaciers of the Carpathians and all the way across the Russian steppe to Irkutsk. The Gravettian hunting people lived about 30,000 years ago.

Hammamat Wadi (now Bir el Hammamat) a valley extending from ancient Coptos (now Ous) to the Red sea (at Quesir). The valley played an important part in early Egyptian industry because of the gold deposits and fine building stone found there. Most likely the early invaders and conquerors of the neolithic peasants living along the bend-of the Nile came up the valley from the Red sea.

Helladic of or pertaining to Hellas, as opposite to Asia, also the divisions of prehistory of the Greek islands, principally Crete, from 2700 B.C. onward.

Hellenistic of or pertaining to Greek history and culture after Alexander the Great when Greece itself had lost its political independence but when Greek civilization and culture reached its maximum extension and influence.

Herodotus (484–425 B.C.) Greek historian, born in Helikarnossos in Asia Minor. He spent 17 years traveling before at the age of 37 he began to write his *Historia* which besides a lot of legends contains a great deal of solid information on Egypt and the East.

Hisarlik a 200×160-meter tell on the Anatolian coast ca. 6 km southeast of the Dardanelles where the six cities of Troy (the first one from Neolithic times) remained hidden until Heinrich Schliemann began digging there in 1870.

Hittites an Indo-European people who settled on the Anatolian highlands ca. 2000 B.C. and founded an empire with Harrusa (Boghazköy) as its capital. At the end of the second millenium the Hittites conquered Syria, then an Egyptian sphere of interest. The two superpowers clashed head on at the battle of Kadesh in 1286 B.C. There is ample evidence to believe that it was Hittite smiths who hit upon the discovery of extracting iron and forging it into tools and weapons. This appears to have taken place around the Van lake some time prior to 1300 B.C.

Horus clan or the "Followers of Horus" were most likely the leaders of the immigrants coming from the Red Sea who in predynastic times made themselves masters of Upper Egypt and eventually also of Lower Egypt and the delta. Their totem was the falcon which later was turned into the most prominent god in Egypt. Each pharao, had besides his own name also an official Horus name.

hydraulic of or pertaining to fluids in motion, conveying and use of water, hence also the building of such structures as dams, locks and other regulating structures. Hydraulic cement sets under water, also used about machines and devices operating by the resistance offered by water or oil.

Hyksos (Eg. *Hig Shasu* ruler of nomadic people) invaders and rulers of Egypt ca. 1650 to 1580 B.C.

hypocaust (Gr. *hypo* under + *kaiein* to burn) Roman central heating system whereby the combustion gases from a furnace were conducted through flues in the floors and walls of houses, used mostly for domestic heating north of the Alps.

hypostyle (Gr. *hypo* under + *stylos* column) constructed by means of columns, applied to such great halls as that in the Karnak temple.

ingeniatore, also engignier, engigneor, engynour, etc., medieval writings of engineer.

ingot a mass of metal cast into a convenient shape and later remelted and finished by rolling or forging.

innovation introduction of something new. A *technical innovation* is a significant improvement of something that exists, a *technological innovation* is an invention or discovery that breaks with the past and sets technology on a new track.

irrigation act or process of wetting or moistening, in farming artificial watering by canals and dtiches. Ancient practice made use of *perennial irrigation* (as in Sumer) or *basin irrigation* (as in Egypt).

joist any small rectangular timber or rolled steel beam directly supporting a floor.

Larsa important city on the Euphrates 25 miles from Ur.

legion (L. *legere* to collect) in Rome a body of soldiers forming principal unit of the army and varying from 3,000 foot and 300 horse in early times to 5,000–6,000 foot under the Empire.

Lias limestone rock, oldest division of the Jurassic system.

libra a Roman pound weighing 0.722 lb.

limes a boundary, specifically a fortified frontier such as *Limes Germanicus* and *Limes Britannicus* or the wall of Hadrian.

loess unstratified deposit of loam covering extensive areas of central Europe, Russia, North America in parts, and eastern China, believed to have been transported and deposited by the wind.

Madeleine name of an archeological site on the Vèzère, a tributary of Dordogne, used by the gifted Cro-Magnon people about 15,000 years ago.

magister originally a republican Roman office title, for example *magister equition* — master of horse.

malachite a green basic carbonate of copper $Cu_2CO_3(OH)_2$, a copper ore.

mansio, mansione house for temporary stay, a post station.

Mari culturally advanced city and state around the upper Euphrates contemporary with Sumer.

mastaba (Ar. *mactabah* bench) a type of tomb oblong in shape with sloping sides from the time of the Memphis dynasties.

megalith huge boulder of stone used in prehistoric monuments, particularly from the Bronze Age.

megaron (Gr. *megas* great) the central rectangular hall of a Mycenaean house.

merlon one of the solid members of a battlement.

Merthyr Tydfil ironworks in Wales that rapidly grew to the leading producer of puddle iron in the world in the early 19th century.

Mesolithic stage intermediate between Paleolithic and Neolithic stone ages.

microliths small stone tools made of flint chips invented by the Aurignacians and further improved by the Magdalenians after which the technique declined in Mesolithic times. The tools were rhomboid, triangular, semicircular, etc., and some were fitted to spears used for fishing and bird hunting.

Moeris (now Birket Garum) lake ca. 40 miles west of the Nile at the northern end of the Fayum oasis. The lake is 475 feet below sea level and could have been used as a storage of water, but that would have required a tunnel through the escarp. The so-called Labyrinth, headquarters of the Egyptian water administration, was situated 20 km west of the Nile at the end of a much larger lake, as described by Herodotus. He also speaks of a tunnel through the mountains to the west of Memphis and one to Syrtis in Libya. From Herodotus it appears as if the low Fayum oasis was used as a reservoir, and nothing in the topography contradicts him.

monolith a single stone block shaped into a pillar, statue or monument.

Muggatan limestone quarries to the south of Cairo, now actually on the outskirts of the city where the Tura limestone was quarried.

municipium a privileged class of Roman towns, a free city.

Mycenae city of Agamemnon and the first major political center in Europe. The site was occupied in 2500 B.C. and fortified in 1900 B.C. It began to flourish in 1500 B.C. and became Europe's first commercial and cultural center after Knossos on Crete was destroyed. The city retained its importance to about 1100 B.C. when a catastrophy of some kind put an end to its Homeric splendor.

Natufians a Mesolithic people living in Palestine and regarded as the ancestors of the Hamites and Semites. The Natufian women are thought to be the first cultivators of the grain grasses growing wild in Palestine. Evidence points to the Natufians having brought farming to the Nile delta.

Neolithic pertaining to a stage of culture characterized by the use of polished stone implements. In Europe, it was associated with farming and can be considered to have begun ca. 6000 B.C.; in America, the Indians used Neolithic tools at the time of Columbus.

nilometer (Gr. *Neilos* the Nile + *metron* measure) gage for measuring the water level of the Nile, usually a scale cut into the rock.

nitroglycerin an oily explosive liquid obtained by treating glycerol with a mixture of nitric and sulfuric acids. It was invented by Ascanio Sobrero, an Italian physician, in 1846 and used originally as a heart medicine. Alfred Nobel made a fortune mixing it with kiselguhr and turning it into dynamite.

nome (Gr. *nomos* district) Greek word for a province in ancient Egypt.

obelisk (Gr. *obeliskos* pointed pillar) a four-sided pillar terminating in a pyramid, usually monolithic and decorated with hieroglyphics.

obsidian volcanic glass of a solid compact structure used for cutting tools.

Oman a peninsula in the Persian Gulf with "natural bronze" deposits, that is, a mineral containing metallic copper mixed with tin and nickel, worked by the merchants of Ur and turned into the best tools and weapons in the ancient world. Mining of Oman copper stopped 2200 B.C.

oolite rock consisting of small round grains cemented together, belongs to the upper part of the Jurassic system.

opus signum pozzolana cement mixed with crushed shards used as a waterproof lining of Roman water conduits.

Paleolithic pertaining to the early human culture characterized by rough or simply chipped stone tools. Considered as an industry it may have been begun with the Acheullean, say 380,000 years ago, and ended with the Mesolithic about 8000 B.C. The tools developed from roughly-shaped hand-axes to technically sophisticated pressure flaking, but the tools were never ground or polished.

palisade (L. *palus* a stake) long strong stakes pointed at the top set in the ground in a close row as a means of defense.

parapet (It. *parare* ward off + *petto* breast) a wall, rampart, or elevation to protect soldiers, a breastwork raised above a wall.

passus a step, Roman length measure (= 5 pedes) 4.85 feet.

pelton wheel an impulse turbine with a row of buckets arranged round the rim of the wheel, named after the inventor and used in high-head power developments.

Pentelicus a 330-foot-high mountain near Athens with marble quarries worked since 500 B.C.

peperino a dark-colored tuff found in the Alban hills near Rome and used as a building stone.

perimeter outer boundary of a body or figure, used for closed lines of defense.

pisé (L. *pisare* to pound) a building material consisting of stiff soil or clay mixed with dung or chipped straw, either rammed between shutters or formed into blocks left to dry in the sun.

poras a porous limestone used in early Greek stone buildings because its large pores gave good key to plaster.

porphyry rock consisting of felspar crystals embedded in a compact red or purple grained mass.

portcullis a grating of iron or heavy timbers pointed with iron hung in or over the gateway of a castle.

portland stone a white oolite limestone from the Isle of Portland in England and used as building stone.

postern (L. *posterus* coming after) back door or gate, a way to escape, a subterranean passage between the ditch and interior of a fortification.

pozzolana or **pozzuolana** a loosely compacted siliceous rock of volcanic origin, a mortar used in Roman building made with pozzolana.

Premonstratensian a religious order founded at Premontré near Laon in France in 1119.

puddle to knead or work clay or a mixture of clay and sand to make it impervious to water, also a method of converting pig iron into wrought iron by stirring the heat in a reverberatory furnace.

puteus Roman technical term for tunnel shaft.

pylon (Gr. *pylon* gateway) a gateway building having a truncated pyramidical form.

qanaat a subterranean water tunnel in the Near East.

quinarium Roman measure of capacity of a water conduit.

quicksand sand mixed with water and very dangerous because of the difficulty of extracting anything sinking into it.

quartz a form of silica and the most common of solid materials, essential constituent of granite, sand and gravel.

ravelin a detached work with embankments which makes a salient angle in front of the curtain across the ditch at the top of the counterscarp.

rector fabricae medieval building master in charge of operations, also the designer.

relazione Italian for report, investigation.

rescript written answer of a Roman emperor or pope to an enquiry upon a matter of law or the state, when in reply to a private citizen it was called *annotation*, when to a magistrate *epistle*, and when directed to a city or province a *pragmatic sanction*.

Roma Vecchia Casale a villa property southeast of Rome over which passed five of Rome's largest aqueducts.

Sesklo an archeological site where evidence was found of the introduction of farming into Greece about 5000 B.C.

sesterce a coin equal to 1/4 denarius, originally a silver coin but made of brass during the reign of Augustus.

sheriff originally king's steward over a shire and responsible for the collection of taxes.

siphon a pipe whereby water is conducted over an intermediate elevation to a lower level by the pressure of the atmosphere.

socci team working on a piece-rate contract in medieval construction.

span ancient measure equivalent to the length of the extended hand, equivalent to 9 inches.

specus roofed channel in which water was carried in a Roman aqueduct.

sperone a porous building stone used in Roman buildings during the republic.

stele slab or pillar used as a gravestone or any other similar sculptured monument.

stilting to raise the springing line of an arch or vault.

Strabo, Strabon, Strabone (64–21 B.C.) most competent geographer in classical times. His works comprise 17 books of which 9 exist in medieval copies. Little read by his contemporaries, translated into English 1892 and 1917.

stucco a fine plaster made of lime or gypsum with sand and crushed marble.

Sumer the first populated part of the Euphrates delta and the lower part of the valley.

thanes also **thegns** originally free servants or attendants of Anglo-Saxon kings, also serving as bodyguard, later a military cast which qualified by ownership of five hides of land, and which served as heavily-armed mounted warriors.

tell (Ar. *tall*) hill or mound formed by the disintegration of pisé houses built on the debris of previous ones.

tenaille outwork in the main ditch between two bastions.

therm (L. *therma* hot spring) a hot bath, any bath but particularly Roman public baths.

tholos a circular tomb of beehive shape in the side of a hill.

toise (L. *tensum* to stretch) a French, Swiss, and Dutch fathom, equivalent to 2.13 yards in France and 1.97 yards in Switzerland.

travertine a crystalline calcium carbonate formed by deposits from spring water, used as building stone.

travois primitive vehicle consisting of two trailing poles serving as shafts for an animal and bearing a platform for a load.

trebuchet medieval siege engine for throwing stones.

triforium gallery forming the upper story of the aisle in a church.

truss assemblage of members, such as beams, bars, and rods combined so as to form a rigid framework, properly in the form of a triangle or a combination of triangles.

tympanum the recessed face of a pediment within the frame made by the upper and lower cornices, usually triangular in shape.

Windmill Hill archeological site in southern England where farming was done ca. 3000 B.C.

Würm (after the river of the same name in Bavaria) fourth glaciation during the quarternary epoch when the snow in the Alps remained the year round at an altitude of 3,950 feet. It lasted from 500,000 to 25,000 B.C. and culminated about 40,000 years ago. The ice-free land in Europe constituted an arctic tundra with about the same flora as in northern Scandinavia and Alaska and a fauna with reindeer and such extinct animals as the furred rhinoceros, mammoth and musk ox. In some shielded valleys, such as the Dordogne, Neanderthal homonides struggled in vain against their cruel environment.

weir a dam in a river to stop and raise the water for conducting it to a mill, or to eliminate strong currents and make the river navigable.

Bibliography

Andrews, C. B.: The Railway Age, Country Life, London 1937

Andrews, F. B.: The Medieval Builder and His Methods, University Press, Oxford 1925

Ashby, T.: The Roman Campagna in Classical Times, Benn, London 1937, The Aqueducts of Ancient Rome, Clarendon Press, Oxford 1935

Barois, J.: Irrigations en Egypt, Librairie Polytechnique, Paris 1911

Beadnell, H. J. L.: An Egyptian Oasis, John Murray, London 1909

Bourne, J. C.: London and Birmingham Railway, Acherman & Co., London 1839, History and Descriptions of Great Western RR, David Bogue, London, 1846

Bromehead, C. E. N.: The Early History of Water Supply, The Geographical Journal, London 1942

Calza, G. and G. Becatti: Ostia, Istituto Poligrafico della Stato, Roma 1963

Carcopino, J.: La vie quotidienne à Rome à l'apogée de l'empire, Librairie Hachette, Paris 1960

Childe, Gordon V.: New light on the Most Ancient East, Routledge and Kegan Paul Ltd., London 1958

Clark, S. and Engelbach, R.: Ancient Egyptian Masonry, Oxford University Press, London 1930

Colvin, H. M. et al.: The History of King's Works, Her Majesty's Stationary Office, London 1963

Compressed Air Review: The Story of the Hoover Dam, 1931–1935

Cresy, E.: Encyclopaediea of Civil Engineering, London 1847

Denon, V.: Voyages dans la Basse et La Haute Egypt pendant Les Campagnes du Général Bonapart, P. Didot L'Aîné, Paris 1802

Dinsmoor, W. B.: The Architecture of Ancient Greece, B. T. Batsford Ltd., London 1927

Diodorus Siculus: Library of History, translated by C. H. Oldfather, Heinemann, London 1933

Edwards, I. E. S.: The Pyramids of Egypt, Penguin Books, London 1948

Ellis, Hamilton: British Railway History 1830–1876, George Allen and Unwin Ltd., London 1954

Elton, Arthur: British Railways, Collins, London

Engelbach, R.: The Quarries of the Western Nubian Desert, Annals de Service des Antiquities de l'Egypt, Cairo 1953. The Aswan Obelisk, Cairo 1922

Evans, Sir Arthur: The Palace of Minos, MacMillan, London 1928

Fairservis, W.: The Ancient Kingdoms of the Nile, Mentor Books, New York 1962

Fitzler, K.: Steinbrücke und Bergwerke im Ptolemaischen und Römischen Ägypten, Verlag von Ovelle & Meyer, Leipzig 1910

Fontana, D.: Della transportione dell'obelisco Vaticana et della Fabriche di nostro Signore Papa Sisto V, Dominica Basa, Roma 1590

Frankfort, H.: The Birth of Civilization in the Near East, Williams and Norgate, London 1951

Frontinus: De Aquis Urbis Romae, translated by Charles Bennet, Heinemann, London 1925

Gardiner, A.: Egypt of the Pharaohs, Clarendon Press, Oxford 1961

Gibbon, E.: The Decline and Fall of the Roman Empire, G. Virtue, London

Giedon, S.: The Beginnings of Art, Oxford University Press, London 1962

Glanville, S. R. K.: The Legacy of Egypt, Clarendon Press, Oxford 1941

Goodfield, J.: The Tunnel of Eupalinus, Scientific American, Vol. 210. No. 6, 1964

Hachette: Cairo, Alexandria and Environs, Paris 1963

Hammond, Rolt: Modern Civil Engineering Practice, George Newnes, Ltd., London 1961

Harden, D.: The Phoenicians, Thames & Hudson, London 1962

Hawkes, J. and Sir L. Woolley: Prehistory and the Beginnings of Civilization, George Allen & Unwin Ltd., London 1963

Heer, F.: The Medieval World, Europe from 1100 to 1350, Weidenfeld and Nicolson, London 1961

Hellström, B.: Use of Water in Ancient Civilizations in the Mediterranean Area, Proceedings of the I.U.C.N. Technical Meeting, Athens 1959

Herodotus: History, translated by A. D. Godley. Heinemann, London 1920

The Holy Bible: Newly translated out of the Original Tongues and with the former translations diligently compared. Imprinted by Robert Barker, Printer to the King's most Excellent Majesty, London 1633

Hully de, J. and A. G. Verhoven: Tractaet van Dyjckage (1579), The Hague, 1920

Illustrated London News, 1869

Jacobsen, T. and Lloyd S.: Sennacherib's Aqueduct at Jerwan, University of Chicago Press 1935

Kenyon, K.: Archeology in the Holy Land, Ernest Benn Ltd., London 1965

Laroche, E.: The Enigma of the Hittites, Unesco Courier, Paris 1963

Lauer, J. P.: Les Pyramides de Sakkarah, Cairo 1961

Lazard, P.: Vauban 1633–1707, Librairie Felix Alcan, Paris 1934

Lilienthal, David E.: TVA-Democracy on the March, Harpers & Brothers, New York 1955

Little, T.: The High Dam at Aswan, Methuen & Co. Ltd., London 1965

Lucas, A.: Ancient Egyptian Materials and Industries, Arnold, London 1948

Marlowe, J.: The Making of the Suez Canal, The Cresset Press, London 1964

Moellert, James: Hacilar, A Neolithic Village Site, Scientific American, Vol. 205 No. 2, New York 1961

Meijer, C. J.: L'Arte de restiture a Roma la translasciata Navigatione del suo Tevere, Prt III Del modo di secare Palude Pointine, Roma 1683

Mellart, James: A Neolithic City in Turkey, Scientific American, Vol. 210, No. 4, New York 1964

Mermel, T. W.: Register of Dams in the United States, McGraw-Hill Book Company, New York 1958

Miller, Konrad: Itineria Romanae, Stuttgart 1916

Miller, Konrad: Weltkarte des Catorius, Ravensburg 1888

Moreau, P.: Wissenschaftliche Beschreibung und malersche Ansichten von der Eisenbahn zwischen Liverpool und Manchester, Weimar 1831

Morgan, M.: The Dam, Viking Press, New York 1954

Museum of Modern Art, Twentieth Century Engineering, New York 1964

Needham, Joseph: China and the Invention of the Poundlock, Transactions of the Newcomen Society, Vol. XXXVI, 1963–64

New Horizon: Topmost Dams in the World, Tokyo 1963

Oman, Ahmed Osman: The High Dam. Universitaires d'Egypt, Cairo 1966

Painter, S.: A History of the Middle Ages, MacMillan & Co., London 1964

Pannel, J. P. M.: An Illustrated History of Civil Engineering, Thames and Hudson, London 1964

Pareti, Luigi: The Ancient World, George Allen and Unwin Limited, London 1965

Payne, R.: The Canal Builders, The MacMillan Company, New York 1949

Piggott, S.: Ancient Europe, Edinburgh University Press, Edinburgh 1965

Piranesi: Le Antichita Romane, Roma 1756

Pliny: Natural History, translated by H. Rackham, Heinemann, London 1938

Reisch, Gregor: Margarita philosophica, 1512

Robertson, D. S.: A Handbook of Greek and Roman Architecture, University Press, Cambrigde 1943

Robins, F. W.: The Story of Water Supply, Oxford University Press, London 1946

Rossiter, S.: Rome and Central Italy, The Blue Guides, Ernest Benn, London 1964

Rostovtzeff, M.: The Social and Economic History of the Hellenistic World, Clarendon Press, Oxford 1941, Rome, Oxford University Press, London 1960

Sandström, G.: The History of Tunneling, Barrie & Rockliff, London 1963

Schede, Hartmann: Welt Chronik, Nürnberg 1493

Singer, Charles: A Short History of Scientific Ideas, Clarendon Press, Oxford 1959

Singer, Charles et al.: A Short History of Technology, Clarendon Press, Oxford 1954–58

Strabo: The Geography of Strabo, translated by H. L. Jones, Heinemann, London 1930

Straub, Hans: A History of Civil Engineering, Leonard Hill Ltd., London 1960

Suetonius: Vitae XII Caesarum Caligula, Heinemann, London 1912

Taylor, W.: The Mycenaeans, Thames & Hudson, London 1961

Troy, Sidney: History of Fortification, Heinemann, London 1955

Unger, Eckhard: Babylon, die heilige Stadt, Walter de Gruyter, Berlin 1931

Wace, J. B. and Stubbings F. H.: A Companion to Homer, MacMillan & Co. Ltd., London 1962

Vitruvius: De Architectura, edited from the Harleian manuscript 2767 and translated by Frank Granger, Heinemann, London 1934

Zesen, Filip Jon, Beschreibung der Stadt Amsterdam, Amsterdam, 1664

Sources of illustrations

Most of the black-and-white illustrations were drawn especially for this book by Gyula Buváry (GB below) from originals procured by the author. The sources of these originals, as well as of the engravings and illustrations in color, are listed below; figures refer to page numbers. The author's name is shortened to GS.

Index

Note: Page references to illustrations are in *italics*.

277

278